PRACTICAL RF SYSTEM DESIGN

PRACTICAL RF SYSTEM DESIGN

WILLIAM F. EGAN, Ph.D.

Lecturer in Electrical Engineering
Santa Clara University

The Institute of Electrical and Electronics Engineers, Inc., New York

A JOHN WILEY & SONS, INC., PUBLICATION

Published by John Wiley & Sons, Inc., Hoboken, New Jersey.
Published simultaneously in Canada.

For general information on our other products and services please contact our Customer Care Department within the U.S. at 877-762-2974, outside the U.S. at 317-572-3993 or fax 317-572-4002.

Wiley also publishes its books in a variety of electronic formats. Some content that appears in print, however, may not be available in electronic format.

Library of Congress Cataloging-in-Publication Data is available.

ISBN 0-471-20023-9

Printed in the United States of America

10 9 8 7 6 5 4 3 2 1

To those from whom I have learned:
Teachers, Colleagues, and Students

CONTENTS

PREFACE

This book is about RF system analysis and design at the level that requires an understanding of the interaction between the modules of a system so the ultimate performance can be predicted. It describes concepts that are advanced, that is, beyond those that are more commonly taught, because these are necessary to the understanding of effects encountered in practice. It is about answering questions such as:

- How will the gain of a cascade (a group of modules in series) be affected by the standing-wave ratio (SWR) specifications of its modules?
- How will noise on a local oscillator affect receiver noise figure and desensitization?
- How does the effective noise figure of a mixer depend on the filtering that precedes it?
- How can we determine the linearity of a cascade from specifications on its modules?
- How do we expect intermodulation products (IMs) to change with signal amplitude and why do they sometimes change differently?
- How can modules be combined to reduce certain intermodulation products or to turn bad impedance matches into good matches?
- How can the spurious responses in a conversion scheme be visualized and how can the magnitudes of the spurs be determined? How can this picture be used to ascertain filter requirements?

- How does phase noise affect system performance; what are its sources and how can the effects be predicted?

I will explain methods learned over many years of RF module and system design, with emphasis on those that do not seem to be well understood. Some are available in the literature, some were published in reviewed journals, some have developed with little exposure to peer review, but all have been found to be important in some aspect of RF system engineering.

I would like to thank Eric Unruh and Bill Bearden for reviewing parts of the manuscript. I have also benefited greatly from the opportunity to work with many knowledgeable colleagues during my years at Sylvania-GTE Government Systems and at ESL-TRW in the Santa Clara (Silicon) Valley and would like to thank them, and those excellent companies for which we worked, for that opportunity. I am also grateful for the education that I received at Santa Clara and Stanford Universities, often with the help of those same companies. However, only I bear the blame for errors and imperfections in this work.

William F. Egan

Cupertino, California
February, 2003

GETTING FILES FROM THE WILEY ftp AND INTERNET SITES

To download spreadsheets that are the bases for figures in this book, use an ftp program or a Web browser.

FTP ACCESS

If you are using an ftp program, type the following at your ftp prompt:

```
ftp://ftp.wiley.com
```

Some programs may provide the first “ftp” for you, in which case type

```
ftp.wiley.com
```

Log in as anonymous (e.g., User ID: anonymous). Leave password blank. After you have connected to the Wiley ftp site, navigate through the directory path of:

```
/public/sci_tech_med/rf_system
```

WEB ACCESS

If you are using a standard Web browser, type URL address of:

```
ftp://ftp.wiley.com
```

Navigate through the directory path of:

```
/public/sci_tech_med/rf_system
```

If you need further information about downloading the files, you can call Wiley's technical support at 201-748-6753.

SYMBOLS LIST AND GLOSSARY

The following is a list of terms and symbols used throughout the book. Special meanings that have been assigned to the symbols are given, although the same symbols sometimes have other meanings, which should be apparent from the context of their usage. (For example, A and B can be used for amplitudes of sine waves, in addition to the special meanings given below.)

$\equiv$	is identically equal to, rather than being equal only under some particular condition
$\triangleq$	is defined as
$\sim$	(superscript) indicates rms
$X\|_y$	variable X with the condition y or referring to y
$X\|_{y1}^{y2}$	variable X with y between $y1$ and $y2$
$\angle x$	angle or phase of x
$\thickapprox$	low-pass filter
≋	band-pass filter
acceptance band	band of frequencies beyond the passband where rejection is not required; used to indicate the region between the passband and a rejection band
contaminant	undesired RF power
passband	band of frequencies that pass through a filter with minimal attenuation or with less than a specified attenuation

rejection band	band of frequencies that are rejected or receive a specified attenuation (rejection)
sideband	signal in relation to a larger signal

Generic Symbols (applied to other symbols)

$*$	complex conjugate
$\|x\|$	magnitude or absolute value of x
$\breve{x}$	x is an equivalent noise factor or gain that can be used in standard equations to represent cascades with extreme mismatches (see Section 3.10.4)

Particular Symbols

A	voltage gain in dB. Note that G can as well be used if impedances are the same or the voltage is normalized to R_0.
a	voltage transfer ratio.
$\|a\|$	voltage gain (not in dB)
AM	amplitude modulation
a_n	nth-order transfer coefficient [see Eq. (4.1)]
a_{RT}	round-trip voltage transfer ratio
B	noise bandwidth
B_r	RF bandwidth
B_v	video, or postdetection, bandwidth
BW	bandwidth
$\mathrm{c}(n, j)$	jth binomial coefficient for $(a+b)^n$ (Abromowitz and Stegun, 1964, p. 10)
cas	subscript referring to cascade
CATV	cable television
cbl	subscript referring to cable
CSO	composite second-order distortion (Section 5.2)
CTB	composite triple-beat distortion (Section 5.2)
dB	decibels
DBM	doubly balanced mixer
dBm	decibels referenced to 1 mW
dBc	decibels referenced to carrier
dBV	decibels referenced to 1 V
dBW	decibels referenced to 1 W
e	voltage from an internal generator
F	noise figure, $F = 10\ \mathrm{dB}\log_{10} f$ or fundamental (as opposed to harmonic or IM).
f	noise factor (not in dB) or standard noise factor (measured with standard impedances) or frequency
$\hat{f}$	theoretical noise factor (measured with specified driving impedance) (see Sections 3.1, N.1)

FDM	frequency division multiplex
f_c	center frequency
f_{osc}	oscillator center frequency
f_I or f_{IF}	intermediate frequency, frequency at a mixer's output
f_L or f_{LO}	local oscillator frequency
FM	frequency modulation
f_m	modulation frequency
f_R or f_{RF}	radio frequency, the frequency at a mixer's input
G	power gain, sometimes gain in general, in dB.
g_k	power gain of module k, sometimes gain in general, not in dB.
g_{pk}	power gain preceding module k
H	subscript referring to harmonic
I, IF	intermediate frequency, the result of converting RF using a local oscillator
i	subscript indicating a signal traveling in the direction of the system input
IF	intermediate frequency, frequency at a mixer's output
IIP	input intercept point (IP referred to input levels)
IM	intermodulation product (intermod)
IMn	nth-order intermod or IM for module n
in	subscript indicating a signal entering a module (1) at the port of concern or (2) at the input port
int(x)	integer part of x
IP	intercept point
IPn	intercept point for nth-order nonlinearity or for module n
ISFDR	instantaneous spur-free dynamic range (see Section 5.3)
$\overline{k}$	Boltzmann's constant
$\overline{k}T_0$	approximately 4×10^{-21} W/Hz
L	single-sideband relative power density
L, LO	local oscillator, the generally relatively high-powered, controllable, frequency in a frequency conversion or the oscillator that provides it
L_φ	single-sideband relative power density due to phase noise
$\mathbf{M}$	a matrix (bold format indicates a vector or matrix)
m	modulation index (see Section 8.1)
$\tilde{m}$	rms phase deviation in radians
ma	subscript for "maximum available"
MAX$\{a, b\}$	the larger of a or b
$m \times n$	m refers to the exponent of the LO voltage and n refers to the exponent of the RF voltage in the expression for a spurious product; if written, for example, 3×4, m is 3 and n is 4
N_0	noise power spectral density
N_T	available thermal noise power spectral density at 290 K, $\overline{k}T_0$
o	subscript indicating a signal traveling in the direction of the system output.

OIP	output intercept point (IP referred to output levels)
out	subscript indicating a signal exiting a module (1) at the port of concern or (2) at the output port
P	power in dB.
p	power (not in dB).
$p_{\text{avail},j}$	available power at interface j (preceding module j)
PM	phase modulation
$p_{\text{out},j}$	output power at interface j (preceding module j)
PPSD	phase power spectral density
PSD	power spectral density
R, RF	radio frequency, the frequency at a mixer's input
R_0	agreed-upon interface impedance, a standard impedance (e.g., 50 Ω); characteristic impedance of a transmission line
RT	subscript for "round trip"
S	power spectral density or S parameter (see Section 2.2.1)
$\hat{S}$	sensitivity (see Section 2.5)
S_{ijk}	S parameter of row i and column j in the parameter matrix for module (or element) number k
SF	shape factor, ratio of bandwidth where an attenuation is specified to passband width
SFDR	spur-free dynamic range (see Section 5.3.1)
S/N	signal-to-noise power ratio
SSB	single-sideband; refers to a single signal in relation to a larger signal
SWR	standing wave ratio (see Section F.2)
T	absolute temperature or subscript referring to conditions during test
T_0	temperature of 290 K (16.85°C)
T_{ijk}	T parameter (see Section 2.2.3) of row i and column j in the parameter matrix for module (or element) number k
T_k	noise temperature of module k (see Section 3.2)
UUT	unit under test
$\mathbf{V}$	a vector (bold format indicates a vector or matrix)
v	normalized wave voltage (see Section 2.2.2) or voltage (not in dB.)
V	voltage in dB
$\hat{v}$	phasor representing the wave voltage (see Section 2.2.2)
$\tilde{v}$	phasor whose magnitude is the rms value of the voltage $\tilde{v} = \hat{v}/\sqrt{2}$ (see Section 2.2.2)
v_i, v_{in}, v_o, v_{out}	see Fig. 2.2 and Section 2.2.1
$\Delta_{\pm}$	maximum $\pm$ deviation in dB of cable gain A_{cbl}, from the mean
Δf	peak frequency deviation or frequency offset from spectral center
ρ	reflection coefficient (see Section F.2)
σ	standard deviation

σ^2	variance
τ	voltage transfer ratio of a matched cable (i.e., no reflections at the ends)
$\varphi(t)$	$\omega t + \theta$

PRACTICAL RF SYSTEM DESIGN

CHAPTER 1

INTRODUCTION

This book is about systems that operate at radio frequencies (RF) (including microwaves) where high-frequency techniques, such as impedance matching, are important. It covers the interactions of the RF modules between the antenna output and the signal processors. Its goal is to provide an understanding of how their characteristics combine to determine system performance. This chapter is a general discussion of topics in the book and of the system design process.

1.1 SYSTEM DESIGN PROCESS

We do system design by conceptualizing a set of functional blocks, and their specifications, that will interact in a manner that produces the required system performance. To do this successfully, we require imagination and an understanding of the costs of achieving the various specifications. Of course, we also must understand how the characteristics of the individual blocks affect the performance of the system. This is essentially analysis, analysis at the block level. By this process, we can combine existing blocks with new blocks, using the specifications of the former and creating specifications for the latter in a manner that will achieve the system requirements.

The specifications for a block generally consist of the parameter values we would like it to have plus allowed variations, that is, tolerances. We would like the tolerances to be zero, but that is not feasible so we accept values that are compromises between costs and resulting degradations in system performance. Not until modules have been developed and measured do we know their parameters to a high degree of accuracy (at least for one copy). At that point we might insert the module parameters into a sophisticated simulation program to compute the expected cascade performance (or perhaps just hook them together to see how the cascade works). But it is important in the design process to ascertain the range of performance to be expected from the cascade, given its module specifications. We need this ability so we can write the specifications.

Spreadsheets are used extensively in this book because they can be helpful in improving our understanding, which is our main objective, while also providing tools to aid in the application of that understanding.

1.2 ORGANIZATION OF THE BOOK

It is common practice to list the modules of an RF system on a spreadsheet, along with their gains, noise figures, and intercept points, and to design into that spreadsheet the capability of computing parameters of the cascade from these module parameters. The spreadsheet then serves as a plan for the system. The next three chapters are devoted to that process, one chapter for each of these parameter.

At first it may seem that overall gain can be easily computed from individual gains, but the usual imperfect impedance matches complicate the process. In Chapter 2, we discover how to account for these imperfections, either exactly or, in most cases, by finding the range of system gains that will result from the range of module parameters permitted by their specifications.

The method for computing system noise figure from module noise figures is well known to many RF engineers but some subtleties are not. Ideally, we use noise figure values that were obtained under the same interface conditions as seen in the system. Practically, that information is not generally available, especially at the design concept phase. In Chapter 3, we consider how to use the information that is available to determine system noise figure and what variations are to be expected. We also consider how the effective noise figures of mixers are increased by image noise. Later we will study how the local oscillator (LO) can contribute to the mixer's noise figure.

The concept of intercept points, how to use intercept points to compute intermodulation products, and how to obtain cascade intercept points from those of the modules will be studied in Chapter 4. Anomalous intermods that do not follow the usual rules are also described.

The combined effects of noise and intermodulation products are considered in Chapter 5. One result is the concept of spur-free dynamic range. Another is the portrayal of noise distributions resulting from the intermodulation of bands of noise. The similarity between noise bands and bands of signals both aids the analysis and provides practical applications for it.

Having established the means for computing parameters for cascades of modules connected in series, in Chapter 6 we take a brief journey through various means of connecting modules or components in parallel. We discover the advantages that these various methods provide in suppressing spurious outputs and how their overall parameters are related to the parameters of the individual components.

Then, in Chapter 7, we consider the method for design of frequency converters that uses graphs to give an immediate picture of the spurs and their relationships to the desired signal bands, allowing us to visualize problems and solutions. We also learn how to predict spurious levels and those, along with the relationships between the spurs and the passbands, permit us to ascertain filter requirements.

The processes described in the initial chapters are linear, or almost so, except for the frequency translation inherent in frequency conversion. Some processes, however, are severely nonlinear and, while performance is typically characterized for the one signal that is supposed to be present, we need a method to determine what happens when small, contaminating, signals accompany that desired signal. This is considered in Chapter 8. The most important nonlinearity in many applications is that associated with the mixer's LO; so we emphasize the system effects of contaminants on the LO.

Lastly, in Chapter 9, we will study phase noise: where it comes from, how it passes through a system, and what are its important effects in the RF system.

1.3 APPENDIXES

Material that is not essential to the flow of the main text, but that is nevertheless important, has been organized in 17 appendixes. These are designated by letters, and an attempt has been made to choose a letter that could be associated with the content (e.g., G for gain, M for matrix) as an aid to recalling the location of the material. Some appendixes are tutorial, providing a reference for those who are unfamiliar with certain background material, or who may need their memory refreshed, without holding up other readers. Some appendixes expand upon the material in the chapters, sometimes providing more detailed explanations or backup. Others extend the material.

1.4 SPREADSHEETS

The spreadsheets were created in Microsoft® Excel and can be downloaded as Microsoft Excel 97/98 workbook files (see page xix). This makes them available for the readers' own use and also presents an opportunity for better understanding. One can study the equations being used and view the charts, which appear in black and white in the text, in color on the computer screen. One can also make use of Excel's Trace Precedents feature (see, e.g., Fig. 3.5) to illustrate the composition of various equations.

1.5 TEST AND SIMULATION

Ultimately, we know how a system performs by observing it in operation. We could also observe the results of an accurate simulation, that being one that produces the same results as the system. Under some conditions, it may be easier, quicker, or more economical to simulate a system than to build and test it. Even though the proof of the simulation model is its correspondence to the system, it can be valuable as an initial estimate of the system to be improved as test data becomes available. Once confidence is established, there may be advantages in using the model to estimate system performance under various conditions or to predict the effect of modifications. But modeling and simulating is basically the same as building and testing. They are the means by which system performance is verified. First there must be a system and, before that, a system design.

In the early stages of system design we use a general knowledge of the performance available from various system components. As the design progresses, we get more specific and begin to use the characteristics of particular realizations of the component blocks. We may initially have to estimate certain performance characteristics, possibly based on an understanding of theoretical or typical connections between certain specifications. As the design progresses we will want assurance of important parameter values, and we might ultimately test a number of components of a given type to ascertain the repeatability of characteristics. Finally we will specify the performance required from the system's component blocks to ensure the system meets its performance requirements.

Based on information concerning the likelihood of deviations from desired performance provided by our system design analysis, we may be led to accept a small but nonzero probability of performance outside of the desired bounds. Once the system has been built and tested, it may be possible to use an accurate simulation to show that the results achieved, even with expected component variations, are better than the worst case implied by the combination of the individual block specifications. To base expected performance on simulated or measured results, rather than on functional block specifications, however, requires that we have continuing control over the construction details of the components of various copies of the system, rather than merely ensuring that the blocks meet their specifications. For example, a particular amplifier design may produce a stable phase shift that has a fortuitous effect on system performance, but we would have to control changes in its design and in that of interacting components.

Another important aspect of test is general experimentation, not confined to a particular design, for the purpose of verifying the degree of applicability of theory to various practical components. Examples of reports giving such supporting experimental data can be seen in Egan (2000), relative to the theory in Chapter 8, and in Henderson (1993a), relative to Chapter 7. We can hope that these, and the other, chapters will suggest opportunities for additional worthwhile papers.

1.6 PRACTICAL SKEPTICISM

There is a tendency for engineering students to assume that anything written in a book is accurate. This comes naturally from our struggle just to approach the knowledge of the authors whose books we study (and to be able to show this on exams). With enough experience in using published information, however, we are likely to develop some skepticism, especially if we should spend many hours pursuing a development based on an erroneous parameter value or perhaps on a concept that applies almost universally — but not in our case. Even reviewed journals, which we might expect to be most nearly free of errors, and classic works contain sources of such problems. But the technical literature also contains valuable, even essential, information; so a healthy skepticism is one that leads us to consult it freely and extensively but to continually check what we learn. We check for accuracy in our reference sources, for accuracy in our use of the information, and to ensure that it truly applies to our development. We check by considering how concepts correlate with each other (e.g., does this make sense in terms of what I already know), by verifying agreement between answers obtained by different methods, and by testing as we proceed in our developments. The greater the cost of failure, the more important is verification. Unexpected results can be opportunities to increase our knowledge, but we do not want to pay too high a price for the educational experience.

1.7 REFERENCES

References are included for several reasons: to recognize the sources, to offer alternate presentations of the material, or to provide sources for associated topics that are beyond the scope of this work. The author–date style of referencing is used throughout the book. From these, one can easily find the complete reference descriptions in the References at the end of the text. Some notes are placed at the end of the chapter in which they are referenced.

CHAPTER 2

GAIN

In this chapter, we determine the effect of impedance mismatches (reflections) on system gain. For a simple cascade of linear modules (Fig. 2.1), we could write the overall transfer function or ratio as

$$g = g_1 g_2 \cdots g_N, \tag{2.1}$$

where

$$g_j \stackrel{\Delta}{=} \frac{u_{j+1}}{u_j} \tag{2.2}$$

and u is voltage or current or power. The gain is $|g|$, which is the same as g if u is power. This would require that we measure the values of u in the cascade. If we measure them in some other environment, we could get different gains because of differing impedances at the interfaces. However, it may be difficult to measure u in the cascade, and a gain that must be measured in the final cascade has limited value in predicting or specifying performance. For example, a variation of about ± 1 dB in overall gain can occur for each interface where the standing-wave ratios (SWRs) are 2 and a change as high as 2.5 dB can occur when they are 3. (See Appendix F.1 for a discussion of decibels (dB).)

Here we consider how the expected gain of a cascade of linear modules can be determined, as well as variations in its gain, based on measured or specified parameters of the individual modules. Throughout this book, gains and other parameters are so generally functions of frequency that the functionality is not shown explicitly. Equations whose frequency dependence is not indicated will apply at any given frequency.

We begin with a description, for modules and their cascades, that applies without limitations but which requires detailed knowledge of impedances and

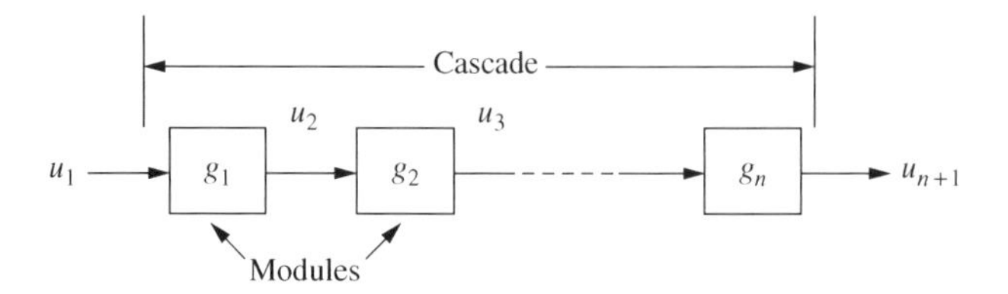

Fig. 2.1 Transfer functions in a simple cascade.

which can be complicated to use. Then we will discover a way to simplify the description of the overall cascade by taking into account special characteristics of some of its parts. This will lead us to a standard cascade, composed of unilateral modules separated by interconnects (e.g., cables) that have well-controlled impedances. The unilateral modules, usually active, have negligible reverse transmission. The passive cables are well matched at the standard impedance (e.g., 50 Ω) of the cascade interfaces; these are the impedances used in characterizing the modules.

It is common to specify the desired performance of each module plus allowed variations from that ideal. The desired performance includes a gain and standard interface impedances. The allowed variations are given by a gain tolerance and the required degree of input and output impedance matches, expressed as maximum SWRs or, equivalently, return losses or reflection coefficient magnitudes (see Appendix F.2). These are the parameters required for determination of the performance of the standard cascade. We will also find ways to fit bilateral modules into this scheme.

We will also consider the case where the modules are specified in terms of their performance with various nonstandard interface impedances (e.g., $2000\ \Omega - j500\ \Omega$), and we will discover how to characterize cascades of these modules. For cases where it may be desirable to include these nonstandard cascades as parts of a standard cascade, we will determine how to describe them in those terms.

Finally, we will study the use of sensitivities in analyzing cascade performance.

Many varieties of power gains are described in Appendix G. If all interfaces were at standard impedance levels (e.g., 50 Ω everywhere), these gains would all be the same, but the usually unintended mismatches lead to differing values for gain, depending on the definitions employed.

2.1 SIMPLE CASES

In some cases these complexities are unimportant. For example, where operational amplifiers (op amps) are used at lower frequencies, measurements of voltages at interfaces can be practical and their low output impedances and high input impedances allow performance in the *voltage-amplifier cascade* to duplicate what was measured during test. However, this luxury is rare at radio frequencies.

In other cases, complexities may be ignored in an effort to get an answer with minimum effort or with the available information. That answer may be adequate for the task at hand; at least it is better than no estimate. Commonly, we simply assume that gains will be the same as when a module or interconnect was tested in a standard-impedance environment. We try to make this so by keeping input and output impedances close to that standard impedance when designing or selecting modules.

While this simplified approach can be useful, we will consider here how to make use of additional information about modules to get a better estimate of cascade performance, one that includes the range of gain values to be expected.

2.2 GENERAL CASE

To characterize the modules so their performance in the system can be predicted, we need more parameters, a set of four (generally called two-port parameters; we are characterizing our modules as having two ports, an input port and an output port) for each module (Gonzalez, 1984, pp. 1–31; Pozar, 2001, pp. 47–55). We begin by considering the parameters that we can use to describe the modules.

2.2.1 *S* Parameters

Individual RF modules are usually defined by their S (scattering) parameters (Pozar, 2001, pp. 50–53; Gonzalez, 1984, pp. 9–10). This can be done with the help of the matrix (see Appendix M for help in using matrices),

$$\begin{bmatrix} v_{\mathrm{out},1} \\ v_{\mathrm{out},2} \end{bmatrix} = \begin{bmatrix} S_{11} & S_{12} \\ S_{21} & S_{22} \end{bmatrix}_1 \begin{bmatrix} v_{\mathrm{in},1} \\ v_{\mathrm{in},2} \end{bmatrix}. \tag{2.3}$$

The subscripts in and out refer to waves propagating[1] into and out of the module at either port (1 or 2). The other subscripts on the vector components indicate the input port 1 or output port 2, whereas the subscript on each matrix element is its row and column, respectively. Subscript 1 on the matrix indicates module 1. We use the same index for the module and for its input port (port 1 here).

We can also write the subscripts in terms of the system with i or o, referring to waves traveling toward the input or toward the output of the system, respectively. Refer to Fig. 2.2. With this notation, Eq. (2.3) becomes

$$\begin{bmatrix} v_{i1} \\ v_{o2} \end{bmatrix} = \begin{bmatrix} S_{11} & S_{12} \\ S_{21} & S_{22} \end{bmatrix}_1 \begin{bmatrix} v_{o1} \\ v_{i2} \end{bmatrix}. \tag{2.4}$$

More generally, for the jth module,

$$\begin{bmatrix} v_{i,j} \\ v_{o,j+1} \end{bmatrix} = \begin{bmatrix} S_{11} & S_{12} \\ S_{21} & S_{22} \end{bmatrix}_j \begin{bmatrix} v_{o,j} \\ v_{i,j+1} \end{bmatrix}. \tag{2.5}$$

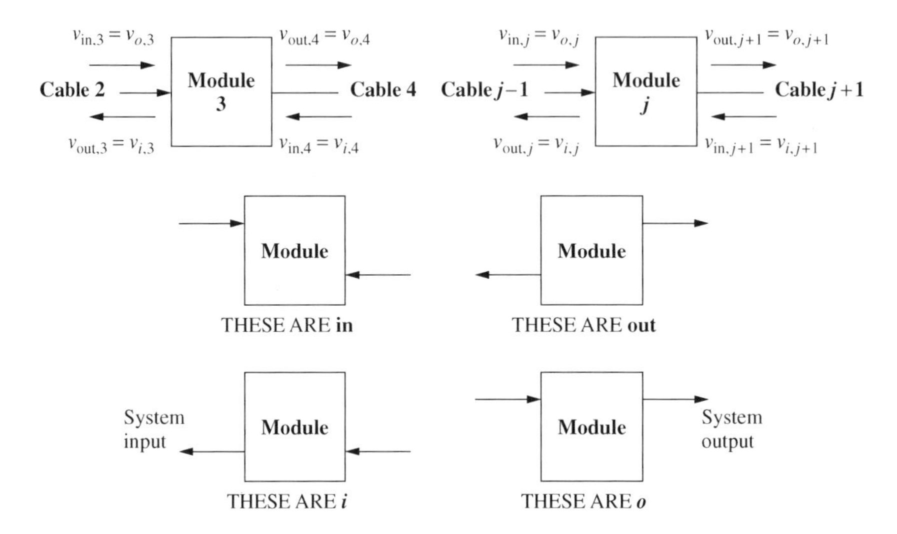

Fig. 2.2 Definitions of wave subscripts.

By normal matrix multiplication then,

$$v_{i,j} = S_{11j}v_{o,j} + S_{12j}v_{i,j+1} \tag{2.6}$$

and

$$v_{o,j+1} = S_{21j}v_{o,j} + S_{22j}v_{i,j+1}. \tag{2.7}$$

This is a convenient form for measurements. It relates signals coming "out" of the module, at either port, to those going "in" at either port. We can control the inputs, ensuring that there is only one by terminating the port to which we do not apply a signal, and measuring the two resulting outputs, one at each port (Fig. 2.3). These give us two of the four parameters and a second measurement, with input to the other port, gives the other two.

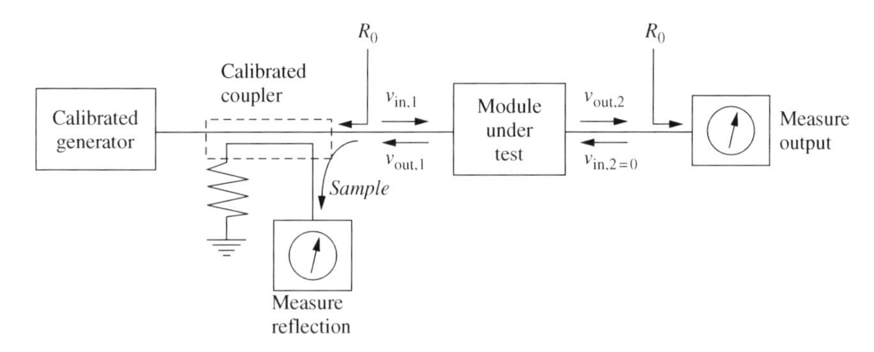

Fig. 2.3 Measurement setup.

Thus, for module 1, with port 2 terminated ($v_{\text{in},2} \equiv v_{i2} = 0$), we measure the reflected signal at port 1 to give the reflection coefficient for that port,

$$S_{11} = \frac{v_{\text{out},1}}{v_{\text{in},1}} \equiv \frac{v_{i1}}{v_{o1}} \tag{2.8}$$

and the transmission coefficient from port 1 to port 2,

$$S_{21} = \frac{v_{\text{out},2}}{v_{\text{in},1}} \equiv \frac{v_{o2}}{v_{o1}}. \tag{2.9}$$

Then we turn the module around and input to port 2 while terminating port 1, giving the reverse transmission coefficient and port 2 reflection coefficient, respectively:

$$S_{12} = \frac{v_{\text{out},1}}{v_{\text{in},2}} \equiv \frac{v_{i1}}{v_{i2}}, \tag{2.10}$$

$$S_{22} = \frac{v_{\text{out},2}}{v_{\text{in},2}} \equiv \frac{v_{o2}}{v_{i2}}. \tag{2.11}$$

(We are using both subscript forms here as an aid in understanding their equivalency.) In each case the S parameter subscripts represent the ports of effect and cause, respectively, $S_{\text{effect cause}}$, where "effect" is the port where "out" occurs and "cause" is the port where "in" occurs.

2.2.2 Normalized Waves

We have called v_x (i.e., v_o, v_i, v_{out}, or v_{in}) a "wave," but the symbol implies a voltage. It is customary to use normalized voltages with S parameters, and the usual way to normalize them is by division of the root-mean-square (rms) voltage by $\sqrt{R_0}$, where R_0 is the real part of the characteristic impedance Z_0 of the transmission line in which the waves reside. We will assume that Z_0 is real.[2] An RF voltage corresponding to v_x can be represented by

$$V_{mx}\cos(\omega t + \theta) = \text{Re}\, V_{mx} e^{j(\omega t + \theta)}. \tag{2.12}$$

This can be abbreviated

$$\hat{v}_x(t) = \hat{v}_x e^{j\omega t}, \tag{2.13}$$

where

$$\hat{v}_x = V_{mx} e^{j\theta}. \tag{2.14}$$

Sometimes a phasor is employed whose magnitude is the effective (rms) value (Hewlett-Packard, 1996; Yola, 1961; Kurokawa, 1965):

$$\tilde{v}_x = (V_{mx}/\sqrt{2}) e^{j\theta}. \tag{2.15}$$

Our normalized voltage,

$$v_x = \tilde{v}_x/\sqrt{R_0}, \tag{2.16}$$

uses this form, which has the advantage that the available power in the traveling wave can be expressed simply as

$$p_x = |v_x|^2. \tag{2.17}$$

Traditionally, the symbol a_n is used for $v_{\text{in},n}$ and b_n is used for $v_{\text{out},n}$.

If, on the other hand, the phasor employed in Eq. (2.16) is $\hat{v}_x$ rather than $\tilde{v}_x$ (Pozar, 1990, p. 229, 1998, p. 204), the power will be $|v_x|^2/2$. In most cases the module parameters are ratios of two waves at the same impedance; so it makes no difference whether they are ratios of v_x or of $\hat{v}_x$ or of $\tilde{v}_x$.

2.2.3 *T* Parameters

Unfortunately, we cannot use S matrices conveniently for determining overall response because we cannot multiply them together to produce anything useful. We require a matrix equation for overall transfer function of the form

$$\mathbf{V}_1 = \mathbf{M}\mathbf{V}_{n+1} = \mathbf{M}_1\mathbf{M}_2\mathbf{M}_3 \cdots \mathbf{M}_n\mathbf{V}_{n+1}. \tag{2.18}$$

Here the vector $\mathbf{V}_j$, representing a module input, has the same identifying number (subscript) as the matrix $\mathbf{M}_j$, representing the module. Note that we are operating on outputs to give inputs. This is nice in that the matrices are then written in the same order in which the modules are traditionally arrayed in a drawing (left to right from input to output, as in Fig. 2.1). There is also an even better reason. The vector on which the matrix operates (multiplies) must contain the information needed to produce the resulting product. Unilateral modules that have little or no reverse transmission do not provide significant information about the output to the input; thus a mathematical representation in which the matrix operated on that input would not work well. On the other hand, all modules of interest produce outputs that are functions of their inputs; so there is sufficient information in the vector representing the output to form the input.[3]

Equation (2.18) implies

$$\mathbf{V}_1 = \mathbf{M}_1\mathbf{V}_2 \tag{2.19}$$

and

$$\mathbf{V}_2 = \mathbf{M}_2\mathbf{V}_3 \tag{2.20}$$

in order that

$$\mathbf{V}_1 = \mathbf{M}_1(\mathbf{M}_2\mathbf{V}_3) = \mathbf{M}_1\mathbf{M}_2\mathbf{V}_3 \tag{2.21}$$

and so on. All this implies that $\mathbf{V}_2$ represents the state between modules 1 and 2 so we define the vector

$$\mathbf{V}_j = \begin{bmatrix} v_o \\ v_i \end{bmatrix}_j = \begin{bmatrix} v_{oj} \\ v_{ij} \end{bmatrix}, \tag{2.22}$$

where j represents the port and o and i indicate the voltage wave moving right toward the system output or left toward its input, respectively. Thus the matrix connecting such vectors has the form (Dechamps and Dyson, 1986; Gonzalez, 1984, pp. 11–12)

$$\begin{bmatrix} v_o \\ v_i \end{bmatrix}_1 = \begin{bmatrix} T_{11} & T_{12} \\ T_{21} & T_{22} \end{bmatrix}_1 \begin{bmatrix} v_o \\ v_i \end{bmatrix}_2 . \tag{2.23}$$

As before, the module and its input have the same subscript. In many cases it will be more convenient to move the subscript from the vector or matrix to its individual elements, adding the port number as the last subscript:

$$\begin{bmatrix} v_{o1} \\ v_{i1} \end{bmatrix} = \begin{bmatrix} T_{111} & T_{121} \\ T_{211} & T_{221} \end{bmatrix} \begin{bmatrix} v_{o2} \\ v_{i2} \end{bmatrix} . \tag{2.24}$$

Each vector, in this representation, describes two waves that occur at a single point in the system whereas, for the S parameters, the vector elements represented waves from different ports.[4] However, S-parameter measurements are simpler than T-parameter measurements. Consider that T_{121} is the ratio between a wave entering the module at port 1, v_{o1}, and one entering it at port 2, v_{i2}, while the wave leaving it at port 2, v_{o2}, is set to zero. To measure this directly, we would require two phase-coherent generators, one driving each port, that would be adjusted so the outputs due to each at port 2 would cancel.

2.2.4 Relationships Between *S* and *T* Parameters

It is simpler to measure the S parameters and obtain the T parameters from them. For example, T_{22} for module 1 is

$$T_{22} = \left. \frac{v_{i1}}{v_{i2}} \right|_{v_{o2}=0} . \tag{2.25}$$

Equation (2.7) indicates that the condition $v_{o2} = 0$ requires

$$S_{21} v_{o1} = -S_{22} v_{i2} . \tag{2.26}$$

Combining this with Eq. (2.6) we obtain

$$v_{i1} = -\frac{S_{11} S_{22}}{S_{21}} v_{i2} + S_{12} v_{i2} = \left[S_{12} - \frac{S_{11} S_{22}}{S_{21}} \right] v_{i2} \tag{2.27}$$

from which we obtain the T parameter in terms of S parameters,

$$T_{22} = S_{12} - \frac{S_{11} S_{22}}{S_{21}} . \tag{2.28}$$

By a similar process we can obtain the other values of T_{ij} in terms of the S_{ij}:

$$\begin{bmatrix} T_{11} & T_{12} \\ T_{21} & T_{22} \end{bmatrix} = \begin{bmatrix} \dfrac{1}{S_{21}} & -\dfrac{S_{22}}{S_{21}} \\ \dfrac{S_{11}}{S_{21}} & S_{12} - \dfrac{S_{11}S_{22}}{S_{21}} \end{bmatrix} \tag{2.29}$$

$$= \frac{1}{S_{21}} \begin{bmatrix} 1 & -S_{22} \\ S_{11} & S_{12}S_{21} - S_{11}S_{22} \end{bmatrix}, \tag{2.30}$$

and of S_{ij} in terms of T_{ij},

$$\begin{bmatrix} S_{11} & S_{12} \\ S_{21} & S_{22} \end{bmatrix} = \begin{bmatrix} \dfrac{T_{21}}{T_{11}} & T_{22} - \dfrac{T_{12}T_{21}}{T_{11}} \\ \dfrac{1}{T_{11}} & -\dfrac{T_{12}}{T_{11}} \end{bmatrix} \tag{2.31}$$

$$= \frac{1}{T_{11}} \begin{bmatrix} T_{21} & T_{11}T_{22} - T_{12}T_{21} \\ 1 & -T_{12} \end{bmatrix}. \tag{2.32}$$

2.2.5 Restrictions on *T* Parameters

We can now show more specifically why the T matrix was designed to give input as a function of output, rather than the converse. For unilateral gain in the forward direction, $S_{12} = 0$. This simplifies T_{22} in Eq. (2.30). On the other hand, unilateral gain in the reverse direction, $S_{21} = 0$, causes the elements in Eq. (2.30) to become infinite. As S_{21} approaches 0, $\mathbf{V}_2$ becomes a weak function of $\mathbf{V}_1$, so a large number is required to give $\mathbf{V}_1$ in terms of $\mathbf{V}_2$. Moreover, if forward transmission is small, v_{o2} may become a stronger function of v_{i2} than of v_{o1}, in which case $\mathbf{V}_1$ becomes dependent on the difference between the two components of $\mathbf{V}_2$ and subject to error due to small inaccuracies in $\mathbf{M}$. As a result, $\mathbf{M}$ should not represent a process where transmission from $\mathbf{V}_1$ to $\mathbf{V}_2$, as defined by Eq. (2.9), is small or zero. For this reason, Eq. (2.19) is written as it is, since transmission toward the system output S_{21} is a purpose of a system, and thus is expected to be appreciable, whereas reverse transmission S_{12} is often minimized.

2.2.6 Cascade Response

Now we can obtain the overall response of a series of modules (a cascade) by multiplying their individual T matrices. The sequence in which the matrices are arrayed must be the same as the sequence, from input to output, of the elements in the cascade and the interface (standard) impedances must be those in which the S or T parameters were measured. If the parameters of adjacent modules are defined for different standard impedances at the same interface, one of them must be recharacterized. This can be done by inserting a T matrix representing the impedance transition, as described in Appendix I.

The process can be aided by a mathematical program (e.g., MATLAB®), or perhaps done implicitly using a network analysis program, if we have values for all the parameters in all the modules. However, we will often not have values for all the parameters and, generally, when we do have such information, it will be in terms of ranges of parameters, maximums and minimums or expected distributions. We could estimate the distribution of all the parameters and do a Monte Carlo analysis, obtaining a distribution of solutions based on trials with various parameter values drawn according to their distributions. Both the complexity of such a process and the desire for a better understanding of the results suggest that simpler methods are desirable.

2.3 SIMPLIFICATION: UNILATERAL MODULES

In general, the reflection at any module input port in a cascade depends on the part of the cascade that follows. Looking into a given module, we see an impedance that is affected by every following module. That is why we must multiply T matrices.

When a module has zero reverse transmission ($S_{12j} = 0$), Eq. (2.6) shows that the forward and reverse waves at the input port are related just by the module parameter S_{11j}. Nothing that occurs at the output port can influence this relationship so the reflection at the input port is independent of the impedance seen at the module output. This greatly simplifies the determination of the reflection at the input port, making it dependent on the parameters of just that one module. Similarly, since the reverse wave at the module output does not influence the input, the output reflection is independent of the parameters of preceding modules.

As a result, if the modules are unilateral, the gain of the cascade can be determined from the parameters of the individual modules, rather than by matrix multiplication. Therefore, it is important to consider what kinds of modules (or combinations of modules) can be treated as unilateral and, then, how cascades of unilateral modules can be analyzed.

Some modules tend to be unilateral, to transmit information from input to output but not in the reverse direction, or only weakly in the reverse direction. Complex modules [e.g., frequency converters, modules with digital signal processing (DSP) between input and output] often fit this category. Even amplifiers, if they are unconditionally stable, have

$$|S_{21}S_{12}| < 1; \tag{2.33}$$

so, when they are well terminated, the reverse transmission is small.

2.3.1 Module Gain

For module gain we will use the commonly specified transducer power gain (Appendix G) with given interface impedances (usually 50 Ω for RF). This is the ratio of output power into the nominal load resistance to the power available from a source that has nominal input resistance. It differs from available gain, for which the load would be the conjugate of the actual module output impedance rather than a standardized nominal resistance.

In testing a module with index j, the output power can be read from a power meter or spectrum analyzer, one with impedance equal to the nominal impedance of the output port, R_L. It is related to the forward output voltage during the test $v_{o,j+1,T}$ by

$$p_{\text{out},j+1} = |v_{o,j+1,T}|^2 = |\tilde{v}_{o,j+1,T}|^2/R_L. \tag{2.34}$$

The input power can be read from a signal generator that is, as is usual, calibrated in terms of its available power. It is related to the forward input voltage v_{oj} by

$$p_{\text{avail},j} = |v_{o,j}|^2 = |\tilde{v}_{o,j}|^2/R_S, \tag{2.35}$$

where R_S is the source resistance. Therefore, the transducer power gain given for module j is

$$g_j = \left|\frac{v_{o,j+1,T}}{v_{oj}}\right|^2 = \left|\frac{v_{o,j+1}}{v_{oj}}\right|^2_{v_{i,j+1}=0} = |S_{21j}|^2 \tag{2.36}$$

$$= \left|\frac{\hat{v}_{o,j+1,T}}{\hat{v}_{oj}}\right|^2 \frac{R_s}{R_L} = \left|\frac{\hat{v}_{o,j+1}}{\hat{v}_{oj}}\right|^2_{\hat{v}_{i,j+1}=0} \frac{R_s}{R_L}. \tag{2.37}$$

Note that $v_{o,j+1,T}$ is equivalent to $v_{o,j+1}$ with $v_{i,j+1} = 0$ because the module is tested with a load that equals the impedance of the interconnect and of the device in which the waves are measured so there is no measured reflection during test.

Usually $R_s = R_L$ and the last resistor ratio disappears. In any case, $|S_{21}|$ can be related to the transducer power gain by Eq. (2.36).

The variables that form the ratio g_j during the test must also be those to which g_j refers in the cascade. These are the wave induced by the module in its output cable (excluding any wave reflected from the output of the module) and the forward wave impinging on the module input.

2.3.2 Transmission Line Interconnections

Now we determine the gain of a cascade of unilateral elements interconnected by cables (transmission lines) whose characteristic impedances are the same as those used in characterizing the modules. We will call this a *standard cascade*. Because they are unilateral, we look at each pair of interconnected modules as a source and a load with all interaction between them being independent of anything that precedes the source (excepting its driving voltage) or follows the load (Fig. 2.4). We require a means to account for the effects of mismatches at the source output and the load input on the performance of the combined pair. Direct connection of the modules is a degenerate case where the cable length goes to zero.

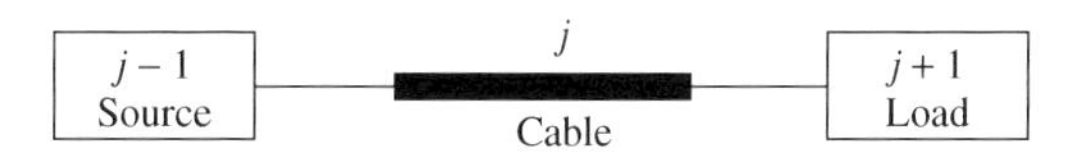

Fig. 2.4 Source and load connected.

Since we use the variables v_{ojT} and $v_{o,j+1}$ in defining the source $(j-1)$ and load $(j+1)$ module gains, respectively, the gain of cable j that connects them must be the ratio of $v_{o,j+1}$ to v_{ojT}. Then we will be able to write a cascade voltage transfer function as

$$a_{\text{cas}} = a_{m1}a_{\text{cbl},2}a_{m3}a_{\text{cbl},4}\cdots a_{mN}, \tag{2.38}$$

where the first subscript indicates module m or cable, cbl,

$$a_{mj} = \frac{v_{o,j+1,T}}{v_{o,j}} \tag{2.39}$$

and

$$a_{\text{cbl},j} = \frac{v_{o,j+1}}{v_{ojT}}. \tag{2.40}$$

Then the overall transfer function will be

$$a_{\text{cas}} = \frac{v_{o2T}}{v_{o1}}\frac{v_{o3}}{v_{o2T}}\frac{v_{o4T}}{v_{o3}}\frac{v_{o5}}{v_{o4T}}\cdots\frac{v_{o,N+1,T}}{v_{o,N}} = \frac{v_{o,N+1,T}}{v_{o1}}. \tag{2.41}$$

We assume for now that the final module drives a matched load so $v_{o,N+1,T} = v_{o,N+1}$ and $a_{\text{cas}} = v_{o,N+1}/v_{o1}$, as desired. (Other cases will also be handled.)

When the source is tested, it sends a forward wave v_{ojT} into a cable and load that have nominal real impedances (Fig. 2.5). This produces, at the test cable output,

$$v_{o,j+1,T} = \tau v_{ojT}, \tag{2.42}$$

where the factor τ is the voltage transfer ratio representing the time delay and attenuation in the cable.

During test, the output $v_{o,j+1,T}$ is absorbed in, and measured at, the load. In the cascade, the value of the forward wave $v_{o,j+1}$ is the value that appears during test $(v_{o,j+1,T})$ plus waves reflected in sequence from the load $(S_{11,j+1})$

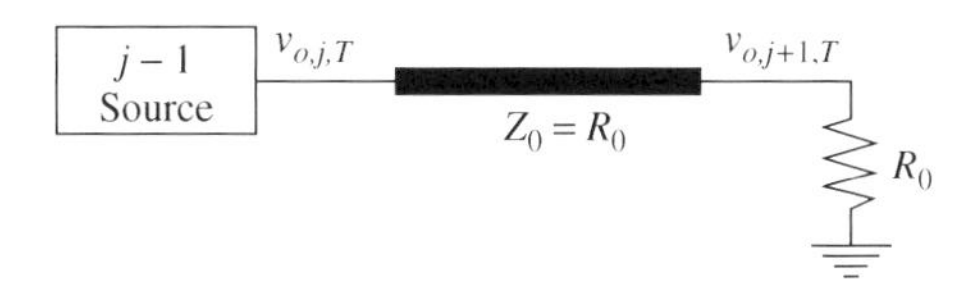

Fig. 2.5 Forward wave from source.

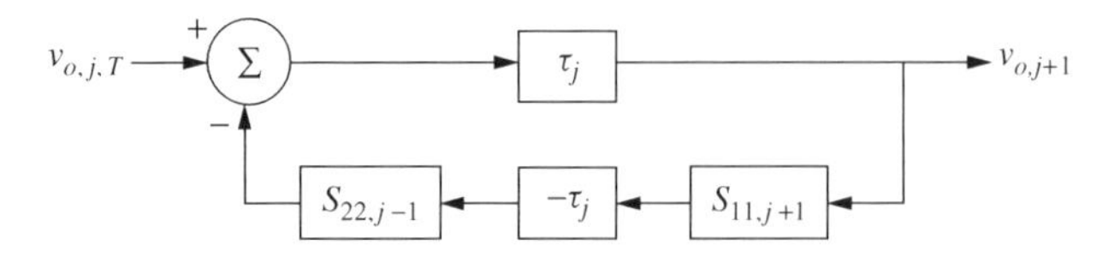

Fig. 2.6 Multiple reflections in cascade.

and the source ($S_{22,j-1}$). Refer to Fig. 2.6. We must determine the value of that net forward wave $v_{o,j+1}$ since this is what drives the load module $j+1$ and determines the output from that module. The load module will respond as if it were sent a signal $v_{o,j+1}$ from a matched source during test.

The primary state variables in the standard cascade are:

- The forward wave at the output of each interconnect
- The induced wave at the input of each interconnect

The latter would be the forward wave at the input if the interconnect were properly terminated at its output. Otherwise, however, the forward wave also includes double reflections from the input of the driven module and the output of the driving module.

The ratio $a_{\text{cbl},j}$ of the closed-loop output in Fig. 2.6 to the forward wave that drives its input during test (when there is no reflected wave in the cable) we call the cable gain. It is given by the normal equation for closed-loop transfer function:

$$a_{\text{cbl},j} = \frac{v_{o,j+1}}{v_{ojT}} = \frac{\tau_j}{1 - S_{22,j-1}S_{11,j+1}\tau_j^2}, \tag{2.43}$$

where

$$\tau_j = \exp(h - jb), \tag{2.44}$$

where $-h = \alpha d$ is loss in nepers[5] and $b = \beta d$ is the phase lag in the cable of length d. A minus has been used in the feedback path to cancel the minus at the summer of the customary feedback configuration.

The corresponding gain in forward power (or squared voltage if the input and output impedances differ) is

$$g_{\text{cbl},j} = |a_{\text{cbl},j}|^2 \tag{2.45}$$

$$= \frac{|\tau_j|^2}{(1 - S_{22,j-1}S_{11,j+1}\tau_j^2)[1 - S_{22,j-1}^* S_{11,j+1}^* (\tau_j^2)^*]} \tag{2.46}$$

$$= \frac{|\tau_j|^2}{1 - 2|S_{22,j-1}S_{11,j+1}\tau_j^2|\cos\theta + |S_{22,j-1}S_{11,j+1}\tau_j^2|^2} \tag{2.47}$$

$$= \frac{1}{e^{-2h} - 2|S_{22,j-1}||S_{11,j+1}|\cos\theta + |S_{22,j-1}|^2|S_{11,j+1}|^2 e^{2h}}, \tag{2.48}$$

where

$$\theta = -2b + \varphi_{j-1} + \varphi_{j+1}, \tag{2.49}$$

$$\varphi_{j-1} = \angle S_{22,j-1}, \tag{2.50}$$

and

$$\varphi_{j+1} = \angle S_{11,j+1}. \tag{2.51}$$

We can see here that, if the attenuation is high ($h \ll 1$), the power gain is just the interconnect loss, e^{2h}.

We define the round-trip, or open-loop, voltage gain,

$$|a_{\text{RT}j}| \triangleq |\tau_j|^2 |S_{22,j-1}||S_{11,j+1}| \tag{2.52}$$

$$= |\tau_j|^2 \frac{\text{SWR}_j - 1}{\text{SWR}_j + 1} \frac{\text{SWR}_{j+1} - 1}{\text{SWR}_{j+1} + 1}, \tag{2.53}$$

where $|\tau_j| = \exp(h_j)$ and SWR_j and SWR_{j+1} are standing-wave ratios associated with the reflections. We have given the SWR a subscript corresponding to the interface where it occurs (as we do for the voltage vector there). We can do this because the cable is assumed to have SWR $= 1$ so only the module's SWR requires a value at each interface.

Using Eq. (2.52), we can write Eq. (2.47) as

$$g_{\text{cbl},j} = \frac{|\tau_j|^2}{1 - 2|a_{\text{RT}j}|\cos\theta + |a_{\text{RT}j}|^2}. \tag{2.54}$$

2.3.2.1 *Effective Power Gain* We now compute the mean and peak values of the gain in forward power (the square of the voltage magnitude if impedances differ), in the cascade relative to that in test, over all values of θ. These can be considered to be the values expected over a random distribution of phases of the reflections or the values that will be seen as frequency changes in a cable that is many wavelengths long (thus changing the phase shift through the cable). From Eq. (2.54) (dropping the subscript j for simplicity), the minimum and maximum gains in the cable are

$$|a_{\text{cbl}}|_{\max} = \frac{|\tau|}{\sqrt{1 - 2|a_{\text{RT}j}| + |a_{\text{RT}j}|^2}} = \frac{|\tau|}{1 - |a_{\text{RT}}|} \tag{2.55}$$

and

$$|a_{\text{cbl}}|_{\min} = \frac{|\tau|}{\sqrt{1 + 2|a_{\text{RT}j}| + |a_{\text{RT}j}|^2}} = \frac{|\tau|}{1 + |a_{\text{RT}}|}. \tag{2.56}$$

The average gain as the frequency varies is the same as the average as θ varies since Eq. (2.49) can be written

$$\theta = \varphi_{j-1} + \varphi_{j+1} - 2\omega d/v, \tag{2.57}$$

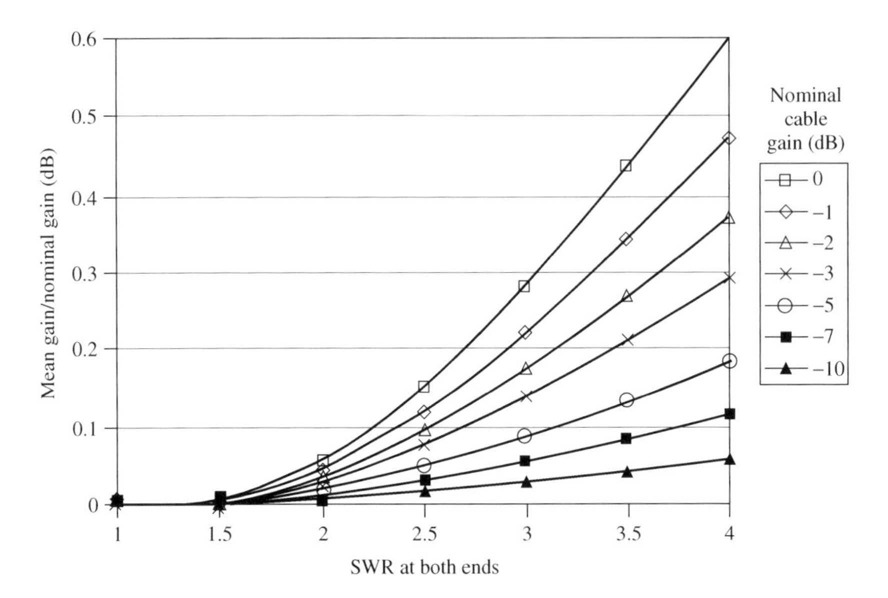

Fig. 2.7 Excess of mean cable gain over nominal cable gain due to reflections.

where v is the velocity in the cable and d is its length. This average is obtained from

$$\overline{g_{\text{cbl}}} = \frac{|\tau|^2}{2\pi}\int_0^{2\pi} \frac{d\theta}{1 - 2|a_{\text{RT}}|\cos\theta + |a_{\text{RT}}|^2} \tag{2.58}$$

$$= \frac{|\tau|^2}{1 - |a_{\text{RT}}|^2}. \tag{2.59}$$

This indicates that the average cable loss is reduced by the reflections. The relationship is plotted in Fig. 2.7. From this we can see that the mean cable gain differs little from the nominal value, $|\tau|^2$, in many practical cases.

It is apparent, from Eqs. (2.59), (2.55), and (2.56), that the average value of power gain is the geometric mean of the maximum and the minimum,

$$\overline{g_{\text{cbl},j}} = |a_{\text{cbl}}|_{\text{max}}|a_{\text{cbl}}|_{\text{min}}, \tag{2.60}$$

and it follows that, in dB, it is the arithmetic mean,

$$\overline{G_{\text{cbl}}} = \frac{G_{\text{cbl,max}} + G_{\text{cbl,min}}}{2}. \tag{2.61}$$

The maximum deviation from the mean is, in dB,

$$\Delta_+ \triangleq G_{\text{cbl,max}} - \overline{G_{\text{cbl}}} \tag{2.62}$$

$$= 10 \text{ dB} \log_{10} \frac{|\tau|}{1 - |a_{\text{RT}}|} - 10 \text{ dB} \log_{10} \frac{|\tau|}{1 + |a_{\text{RT}}|} \tag{2.63}$$

$$= 10 \text{ dB} \log_{10} \left(\frac{1 + |a_{\text{RT}}|}{1 - |a_{\text{RT}}} \right). \tag{2.64}$$

It is also quickly apparent that $\Delta_+ = -\Delta_-$. That is, the deviation from mean, in dB, at the maximum, is the same as at the minimum.

Since $\log_{10}(x) = 0.434 \ln(x)$ and $\ln[(1 + |a_{\text{RT}}|)/(1 - |a_{\text{RT}}|)] = 2[|a_{\text{RT}}| + |a_{\text{RT}}|^3/3 + |a_{\text{RT}}|^5/5 + \ldots]$,

$$\Delta_+ \approx 8.7 \text{ dB } |a_{\text{RT}}| \quad \text{for} \quad |a_{\text{RT}}| \ll 1. \tag{2.65}$$

Example 2.1 Cable Gain Find the minimum, maximum, and mean cable gains for a cable that has a loss of 2 dB in a matched environment (its nominal loss) but is operating with a SWR of 2 looking into the driving module and a SWR of 3 looking into the load.

We obtain the magnitude of the voltage transfer ratio for the matched cable,

$$|\tau| = 10^{(-2 \text{ dB}/20 \text{ dB})} = 0.7943. \tag{2.66}$$

The round-trip voltage gain, from Eq. (2.53), is

$$|a_{\text{RT}}| = (0.7943)^2 \frac{2-1}{2+1} \frac{3-1}{3+1} = 0.631 \times \frac{1}{3} \times \frac{1}{2} = 0.1052. \tag{2.67}$$

From Eqs. (2.55) and (2.56) the extremes of the cable voltage gain are

$$|a_{\text{cbl}}|_{\max} = \frac{0.7943}{1 - 0.1052} = 0.8876 \Rightarrow -1.035 \text{ dB} \tag{2.68}$$

and

$$|a_{\text{cbl}}|_{\min} = \frac{0.7943}{1 + 0.1052} = 0.7187 \Rightarrow -2.869 \text{ dB}. \tag{2.69}$$

The mean power gain is obtained from Eq. (2.59) as

$$\overline{g_{\text{cbl}}} = \frac{0.7943^2}{1 - 0.105^2} = 0.6380 \Rightarrow -1.952 \text{ dB}, \tag{2.70}$$

which is also the average of the maximum and minimum gains in dB, Eqs. (2.68) and (2.69).

Alternatively, we can find the values in Eqs. (2.68) and (2.69) approximately using Eq. (2.65). The deviation of the maximum and minimum gains in dB from their mean is

$$\Delta \approx 8.7 \text{ dB} \times 0.1052 = 0.915 \text{ dB}. \tag{2.71}$$

This approximation along with Eq. (2.70) implies

$$A_{\text{cbl,max}} \equiv G_{\text{cbl,max}} \approx -1.952 \text{ dB} + 0.915 \text{ dB} = -1.037 \text{ dB} \tag{2.72}$$

and

$$A_{\text{cbl,min}} \equiv G_{\text{cbl,min}} \approx -1.952 \text{ dB} - 0.915 \text{ dB} = -2.867 \text{ dB}, \tag{2.73}$$

which are approximately the values obtained in Eqs. (2.68) and (2.69).

Example 2.2 Effect of Mismatch The gain of a cascade is estimated by adding (in dB) the transducer gains of all its modules and subtracting the nominal losses of the cables. If we accept an SWR specification of 2 at the output of one of the modules and 3 at the input to the following module, and if these modules are connected by a cable with 2 dB of nominal loss, how will this affect the gain of the cascade.

Based on Example 2.1, we know that the gain of the cascade can vary about ± 0.92 dB [Eq. (2.71)] due to such an interface. There would also be an increase in mean gain of about 0.05 dB [Eq. (2.70)] under any conditions where the specified SWRs actually occurred. This is the mean over all possible phases due to the reflections and cable delay. It is small compared to the maximum and minimum gain changes and would be even smaller if averaged over the various actual values of SWR so the main effect is the ± 0.92 dB uncertainty introduced into the cascade gain. This amount of variation requires that the worst-case phase relationships occur when both SWRs are at their maximum allowed values.

The variance of G, σ_G^2, is also important since these variances will add for all of the modules and interconnects to give an overall variance for the cascade. The variance may provide a more useful estimate of the range of gains to be expected if the maximum and minimum are considered too extreme for an application, especially as the number of modules and interconnects grow. The deviation of $G_{\text{cbl}} = 10 \text{ dB} \log(|g_{\text{cbl}}|)$ from its mean, Eq. (2.62), is plotted, for various $|a_{\text{RT}}|$, as a function of θ in Fig. 2.8. From the data represented there, the variance can be computed (summing 40 data points over half a cycle of θ), giving a standard deviation σ_G as plotted in Fig. 2.9. This relationship can be well approximated as

$$\sigma_G \approx 0.7\Delta_+ \tag{2.74}$$

for

$$|a_{\text{RT}}| < 0.7. \tag{2.75}$$

The inequality $|a_{\text{RT}}| < 0.7$ corresponds to SWRs less than 11 at both ends of the cable and should therefore cover most cases.

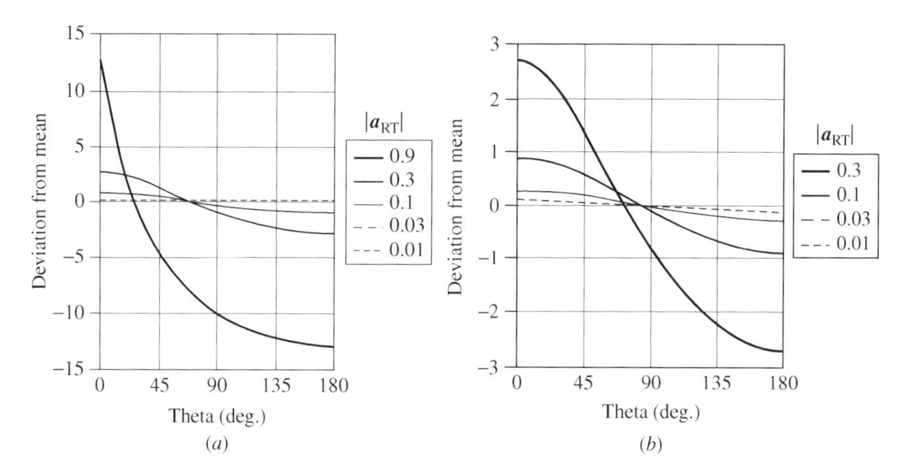

Fig. 2.8 Effective interconnect gain, deviation from mean. [Part (*a*) is expanded at (*b*).]

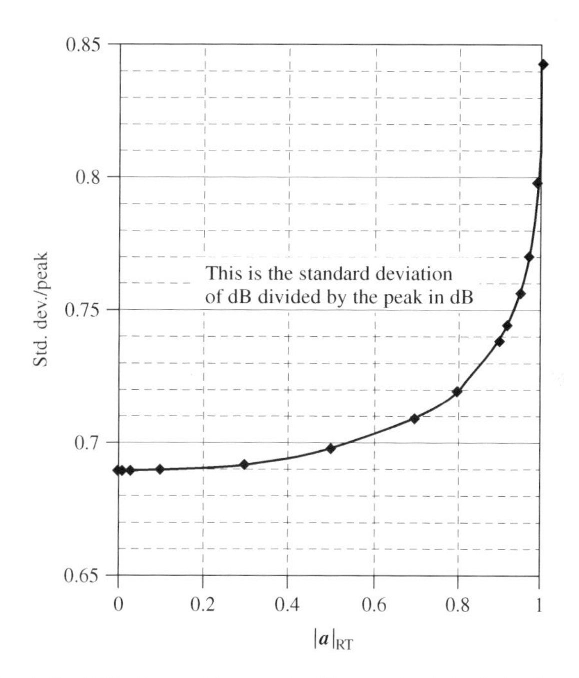

Fig. 2.9 Effective cable gain in dB, standard-deviation/peak.

2.3.2.2 Power Delivered to the Load We briefly consider how much power is delivered by the cable to its load in Appendix L. This is not an important parameter in our cascade since module gains are relative to the forward power at the cable output rather than the absorbed power, but it can be useful for other purposes and it may help to clarify the meaning of the effective gain of the cable.

2.3.2.3 *Phase Variation Due to Reflection* In some cases we may need to know how much the phase delay can vary due to mismatches at the ends of a (possibly calibrated) interconnect. We rewrite Eq. (2.43), using (2.49) and (2.52), as

$$a_{\mathrm{cbl}} = \frac{\exp(h - jb)}{1 - |a_{\mathrm{RT}}|\exp(j\theta)} = \frac{e^h e^{-jb}}{(1 - |a_{\mathrm{RT}}|\cos\theta) - j|a_{\mathrm{RT}}|\sin\theta} \tag{2.76}$$

to make clear that the phase of a_{cbl} is $\gamma - b$, where b is the phase lag due to one-way transmission through the cable, and

$$\gamma = \arctan\frac{|a_{\mathrm{RT}}|\sin\theta}{1 - |a_{\mathrm{RT}}|\cos\theta} \tag{2.77}$$

is the additional phase shift due to the reflections. To find the extreme values of γ as θ varies over 360°, we set the derivative,

$$\frac{d\gamma}{d\theta} = \frac{|a_{\mathrm{RT}}|\cos\theta(1 - |a_{\mathrm{RT}}|\cos\theta) - (|a_{\mathrm{RT}}|\sin\theta)^2}{(1 - |a_{\mathrm{RT}}|\cos\theta)^2 + (|a_{\mathrm{RT}}|\sin\theta)^2} = \frac{|a_{\mathrm{RT}}|(\cos\theta - |a_{\mathrm{RT}}|)}{1 - 2|a_{\mathrm{RT}}|\cos\theta + |a_{\mathrm{RT}}|^2}, \tag{2.78}$$

to zero, obtaining

$$\cos\theta = |a_{\mathrm{RT}}| \quad \text{at} \quad \frac{d\gamma}{d\theta} = 0. \tag{2.79}$$

Using that value of θ in Eq. (2.77), we obtain

$$\gamma_{\mathrm{max,min}} = \arctan\frac{\pm|a_{\mathrm{RT}}|\sqrt{1 - |a_{\mathrm{RT}}|^2}}{1 - |a_{\mathrm{RT}}|^2} = \pm\arctan\frac{|a_{\mathrm{RT}}|}{\sqrt{1 - |a_{\mathrm{RT}}|^2}} \tag{2.80}$$

$$= \pm\arcsin|a_{\mathrm{RT}}|. \tag{2.81}$$

In addition, calculation of γ from Eq. (2.77) for 40 points over one cycle of θ indicates that γ has zero mean and a standard deviation as plotted versus $|a_{\mathrm{RT}}|$ in Fig. 2.10. As was the case for gain variation, the standard deviation can be approximated as 70% of the peak,

$$\sigma_\gamma \approx 0.7\gamma_{\mathrm{max}}, \tag{2.82}$$

with good accuracy for SWRs less than 10.

2.3.2.4 *Generalization to Bilateral Modules* We have written the expressions in this section (2.3) for unilateral modules, but they generally can be applied also to bilateral modules with an appropriate interpretation of the parameters. That requires that $S_{11,j+1}$ and $S_{22,j-1}$ in the expressions for a_{cbl} be changed to the reflection coefficients of the preceding and succeeding cascade sections, respectively. We might give them symbols $\rho_{11,j+1}$ and $\rho_{11,j-1}$ or $S_{11,(j+1)-}$ and $S_{22,(j-1)+}$. This generalization might be useful for some simple problems, but the

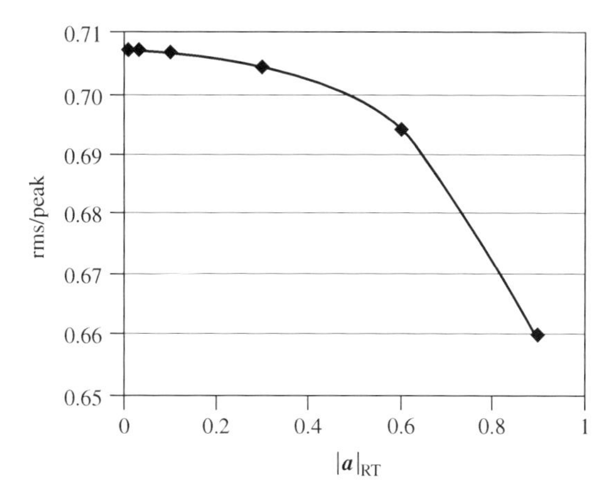

Fig. 2.10 Phase deviation, standard-deviation/peak.

complexity of computing the reflection from two cascades of modules for each a_{cbl} in a cascade shows why unilateral modules are needed for simplicity.

2.3.3 Overall Response, Standard Cascade

2.3.3.1 Gain The total power gain of a standard cascade is the sum of the (dB) module power gains, as measured in an environment of nominal interface impedances, plus the effective gains of the interconnections. For each module we can estimate a mean value and a peak deviation from the mean as well as a standard deviation. From these we can compute the overall cascade gain,

$$G_{\text{cas}} = \sum_{j=1}^{N} G_j, \tag{2.83}$$

where j is the index of either a module or an interconnection, of which there are N total, and G represents mean, maximum, or minimum gain in dB. This is basically the same as Eq. (2.38).

Similarly, the variance of the gain can be computed from

$$\sigma_{\text{cas}}^2 = \sum_{j=1}^{N} \sigma_j^2, \tag{2.84}$$

where σ_j is the estimated standard deviation of gain for a module or of effective gain for an interconnection.

If adjacent modules are connected directly, without a cable, we can still conceive of a zero-length cable between them. That gives us a place in which to define the waves and allows us to use module transducer gains in our standard-impedance framework. Both modules must be characterized using the impedance of the chosen cable at their interface. (In the design phase, characterization may consist of estimates based on expected module designs.) If the output and input impedances of the modules at the interface differ, the impedance of the zero-length cable should be set equal to one of them, preferably the one that can be matched with the smallest SWR, in order to minimize superfluous reflections and the resulting variations in calculated cascade performance. Then the other module must be recharacterized for that interface impedance.

2.3.3.2 *End Elements in the Cascade* The gain given by Eq. (2.83) is the cascade's transducer gain where the impedance of the source is the same as the standard impedance that is defined for the input of the first module and that of the load is the same as for the last module (Fig. 2.11). However, other sources and loads can be accommodated.

The last element N may be a module that drives a load at the nominal impedance or one that drives no load at all. In the latter case, the module can be given a convenient transfer function that represents the ratio of a desired observed variable (e.g., a meter reading) to the driving signal, $v_{o,N}$, the same ratio that is used in characterizing the module. In the former case, output conditions will be the same as during measurement so the measured gain of module N will apply. (If the load is separated from the module by a cable of nominal impedance, the power dissipated in that load can easily be related to the power at the module output.)

A load that is not at nominal impedance can be treated like the final module in the cascade. For example, a 10-Ω resistive load connected to a 50-Ω output cable provides an SWR of 5 at the cable output. The power dissipated in the load will be 0.556 times the power in the forward wave in the cable[6], so the last module can be characterized by a SWR of 5 and a power gain of 0.556. The computed cascade output will then be the power delivered to the 10-Ω load.

If the cascade source impedance is not matched to the standard impedance of the cable to which it is connected, that cable becomes the first element in the cascade and has the source SWR at its input. The cascade gain is then relative to the power that the source delivers to that cable (in v_{ojT} of Fig. 2.6). For example, an antenna might be designed to match 50 Ω and its SWR and output power into 50 Ω specified. That specified power would be the power induced into the cable, and the forward power at the cable output would depend on that induced power and on the SWRs at the antenna and at the cable output, just as if the cable were

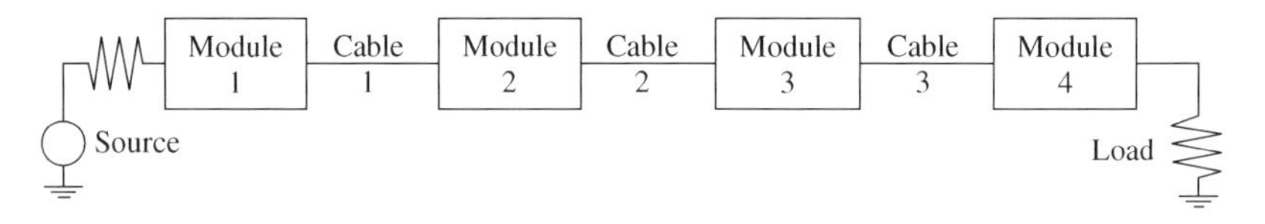

Fig. 2.11 Cascade of unilateral modules.

driven by a module. The cascade gain would be relative to the power that the antenna would deliver to a 50-Ω resistor.

2.3.3.3 Phase Since phase shifts of the modules and effective phase shifts of the interconnections add to give the cascade phase shift, these can also be summed, based on specifications or estimates for the modules and the expected phase shift due to cable length ($-b$) plus γ [Eq. (2.77)]. Maximum variations can be estimated for the modules and added to those given by the extremes for the interconnections in Eq. (2.80). Variances can also be estimated and added, as in Eq. (2.84), for each of the series elements, using Eq. (2.82) for interconnections.

2.3.3.4 Cascade Calculations

Example 2.3 Figure 2.11 shows a cascade of unilateral modules separated by cables at the nominal impedance for the system, the impedance at which the module parameters are characterized (say 50 Ω). Figure 2.12 is a spreadsheet used in calculating the characteristics of the overall cascade. (This should be downloaded so the underlying equations can be read.)

	A	B	C	D	E	F	G	H
2		Gain	Gain	SWR				
3		nom	+/–	at out			$\lvert a_{RT}\rvert$	
4	Module 1	12.0 dB	1.0 dB	1.5				
5	Cable 1	–1.5 dB		1.5			0.028318	
6	Module 2	8.0 dB	2.0 dB	2				
7	Cable 2	–1.0 dB		2			0.088259	
8	Module 3	2.0 dB	2.0 dB	2.8				
9	Cable 3	–0.8 dB		3.2			0.206377	
10	Module 4	30.0 dB	2.0 dB					
11				DERIVED				
12		Gain	Gain	Gain	Gain	Gain	phase	phase
13		mean	max	min	±	σ	±	σ
14	Module 1	12.00 dB	13.00 dB	11.00 dB	1.00 dB	0.50 dB		
15	Cable 1	–1.50 dB	–1.25 dB	–1.74 dB	0.25 dB	0.17 dB	1.6227°	1.1359°
16	Module 2	8.00 dB	10.00 dB	6.00 dB	2.00 dB	1.25 dB		
17	Cable 2	–0.97 dB	–0.20 dB	–1.73 dB	0.77 dB	0.54 dB	5.0634°	3.5444°
18	Module 3	2.00 dB	4.00 dB	0.00 dB	2.00 dB	0.80 dB		
19	Cable 3	–0.61 dB	1.21 dB	–2.43 dB	1.82 dB	1.27 dB	11.9101°	8.3371°
20	Module 4	30.00 dB	32.00 dB	28.00 dB	2.00 dB	1.30 dB		
21				CUMULATIVE				
22	at output of							
23	Module 1	12.00 dB	13.00 dB	11.00 dB	1.00 dB	0.50 dB	0.0000°	0.0000°
24	Cable 1	10.50 dB	11.75 dB	9.26 dB	1.25 dB	0.53 dB	1.6227°	1.1359°
25	Module 2	18.50 dB	21.75 dB	15.26 dB	3.25 dB	1.36 dB	1.6227°	1.1359°
26	Cable 2	17.54 dB	21.55 dB	13.52 dB	4.01 dB	1.46 dB	6.6861°	3.7220°
27	Module 3	19.54 dB	25.55 dB	13.52 dB	6.01 dB	1.66 dB	6.6861°	3.7220°
28	Cable 3	18.93 dB	26.76 dB	11.09 dB	7.83 dB	2.10 dB	18.5963°	9.1302°
29	Module 4	48.93 dB	58.76 dB	39.09 dB	9.83 dB	2.47 dB	18.5963°	9.1302°

Fig. 2.12 Spreadsheet for cascade of unilateral modules.

Cells B4–D10 (inclusive) are specified module and cable parameters. From these are derived the individual stage parameters in rows 14–20 and from those are computed the cumulative gains and phase shifts in rows 23–29.

Cells D4–D9 give the SWRs at the outputs of each element. These are due to the modules, not to the interconnects, presuming that the latter are much better matched than the former. Thus cell D5 gives the *input* SWR for Module 2, even though it is labeled as the SWR *at the output* of the preceding interconnect. Source and load are 50 Ω so no SWR is shown for them.

Cells G5, G7, and G9 are the values of $|a_{RT}|$ computed from the loss in column B and the SWRs at either end of the cable (column D) according to Eq. (2.53).

In cells E14–E20, maximum variations for the module gains are taken from corresponding values in cells C4–C10. Maximum variations for cable interconnects are taken from Eq. (2.64), based on values for $|a_{RT}|$ in the corresponding cells G5–G9.

Standard deviations σ of gain are estimated for each module (F14–F20), perhaps from data or perhaps based on the specified maximum deviations and expected distribution of variations. Standard deviations for the interconnects are taken as 0.7 times the peak deviations in the column to their left in accordance with Eq. (2.74).

For phase, we have shown only variations, and those only for the interconnects. We could, of course, also give such values for the modules. The effective variations in phase due to interconnections (cells G15–G19) are computed based on $|a_{RT}|$ (cells G5–G9) using Eq. (2.81). Standard deviations (H15–H19) are computed as 0.7 times these peak variations in accordance with Eq. (2.82).

Maximum and minimum gains (cells C14–D20) are computed from the mean values (cells B14–B20) and peak variations (cells E14–E20).

Cumulative gains and peak variations (cells B23–E29) are obtained by adding the value for that element, given in rows 14–20 of the same column, to the sum in the cell just above. The cumulative standard deviations (cells F23–F29) are obtained similarly except they are squared before adding (and then the root is taken). Cumulative phase peak variations and standard deviations (G23–H29) are similarly computed.

Row 29 gives cumulative values for the cascade. Note that, while the sum of module peak gain variations (cells C4–C10) is ± 7 dB, the cumulative peak variation (cell E29) is ± 9.83 dB, the difference being due to mismatches.

2.3.4 Combined with Bilateral Modules

Modules that are not, or cannot be approximated as, unilateral require a representation such as the T parameters when they are in cascade. A cascade of such modules can then be represented as a single module with parameters obtained by multiplying the T matrices together. The inclusion of any unilateral module in a cascade of otherwise bilateral modules causes the entire cascade to become unilateral. This must be so because the unilateral module prevents reverse transmission through the cascade. We show this mathematically (and obtain some useful expressions in the process) as follows.

The S-parameter matrix for a cascade of two modules is given by (see Appendix S)

$$\mathbf{S}_{\text{comp}} \equiv \begin{pmatrix} S_{11\text{comp}} & S_{12\text{comp}} \\ S_{21\text{comp}} & S_{22\text{comp}} \end{pmatrix} = \begin{pmatrix} S_{111} + \dfrac{S_{112}S_{121}S_{211}}{1 - S_{112}S_{221}} & \dfrac{S_{121}S_{122}}{1 - S_{112}S_{221}} \\ \dfrac{S_{212}S_{211}}{1 - S_{112}S_{221}} & S_{222} + \dfrac{S_{122}S_{212}S_{221}}{1 - S_{112}S_{221}} \end{pmatrix}, \tag{2.85}$$

where the third subscript is the module number and module 1 drives module 2.

If module 1 is unilateral ($S_{121} = 0$, Fig. 2.13*a*), this becomes

$$\mathbf{S}_{\text{comp}|1\,\text{unilateral}} = \begin{pmatrix} S_{111} & 0 \\ \dfrac{S_{212}S_{211}}{1 - S_{112}S_{221}} & S_{222} + \dfrac{S_{122}S_{212}S_{221}}{1 - S_{112}S_{221}} \end{pmatrix}. \tag{2.86}$$

If module 2 is unilateral ($S_{122} = 0$, Fig. 2.13*b*), this becomes

$$\mathbf{S}_{\text{comp}|2\,\text{unilateral}} = \begin{pmatrix} S_{111} + \dfrac{S_{112}S_{121}S_{211}}{1 - S_{112}S_{221}} & 0 \\ \dfrac{S_{212}S_{211}}{1 - S_{112}S_{221}} & S_{222} \end{pmatrix}. \tag{2.87}$$

In each case we see that the composite is unilateral, since $S_{12,\text{comp}} = 0$. If either of these composites is combined with another bilateral module, either after or

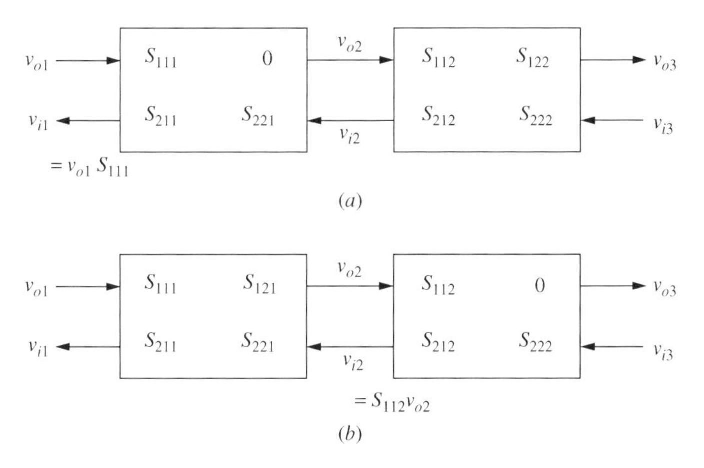

Fig. 2.13 Bilateral module combined with unilateral module.

before it, the composite parameters will be given either by Eq. (2.86) or by Eq. (2.87) with the S parameters of the original pair taken from Eq. (2.86) or from Eq. (2.87) as appropriate. Therefore, the addition of a bilateral module will produce another unilateral composite, and so forth.

These composites can then be used as elements in a cascade of unilateral modules. This will be illustrated in the following example.

Example 2.4 Composite Module from Bilateral and Unilateral Modules Figure 2.14 shows a cascade consisting of two bilateral modules followed by a unilateral module interconnected with cables matched to the nominal system impedance. The S parameters of the cascade elements are shown in the spreadsheet of Fig. 2.15 in cells C3–F12. Note that the last module, E, has $S_{12} = 0$, defining it as unilateral, whereas the other two modules have finite S_{12} and thus are bilateral.

Cells C15–F24 contain the equivalent T parameters, obtained from the S parameters according to Eq. (2.29). These are automatically (i.e., by formulas in the spreadsheet) converted from polar to rectangular form in cells C27–G36. These rows are copied into a MATLAB script (Fig. 2.16). The semicolon required to mark the end of the matrix row in MATLAB is included in cells E27–E36 to facilitate the paste operation. The real and imaginary parts are transferred separately and combined in the script. (Matrix B is shown in rectangular form to illustrate an alternate, if less convenient, way to enter the data.)

After all the T matrices have been filled in the script, it is executed and computes the product of the T matrices. The output from the script is shown at the bottom of Fig. 2.16. (In MATLAB, results of command lines that are not terminated by semicolons are printed, so the various matrices appear in the output.) Only the E matrix and the product T matrix are visible in the figure. The magnitude $\mathbf{T}_m$ and angle $\mathbf{T}_a$ of the product matrix $\mathbf{T}$ are also created to facilitate conversion to S parameters.

The resulting product is converted from T-matrix form to S-matrix form according to Eq. (2.31) and entered into cells C39–F40 (Fig. 2.15). The SWR and dB gains corresponding to the S parameters are automatically computed and entered in rows 41 and 42. Note that S_{12} for the composite is essentially zero, signifying a composite unilateral module.

The conversions from S to T parameters and visa versa were facilitated by an ST-Conversion Calculator spreadsheet, shown in Fig. 2.17. (The second page of

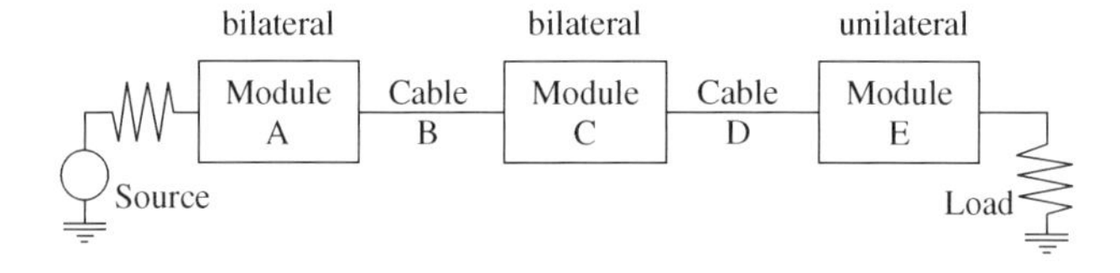

Fig. 2.14 Cascade of bilateral modules and one unilateral module.

	A	B	C	D	E	F	G
2			**S11**	**S12**	**S21**	**S22**	
3	Module A	magnitude	0.224	0.2	1.8	0.4	
4		degrees	−30	0	−45	180	
5	Cable B	magnitude	0	0.9	0.9	0	
6		degrees	0	−60	−60	0	
7	Module C	magnitude	0.2	0.15	1.78	0.25	
8		degrees	0	−30	−30	0	
9	Cable D	magnitude	0	0.9	0.9	0	
10		degrees	0	−60	−60	0	
11	Module E	magnitude	0.2	0	2.2	0.3333	
12		degrees	−30	0	−60	−30	
13							
14			**T11**	**T12**	**T21**	**T22**	
15	Module A	magnitude	0.5555556	−0.2222222	0.1244444	0.2484159	
16		degrees	45.00°	225.00°	15.00°	2.97°	
17	Cable B	magnitude	1.1111111	0	0	0.9	
18		degrees	60	60	60	−60	
19	Module C	magnitude	0.5617978	−0.1404494	0.1123596	0.1381143	
20		degrees	30	30	30	−40.144628	
21	Cable D	magnitude	1.1111111	0	0	0.9	
22		degrees	60	60	60	−60	
23	Module E	magnitude	0.4545455	−0.1515	0.0909091	0.0303	
24		degrees	60	30	30	180	
25					[rad/deg = 0.0174533]		
26			**T11**	**T12**		**T21**	**T22**
27	Module A	real	0.3928371	0.1571348	:	0.1202041	0.24808
28		imaginary	0.3928371	0.1571348	:	0.0322086	0.01288
29	Cable B	real	0.5555556	0	:	0	0.45
30		imaginary	0.9622504	0	:	0	−0.7794
31	Module C	real	0.4865311	−0.1216328	:	0.0973062	0.10558
32		imaginary	0.2808989	−0.0702247	:	0.0561798	−0.089
33	Cable D	real	0.5555556	0	:	0	0.45
34		imaginary	0.9622504	0	:	0	−0.7794
35	Module E	real	0.2272727	−0.1312028	:	0.0787296	−0.0303
36		imaginary	0.3936479	−0.07575	:	0.0454545	3.7E−18
37							
38			**S11**	**S12**	**S21**	**S22**	
39	**Total**	magnitude	0.20022	0.00003	5.62430	0.33352	
40		degrees	−41.6826	1.5398	106.7776	0	
41		**SWR**	**1.50**			**2.00**	
42		**gain**		**−91.48 dB**	**15.00 dB**		

Fig. 2.15 Spreadsheet for composite parameters.

this spreadsheet is an aid to facilitate copying from matrix-shaped format of the script output to the linear-shaped format of the spreadsheet.)

The gain and SWRs for the composite module can now be entered as those of a unilateral module in a cascade, such as that represented by Fig. 2.18 and the spreadsheet in Fig. 2.19 where the composite in Fig. 2.14 becomes Module 2. (Compare its gain and SWR to the values in lines 41 and 42 of Fig. 2.15.)

MATLAB SCRIPT

```
 2 Ar = [0.392837101   0.15713484  ;   0.120204103 0.248081641]
 3 Ai = [0.392837101   0.15713484  ;   0.032208592 0.012883437]
 4 A = Ar + i*Ai
 5
 6 B =[0.5556 + 0.9623i        0
 7          0              0.4500 - 0.7794i]
 8
 9 Cr = [0.486531126   -0.121632781    ;   0.097306225 0.105577254];
10 Ci = [0.280898876   -0.070224719    ;   0.056179775 -0.089044944];
11 C = Cr + i*Ci
12
13 Dr = [0.555555556   0   ;0  0.45];
14 Di = [0.962250449   0   ;0  -0.779422863];
15 D = Dr + i*Di
16
17 Er = [0.227272727   -0.131202849    ;   0.078729582 -0.0303];
18 Ei = [0.393647911   -0.07575    ;   0.045454545 3.71056E-18];
19 E = Er + i*Ei
20
21 T = A*B*C*D*E
22
23 Tm = abs(T)
24 Ta = angle(T)*180/pi
```

RESPONSE WHEN RUN, lower part

```
E =

  2.2727e-01 +  3.9365e-01i -1.3120e-01 -  7.5750e-02i
  7.8730e-02 +  4.5455e-02i -3.0300e-02 +  3.7106e-18i


T =

 -5.1334e-02 -  1.7027e-01i  4.3192e-02 +  4.0592e-02i
 -3.0340e-02 -  1.8621e-02i  1.1861e-02 +  3.1883e-04i


Tm =

   1.7784e-01   5.9273e-02
   3.5598e-02   1.1865e-02


Ta =

  -1.0678e+02   4.3222e+01
  -1.4846e+02   1.5398e+00
```

Fig. 2.16 MATLAB script and response, multiplication of T matrices.

2.3.5 Lossy Interconnections

Well-matched but lossy elements, attenuators or isolators, reduce the interactions between the modules on either side and can cause them to act as if they were unilateral.

ENTER	S11	S12	S21	S22
magnitude	0	0	3.981	−0.524
degrees	0	0	0	0
EQUIVALENT	T11	T12	T21	T22
magnitude	0.25118864	0.13162285	0	0
degrees	0.00°	0.00°	0.00°	0.00°

ENTER	T11	T12	T21	T22
magnitude	4.75E−03	0.00E+00	0	0
degrees	−1.51E+02	0	0	0
EQUIVALENT	S11	S12	S21	S22
magnitude	0	0	210.322635	0
degrees	151.36°	0.00°	151.36°	151.36°

Fig. 2.17 S–T conversion calculator.

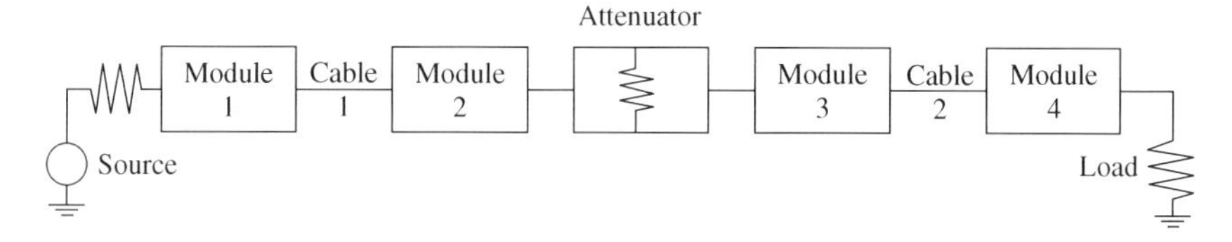

Fig. 2.18 Cascade with attenuator.

Figure 2.20 shows a cascade of three bilateral modules where the middle module (index 2) is reflectionless but lossy. We will treat it as a lossy interconnect. The source might represent all the previous modules, and the load might represent all of the subsequent modules, in the cascade. The reverse wave at port 2, v_{i2}, equals v_{o2} multiplied by the round-trip loss of the following element, 2, times the reflection coefficient at the input to 3. This is reflected at the output of module 1 and combines with the wave transmitted through module 1 to give

$$v_{o2} = S_{211} v_{o1} (1 + a_{\mathrm{RT}} + a_{\mathrm{RT}}^2 + \ldots), \tag{2.88}$$

where the (total) round-trip loss is

$$a_{\mathrm{RT}} = S_{212} \rho_3 S_{122} \rho_1. \tag{2.89}$$

The four parameters in a_{RT} represent the forward transfer function in the lossy element 2, the reflection at the input to element 3, the reverse transmission in the lossy element, and the reflection at the output of element 1, respectively. Here ρ_1 includes reflections due to module 1 directly as well as all previous modules. Likewise, ρ_3 includes reflections from the first and all subsequent modules within the load. All of these parameters can be small so the product a_{RT} can be much less than one, in which case it can be ignored in Eq. (2.88). This condition can be true regardless of ρ_1 and ρ_3 (which are always less than 1) if there is enough

	A	B	C	D	E	F	G	H
2		Gain	Gain	SWR				
3		nom	+/–	at out			$\|a_{RT}\|$	
4	Module 1	12.0 dB	1.0 dB	1.5				
5	Cable 1	–1.5 dB		1.5			0.02832	
6	Module 2	15.0 dB	2.0 dB	2				
7	Attenuator	–8.0 dB		2			0.01761	
8	Module 3	2.0 dB	2.0 dB	2.8				
9	Cable 2	–0.8 dB		3.2			0.20638	
10	Module 4	30.0 dB	2.0 dB					
11				DERIVED				
12		Gain	Gain	Gain	Gain	Gain	phase	phase
13		mean	max	min	±	σ	±	σ
14	Module 1	12.00 dB	13.00 dB	11.00 dB	1.00 dB	0.50 dB		
15	Cable 1	–1.50 dB	–1.25 dB	–1.74 dB	0.25 dB	0.17 dB	1.6227°	1.1359°
16	Module 2	15.00 dB	17.00 dB	13.00 dB	2.00 dB	1.00 dB		
17	Attenuator	–8.00 dB	–7.85 dB	–8.15 dB	0.15 dB	0.11 dB	1.0090°	0.7063°
18	Module 3	2.00 dB	4.00 dB	0.00 dB	2.00 dB	0.80 dB		
19	Cable 2	–0.61 dB	1.21 dB	–2.43 dB	1.82 dB	1.27 dB	11.9101°	8.3371°
20	Module 4	30.00 dB	32.00 dB	28.00 dB	2.00 dB	1.30 dB		
21			CUMULATIVE					
22	at output of							
23	Module 1	12.00 dB	13.00 dB	11.00 dB	1.00 dB	0.50 dB	0.0000°	0.0000°
24	Cable 1	10.50 dB	11.75 dB	9.26 dB	1.25 dB	0.53 dB	1.6227°	1.1359°
25	Module 2	25.50 dB	28.75 dB	22.26 dB	3.25 dB	1.13 dB	1.6227°	1.1359°
26	Attenuator	17.50 dB	20.90 dB	14.11 dB	3.40 dB	1.14 dB	2.6317°	1.3376°
27	Module 3	19.50 dB	24.90 dB	14.11 dB	5.40 dB	1.39 dB	2.6317°	1.3376°
28	Cable 2	18.89 dB	26.11 dB	11.68 dB	7.22 dB	1.88 dB	14.5419°	8.4437°
29	Module 4	48.89 dB	58.11 dB	39.68 dB	9.22 dB	2.29 dB	14.5419°	8.4437°

Fig. 2.19 Spreadsheet for cascade with attenuator.

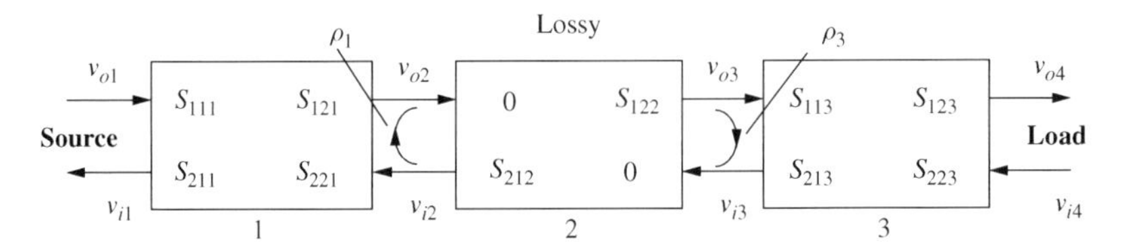

Fig. 2.20 Modules separated by well-matched lossy module.

attenuation in the interconnect. Then the forward wave from the output of module 1 is simply

$$v_{o2} \approx S_{211} v_{o1}, \tag{2.90}$$

and the output from the lossy interconnect is

$$v_{o3} \approx S_{212} S_{211} v_{o1}. \tag{2.91}$$

Thus, transmission through the bilateral module (1) and lossy interconnect (2) is represented by the simple product of S_{21}'s for these two components, as if module 1 were unilateral. Moreover, the wave out of the input of module 1 is (Fig. 2.20)

$$v_{i1} = v_{o1} S_{111} + v_{i2} S_{121} = v_{o1} S_{111} + v_{o2} S_{212} \rho_3 S_{122} S_{121}. \tag{2.92}$$

If we use Eq. (2.90) for v_{o2}, this becomes

$$v_{i1} \approx v_{o1} [S_{111} + S_{211} S_{212} \rho_3 S_{122} S_{121}] \approx v_{o1} S_{111}, \tag{2.93}$$

where the small value of the product of the group of five factors, which includes the round-trip loss of the interconnect ($S_{122}S_{212}$), was used to discard them. We see that v_{i1} is solely due to the reflection at the input of module 1, as if that module were unilateral. Thus module 1 acts like a unilateral module when followed by a sufficiently lossy interconnect. Furthermore, and for similar reasons, the first module following a sufficiently lossy interconnect is effectively unilateral. Any reverse transmission through module 3 is attenuated by the round-trip loss of the interconnect plus the reflection coefficient ρ_1 before reentering module 3. The output of module 3 is, therefore,

$$v_{o4} = v_{o3} S_{213} + v_{i4} S_{223}, \tag{2.94}$$

as if it were unilateral, and we have already shown that v_{i1} is not influenced by v_{i4}, again consistent with unilaterality in module 3.

Example 2.5 Attenuator in Cascade In this example, after first considering the effect of including an attenuator in a cascade of unilateral modules, we will investigate its effectiveness in permitting adjacent bilateral modules to be treated as unilateral.

Figures 2.18 and 2.19 show a cascade that includes an attenuator. These are similar to the cascade discussed in Example 2.3 (Figs. 2.11 and 2.12) except the middle cable has been replaced by an attenuator and the gain of the preceding module has been adjusted to compensate for the added loss. The treatment is not basically different with the attenuator; the interconnect just has more loss. [There could be some additional complexities if the attenuator had a variation in its basic (matched) gain. Then we would have to decide how to combine these variations with the variation due to reflections at the ends of the interconnect (e.g., add them, add their squares, etc.).]

The presence of the attenuator reduces the effects of reflections at that interface by attenuating the reflected waves. Note in Fig. 2.19 the large effective gain variation in cable 2 (cell E19) compared to that for cable 1 (cell E15). This is due to the low attenuation and large SWRs at the ends of the former. Note how the presence of the attenuator has reduced the variations in overall gain between Examples 2.3 and 2.4 (cells E29 in Figs. 2.12 and 2.19).

Now let us test the effectiveness of the attenuator in removing the effects of feedback (S_{12}) in adjacent modules. In these tests we will vary the gain of the attenuator, maintaining constant nominal cascade gain (product of individual element gains) by varying the gain of the final module to compensate. For each setting we will compare the cascade gain when S_{12} is zero (unilateral) in the modules before and after the attenuator to the cascade gain when these modules are bilateral. In the latter case, we will set $S_{12} = 1/S_{21}$ in both modules, the upper limit of reverse gain for unconditional stability.

We will calculate the overall transfer function by multiplying T matrices, using MATLAB to multiply the matrices and Excel spreadsheets for the other calculations. This is similar to what was done in Example 2.4, but this time we will include the S–T matrix conversions on the spreadsheet, rather than using a separate conversion spreadsheet.

First, we must specify the module parameters more completely than given in Fig. 2.19. We must add a phase for each of the S parameters since Fig. 2.19 only gives the magnitude of the transfer functions and the SWRs, which do not reveal the phases of the reflections. We will set all the phases to zero in these experiments, mainly in an attempt to prevent a fortuitous choice of phases from canceling the effects of the reflections. This also reduces the calculation time some since we will not have to copy varying phases into MATLAB.

Excerpts from the spreadsheet are shown in Fig. 2.21. The region of the spreadsheet where we enter S parameters is shown at Fig. 2.21*a*. Note that S_{12} has been set equal to the reciprocal of S_{21} for Modules 2 and 3. This cascade gain will be compared to the cascade gain that occurs when these two values are set equal to zero. The attenuator gain is entered in dB (right column) and $S_{12} = S_{21}$ is automatically set to give that value. The spreadsheet also automatically sets S_{21} of Module 4 to maintain a total nominal (not considering reflections) gain of 48.7 dB.

MATLAB is used, as it was in Example 2.4, to multiply the matrices, but here the spreadsheet includes the conversions between S and T parameters, which employed a separate calculator spreadsheet before. The T parameters of the units (modules and cables) are copied from the spreadsheet into MATLAB, which then computes their product, which is the T matrix for the cascade. This is entered into the spreadsheet (Fig. 2.21*c*), with some help from Excel's Text to Columns feature. The spreadsheet then converts these T parameters to S parameters, as shown in Fig. 2.21*b*. Parts *b* and *c* show portions of the spreadsheet for two attenuator settings. As before, gain in dB and SWR are computed from the S parameters. Note that the overall S_{12} is $-\infty$ dB due to the presence of unilateral modules in the chain.

Test 1: Cascade of Fig. 2.21 Gain is plotted against the attenuator value in Fig. 2.22. Note that the difference between the gain when true unilateral modules are used and that when severely bilateral modules are used, on both sides of the attenuator, goes from 3.7 dB with zero attenuation to only 0.25 dB with 12 dB of attenuation. This confirms that unilateral modules can replace the bilateral modules if the adjacent attenuation is high enough. The gain varies with attenuation, even with unilateral modules, because of the reflections at the interfaces at either end of the attenuator.

		S11	S12	S21	S22	
Module 1	magnitude	0	0	3.981	0.2	12.00 dB
	degrees	0	0	0	0	
Cable 1	magnitude	0	0.841	0.841	0	−1.50 dB
	degrees	0	0	0	0	
Module 2	magnitude	0.2	0.17782794	5.623	0.333	15.00 dB
	degrees	0	0	0	0	
Attenuator	magnitude	0	0.251	0.251	0	−12.00 dB
	degrees	0	0	0	0	
Module 3	magnitude	0.333	0.79432823	1.259	0.474	2.00 dB
	degrees	0	0	0	0	
Cable 2	magnitude	0	0.912	0.912	0	−0.80 dB
	degrees	0	0	0	0	
Module 4	magnitude	0.524	0	50.119	0	34.00 dB
	degrees	0	0	0	0	
						48.70 dB

(*a*) Module *S* parameter input

		S11	S12	S21	S22
M2 AND M3 CONDITIONALLY UNSTABLE (VERGE)					
Total	magnitude	0.00E + 00	0	6.11E + 02	0.00E + 00
Attenuator	degrees	0.00°	0.00°	0.00°	0.00°
0 dB	**SWR**	**1.00**			**1.00**
	gain		**–inf**	**55.73 dB**	
Total	magnitude	0.00E + 00	0	5.31E + 02	0.00E + 00
Attenuator	degrees	0.00°	0.00°	0.00°	0.00°
−1 dB	**SWR**	**1.00**			**1.00**
	gain		**–inf**	**54.51 dB**	

(*b*) Output for cascade

MATLAB Tm	MATLAB Ta		
1.636E − 03	0	0	0
0.000E + 00	0	0	0
1.882E − 03	0	0	0
0.000E + 00	0	0	0

(*c*) Magnitude and phase of four *T*-matrix elements are entered here from MATLAB. This part of the spreadsheet is to the right of (*b*) above. Data for two runs are shown; more can be accommodated.

Fig. 2.21 Spreadsheet for computing cascade gain with bilateral modules.

Test 2: No Reflections at the Attenuator In this test the reflections are removed from the modules at the ends of the attenuator to prevent any variations with attenuation in the true unilateral case. All of the other interfaces are given SWRs of 3 (S_{11} or $S_{22} = 0.5$). The input parameters are shown in Fig. 2.23 and results are plotted in Fig. 2.24. Note that the gain is now not a function of attenuation at all when the modules are truly unilateral. The effect with bilateral modules adjacent to the attenuator varied from about 2.2 dB for no attenuation and 0.25 dB for about 9-dB attenuation.

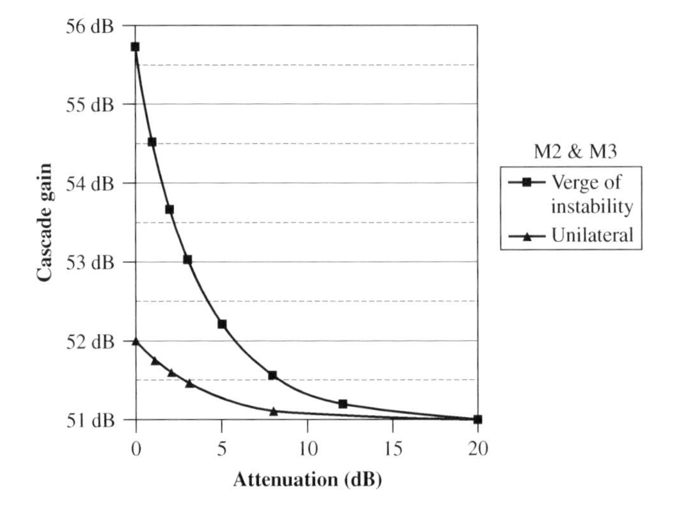

Fig. 2.22 Effect of attenuation and feedback, test 1.

		S11	S12	S21	S22	
Module 1	magnitude	0	0	3.981	0.5	12.00 dB
	degrees	0	0	0	0	
Cable 1	magnitude	0	0.841	0.841	0	−1.50 dB
	degrees	0	0	0	0	
Module 2	magnitude	0.5	0.17782794	5.623	0	15.00 dB
	degrees	0	0	0	0	
Attenuator	magnitude	0	0.398	0.398	0	−8.00 dB
	degrees	0	0	0	0	
Module 3	magnitude	0	0.79432823	1.259	0.5	2.00 dB
	degrees	0	0	0	0	
Cable 2	magnitude	0	0.912	0.912	0	−0.80 dB
	degrees	0	0	0	0	
Module 4	magnitude	0.5	0	31.623	0	30.00 dB
	degrees	0	0	0	0	
						48.70 dB

Fig. 2.23 Parameters for test 2.

These tests show to what degree the attenuator allowed adjacent bilateral modules to be approximated as unilateral. They are only two particular cases (making room for further studies). However, the values of reverse transmission S_{12} were high, at the limit of conditional instability, reflections were relatively high, and phases were all the same to prevent cancellation. We might expect greater effectiveness in many practical cases.

2.3.6 Additional Considerations

2.3.6.1 Variations in SWRs In our examples, we have assumed a fixed SWR for each module in computing variances. If these are maximum SWRs, the

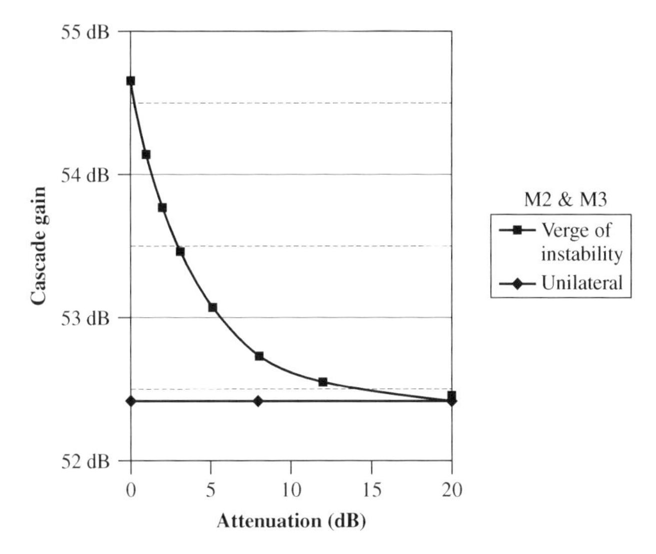

Fig. 2.24 Effect of attenuation and feedback, test 2.

variances will be pessimistic since the variance of the total would be reduced by variations of SWR below its maximum. Figure 2.25 shows variances of gain and phase with SWR in the cascade of Fig. 2.19. These are plotted against a multiplier that was applied simultaneously to each $|\rho|$. The values used in Fig. 2.19 correspond to a multiplier value of one, whereas all SWRs become one when the $|\rho|$ multiplier is zero. In that case, the remaining standard deviation of gain is due to specified gain variations, not SWRs.

2.3.6.2 Reflections at Interconnects We have also neglected the possibility of reflections in the interconnects, including the possibility of some difference in the exact impedances of the interconnects and the measurement system (Egan, 2002, Section R.2). We expect that passive interconnects can be built with relatively good control over interface impedances, but there are bound to be additional reflections. Not surprisingly, they decrease the gain and increase its variability (Egan, 2002, Section R.1). Fortunately, reflections in interconnects and the reduced levels of SWRs that were discussed in the previous paragraph have contrary effects on *gain variation*. Unfortunately, they both tend to decrease mean *gain*.

2.3.6.3 Parameters in Composite Modules While the range of parameters to be expected from individual modules may be available from specifications or test results, it may be more difficult to determine that range for composite modules. These are equivalent unilateral modules composed of one or more bilateral modules plus a unilateral module, as described in Section 2.3.5, or similar composites to be described in the next section. Such composites can be included as equivalent unilateral modules, but it may be necessary to vary some of the

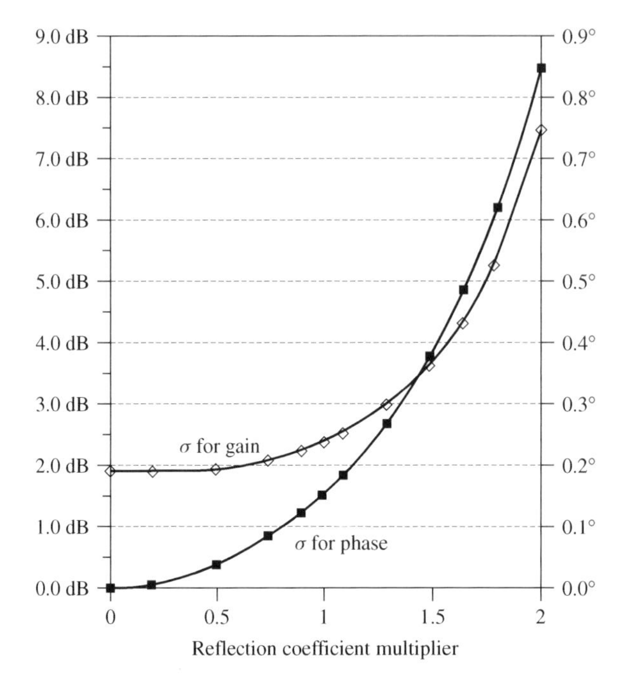

Fig. 2.25 Effect of SWR on gain and phase deviation, cascade of Fig. 2.19.

module parameters (e.g., phases of the S parameters) over their expected ranges to determine the expected range of parameters of the composite.

2.4 NONSTANDARD IMPEDANCES

Some modules may be specified by their input and output impedances, rather than their SWRs. They may also be specified by their maximum available gains, that is, the power delivered to a matched load divided by the power absorbed by the module when it is driven by a matched source (Appendix G).[7] Appendix Z treats unilateral modules that are so specified (we will call them nonstandard modules) and provides formulas and a spreadsheet for computing the response of a cascade of such modules and obtaining the cascade's S parameters. Once that is done, the nonstandard cascade can be included as a module in a standard cascade. (This is also true for a single nonstandard module.)

2.5 USE OF SENSITIVITIES TO FIND VARIATIONS

We have given formulas, in Section 2.3.3, for determining maximums and minimums and variances of cascade gains based on mismatches and on estimates of variances for individual modules. But, if we compose a unilateral module from bilateral modules or nonstandard modules, how are we to determine the range of parameters of the composite module, which is based on many parameters within the individual modules of the composite? One way is to perform a Monte Carlo analysis, but it may be more efficient to determine the sensitivity of the composite parameters to individual parameters and then use these to determine worst-case variations of the composite parameters, perhaps also estimating variances based on the worst cases.

The advantage of the sensitivity analysis is that the individual parameters can be varied one time, whereas in Monte Carlo each of these parameters must be given many values. The disadvantage is that the sensitivity assumes linearity, that the sensitivity is applicable even in the presence of variations of other parameters and for whatever magnitude of parameter changes we ultimately use. Its accuracy declines as the magnitudes of pertinent changes increase, but its relative simplicity may recommend it, at least for initial evaluation.

Sensitivity analysis is more broadly useful than this usage within composite modules, however. It can help us concentrate on module parameters that are most influential in affecting overall cascade performance, and it can help us to quickly estimate the effects of changes in module parameters on cascade parameters.

The basic sensitivity equation gives the change in an overall cascade parameter (e.g., gain) as

$$dy = \sum_{j=1}^{N} \hat{S}_j dx_j, \tag{2.95}$$

where

$$\hat{S}_j \triangleq \frac{\partial y}{\partial x_j} \tag{2.96}$$

is the sensitivity of changes in a scalar quantity y to a change in an individual module parameter x_j, assuming the x_j are independent of each other. We can compute $\hat{S}_j$ by writing an expression for y and performing the differentiation indicated by Eq. (2.96), or we can obtain the derivative by making a small change in x_j and observing the corresponding change in the computed value of y. In some cases we will find the latter easier; we will consider that method here.

We can determine the maximum change in $|y|$ for a given set of changes in $|x|$ from

$$|dy|_{\max} = \sum_{j=1}^{N} |\hat{S}_j dx_j|, \tag{2.97}$$

where dx_j is approximated as the expected change in x_j and dy is approximated as the resulting change in the cascade parameter. (We say "approximate" because this is only strictly true for differential changes.) When the parameter x_j is complex, we include changes of both the real and imaginary parts of x_j in Eq. (2.97). The absolute values of the changes are added to find how dy would change if the signs for the individual dx's were all chosen to cause dy to change in the same direction. This is based on the assumption of linearity, in which case a change in the sign of dx_j causes only a change in the sign of dy.

Example 2.6 Sensitivities Using Spreadsheet Figure 2.26 shows part of the spreadsheet of Fig. 2.19 with some modifications to aid in the computation of sensitivities. In this case, the sensitivity of minimum cascade gain to the SWRs is being computed (the sensitivity of cascade gain to module gain being trivial).

A change of 0.1 has been entered at cell A6. This has caused cell F6 to change by that amount, resulting in a change in the minimum cascade gain in row 29. The value of minimum gain with this change has been copied (by value) from cell E29 to cell C36. This is done for each SWR (using a module or interconnect name to identify the corresponding SWR). Each time that 0.1 is entered into a different cell in A4–A9, we copy (by value) the resulting gain from cell E29 into the appropriate cell in range C34–C39. The value with no modification to the SWR (i.e., with cells A4–A9 blank) is entered in cell C33 for reference. Changes from unmodified to modified gains are given in cells D34–D39. Sensitivities are given in cells E34–E39 for each SWR through division of the changes in cells D34–D39 by the value of the change that was used, which we entered in cell F33.

In creating this spreadsheet from its predecessor (after a new column A was inserted), cells E4–E9 were moved to the right using cut-and-paste (by cell dragging), so the references in the various formulas in the spreadsheet would

	A	B	C	D	E	F	G	H	I
2			Gain	Gain	SWR	SWR			
3	Δ SWR		nom	+/–	at out	modified		$\lvert a_{RT} \rvert$	
4		Module 1	12.0 dB	1.0 dB	1.5	1.5			
5		Cable 1	-1.5 dB		1.5	1.5		0.028318	
6	0.1	Module 2	15.0 dB	2.0 dB	2	2.1			
7		Attenuator	-8.0 dB		2	2		0.018746	
8		Module 3	2.0 dB	2.0 dB	2.8	2.8			
9		Cable 2	-0.8 dB		3.2	3.2		0.206377	
10		Module 4	30.0 dB	2.0 dB					
11					DERIVED				
12			Gain	Gain	Gain	Gain	Gain	phase	phase
13			mean	max	min	±	σ	±	σ
14		Module 1	12.00 dB	13.00 dB	11.00 dB	1.00 dB	0.50 dB		
29		Module 4	48.89 dB	58.12 dB	39.67 dB	9.23 dB	2.29 dB	14.6070°	8.448°
30					⇑				
31			Gain						
32			min	Δ Gain	Sens. At				
33		reference	39.6762		Δ =	0.1			
34		Module 1	39.6394	-0.0367	-0.3672				
35		Cable 1	39.6394	-0.0367	-0.3672				
36		Module 2	39.6665	-0.0097	-0.0969				
37		Attenuator	39.6665	-0.0097	-0.0969				
38		Module 3	39.6339	-0.0422	-0.4223				
39		Cable 2	39.6448	-0.0314	-0.3136				

Fig. 2.26 Spreadsheet with sensitivities.

be changed to the new locations. The values in the cells were then copied back to their former locations. Then the number in cell F4 was replaced with the equation =A4+E4 and this equation was copied to cells F5–F9 (the references will change for each cell as we do so). Thus we can enter new values of SWR into cells E4–E9, and F4–F9 will acquire the changes but will also reflect any change entered in cells A4–A9.

It is tempting to use cut-and-paste (or drag the cell) to move the ΔSWR value down through cells A4–A9 as we observe the effect on the gain. However, that can be disastrous because the spreadsheet equations that reference the dragged cell will change their reference to follow the movement, destroying the integrity of the spreadsheet (the same process that was used in creating F4–F9). This can happen even if the referencing cells are locked. To avoid this we delete the contents of one cell and write the value into the next, or, more conveniently, we can copy-and-paste (not cut-and-paste) the value into a new cell (say by pulling on the cell's lower-right corner) and then delete the original value.

When it is worth the effort, we can create macros using the spreadsheet's built-in capability to do these processes automatically, possibly using other pages in the workbook to hold intermediate data.

Example 2.7 Changes Using Spreadsheet Figure 2.27 is similar to Fig. 2.26, but here we are computing changes in the minimum gain due to specific changes in the SWRs. We proceed as before but we now record, in cells A34–A39, the ΔSWR values used. The sensitivities that were in cells E34–E39 have been replaced with the absolute values of the changes in gains (cells D34–D39). (Since all the changes have the same sign, absolute value is of reduced importance for this case.) The sum of these absolute values is given in cell E40 and below that are the implied minimum and maximum values of minimum gain due to these changes.

Recall that we have not accounted for variations in SWR (Section 2.3.6.1), so we might want to use this process to discern how the gain might be changed when the SWR does vary from the values used in cells E4–E9. If those values are worst case, we might enter expected changes to more typical values as the ΔSWRs. If they are typical, we might use the ΔSWRs to bring them to worst case or to indicate expected variations, sign uncertain. In the latter case, cells E42 and E43 would be pertinent, whereas, in the other cases, cell D41, which retains signs, might be more applicable.

2.6 SUMMARY

- S parameters are a convenient set of two-port parameters for RF modules with standard interface impedances.
- Modules in cascade are represented by T parameters because the T matrices can be multiplied together to produce a representation of the cascade.

	A	B	C	D	E	F	G	H	I
2			Gain	Gain	SWR	SWR			
3	Δ SWR		nom	+/–	at out	modified		$\|a_{RT}\|$	
4		Module 1	12.0 dB	1.0 dB	1.5	1.5			
5		Cable 1	–1.5 dB		1.5	1.5		0.02832	
6		Module 2	15.0 dB	2.0 dB	2	2			
7		Attenuator	–8.0 dB		2	2		0.01761	
8		Module 3	2.0 dB	2.0 dB	2.8	2.8			
9	0.2	Cable 2	–0.8 dB		3.2	3.4		0.21491	
10		Module 4	30.0 dB	2.0 dB					
11					DERIVED				
12			Gain	Gain	Gain	Gain	Gain	phase	phase
13			mean	max	min	±	σ	±	σ
14		Module 1	12.00 dB	13.00 dB	11.00 dB	1.00 dB	0.50 dB		
29		Module 4	48.91 dB	58.21 dB	39.61 dB	9.30 dB	2.32 dB	15.0417°	8.7894°
30					⇑				
31			Gain						
32			min	Δ Gain	|Δ|				
33		reference	39.6762						
34	0.05	Module 1	39.6574	–0.0187	0.01874				
35	0.07	Cable 1	39.6501	–0.026	0.02602				
36	0.1	Module 2	39.6665	–0.0097	0.00969				
37	0.14	Attenuator	39.6628	–0.0134	0.01339				
38	0.14	Module 3	39.6177	–0.0585	0.05847				
39	0.2	Cable 2	39.615	–0.0612	0.06119				
40			sum:	–0.1688	0.16876				
41	changed min Gain:			39.5074					
42		min min Gain:			39.5074				
43		max min Gain:			39.8449				

Fig. 2.27 Spreadsheet with changes.

- Unilateral modules in cascade can be represented by their transducer gains and SWRs without complete knowledge of their impedances.
- The range of expected gains can be obtained for a standard cascade of unilateral modules separated by standard-impedance interconnects.
- Bilateral modules can be combined with a unilateral module to make a composite unilateral module that can be included in a cascade of unilateral modules.
- Lossy interconnects reduce the influence of SWR and sufficiently lossy interconnects allow adjacent bilateral modules to be treated as unilateral.
- Gain can be computed for nonstandard cascades of unilateral modules if module input and output impedances are known.
- Such modules, or cascades of them, can be represented as equivalent standard modules and interfaced with the standard (impedance) modules for analysis.
- Spreadsheets can be used to compute sensitivities of cascade parameters to module parameters.

- Spreadsheets can be used to show the maximum variation in a cascade parameter caused by specified variations in module parameters.

ENDNOTES

[1]Other, nonpropagating, electric and magnetic fields can extend through a module port, decaying along a transmission line (e.g., evanescent fields). If the line is short enough, module performance might then be affected by a structure attached to the other end of the line. We are not considering such effects, which are akin to shielding problems.

[2]Although Z_0 for lossy transmission lines can have an imaginary component (Ramo et al., 1984, pp. 249–251; Pozar, 2001, pp. 31–32), we would normally expect and require it to be small. For example, the properties of a 0.2 inch diameter 50-Ω cable, RG58 (Jordan, 1986, pp. 29-27–29-29), indicate that the imaginary part of Z_0 is less than 2% of total at 10 MHz and less than about 0.2% at 100 MHz, based on formulas for the attenuation constant and characteristic impedance in low-loss cables (Ramo et al., 1984, pp. 250–251). We assume $Z_0 = R_0$ for simplicity, but it appears that complex Z_0 can be accommodated if the traveling waves that we define in Section 2.2 (e.g., $\hat{v}_x$ and $\tilde{v}_x$) are taken across the real part of Z_0 (Kurokawa, 1965; Yola, 1961). The traveling voltage would then be higher than $\hat{v}_x$, but $\hat{v}_x$ would appear across the real part of a reflectionless termination Z_0, and p_x in Eq. (2.17) would give the power delivered to that termination. In addition, p_x would be the available power from a source that is matched to the line, that is, one with output impedance Z_0^*, although the voltage at the input to the line would be higher due to what appears across the reactive component.

[3]Some texts have used the inverse of the T parameters that we use here (Dicke, 1948, pp. 150–151; Ramo et al., 1984, pp. 535–539]. These concentrate on passive microwave circuits that are usually bilateral. Many different names have been used to describe T parameters and their inverse: transmission coefficients, T matrix, scattering transfer parameters, chain scattering parameters.

[4]An alternate type of matrix that can be multiplied to form the representation for a cascade uses the $ABCD$ parameters (Pozar, 2001, pp. 53–55). The state vector used there consists of the voltage and current at a terminal rather than the forward and reverse waves.

[5]There are 8.686 dB per neper, which we can see as follows. Since

$$e^{-h} = 10^{-L/(20\text{ dB})} = (e^{\ln 10})^{-L/(20\text{ dB})}, \qquad h = L\frac{\ln 10}{20\text{ dB}},$$

giving $(8.686\text{ dB})h = L$.

[6]

$$|\rho|^2 = \left(\frac{50\ \Omega - 10\ \Omega}{50\ \Omega + 10\ \Omega}\right)^2 = \frac{4}{9}.$$

The part of the forward power that gets into the load is $1 - \frac{4}{9} = \frac{5}{9} = 0.556$.

[7]Available gain is module output power into a matched load divided by source power into a matched (to the source) load. If the source impedance is the complex conjugate of the module input impedance, the input power in the gain definition will be the power actually absorbed in the module. The module output power will then be maximum so the gain will be the maximum available gain.

CHAPTER 3

NOISE FIGURE

The amount of noise added to a signal that is being processed is of critical importance in most RF systems. This addition of noise by the system is characterized by its noise figure (or, alternatively, noise temperature). In this chapter we consider how the noise figure for a simple cascade of modules can be obtained from individual module noise figures. We then extend the concept to standard cascades, voltage-amplifier cascades, and combinations of the three types. We also learn how to account for image noise in mixers.

3.1 NOISE FACTOR AND NOISE FIGURE

Noise factor (Hewlett-Packard, 1983; Haus et al., 1960a) is the signal-to-noise power ratio at the input (1) of a module or cascade divided by the signal-to-noise power ratio at its output (2):

$$f = \frac{(S/N)_{\text{in}}}{(S/N)_{\text{out}}} \tag{3.1}$$

$$= \frac{p_{\text{signal},1}/p_{\text{noise},1}}{p_{\text{signal},2}/p_{\text{noise},2}} \tag{3.2}$$

$$= \frac{p_{\text{noise},2}/p_{\text{noise},1}}{p_{\text{signal},2}/p_{\text{signal},1}}. \tag{3.3}$$

We will use the term *noise figure* (NF) and symbol F for f expressed in dB:

$$F \triangleq 10 \log_{10} f. \tag{3.4}$$

The input noise power $p_{\text{noise},1}$ is, by definition, the thermal (Johnson) noise power from the source at 290 K (about 17°C) into a matched load, the available noise power at that temperature. This theoretical noise level is $p_{\text{noise},1} = \overline{k}T_0B$, where $\overline{k}$ is Boltzmann's constant, T_0 is 290 K, and B is noise bandwidth. The value of $N_T \triangleq \overline{k}T_0$ is approximately 4×10^{-21} W, or -174 dBm, per Hz bandwidth.[1] [Resistors also have flicker noise, which dominates at low frequencies (Egan, 2000, p. 119).] The input signal power $p_{\text{signal},1}$ is the available source power of the signal.

The output powers are also defined into a matched load. The ratio of output power to input power then meets the definition of available gain (see Appendix G). Figure 3.1 shows a noise figure test setup where some of the variables have circumflexes (hats) to identify them with this theoretical setup. Note that the impedance of the source and load must, in general, be changed for each device under test (DUT), the source impedance to correspond to the specified source and the load impedance to match the impedance at the DUT output.

The noise factor is the factor by which the inherent random noise of the source resistance at 290 K would have to increase to account for the additional output noise that is actually produced by the DUT.

An alternate representation of module noise is noise temperature, which is the increase in source temperature that could have accounted for the module noise contribution. We will include both representations in some of the development that follows.

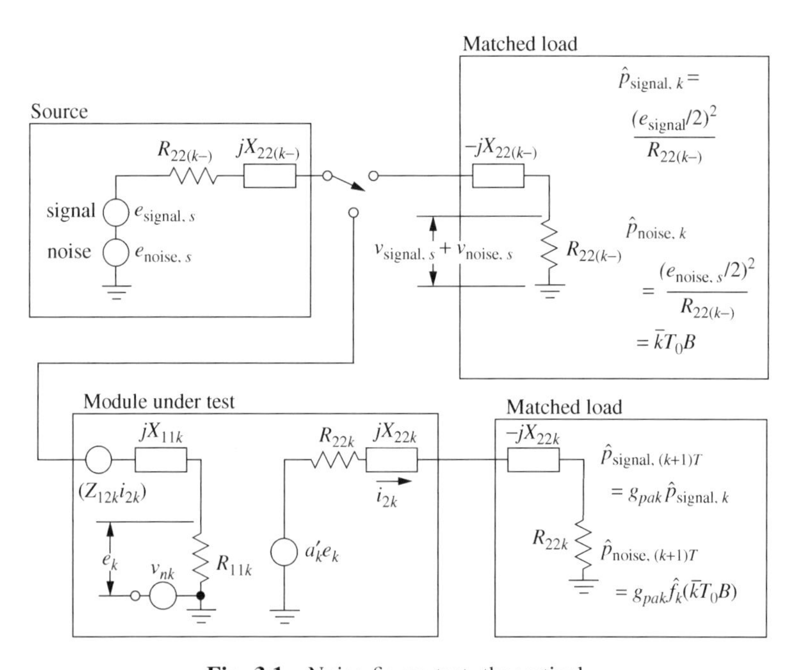

Fig. 3.1 Noise figure test, theoretical.

Noise is usually computed by integrating the noise density N_0 over a frequency band that, by definition of noise bandwidth B, gives the same results as multiplication by the single number B (Egan, 1998, pp. 357–360). This process is accomplished experimentally by measuring the total noise power passing through the passband of the device with two known input noise levels. From these two measurements, the available gain and the noise figure can be computed. (If the lower noise level is the inherent source noise, the higher level can be considered to simulate a broadband signal added to the inherent noise.) Sometimes a narrow filter, centered on the signal frequency, is provided, experimentally or theoretically, and the resulting noise figure is called the *spot noise figure* because it provides information at a particular frequency (spot) rather than averaging it over a wider passband.

We can replace the signal power ratio in Eq. (3.3) with the available power gain g_a and can replace $p_{\text{noise},1}$ with available noise power, giving the theoretical measured noise factor:

$$\hat{f} = \frac{p_{\text{noise},2}/g_a}{\overline{k}T_0 B}. \tag{3.5}$$

This form illustrates that the noise factor is the ratio of actual noise, referenced to the source, to theoretical source noise.

3.2 MODULES IN CASCADE

First we consider a single module with an ideal source and load. Ideally, it would output a noise level that would be the ideal source noise times the gain. Then f would be unity ($F = 0$ dB), and the noise temperature of the module would be absolute zero. Any increase over this amount is due to the module (assuming temperature $T = 290$ K).

The contribution of noise power by module k is the difference between the noise power at its output, $p_{\text{noise},k+1}$, and the ideal source noise, $\overline{k}T_0B$, multiplied by the module gain:

$$\Delta p_{\text{n@out},k} = p_{\text{noise},k+1} - (\overline{k}T_0 B)g_k, \tag{3.6a}$$

which can also be written

$$\Delta p_{\text{n@out},k} = \overline{k}BT_k g_k, \tag{3.6b}$$

where T_k is the noise temperature of module k.

This can be referred to the input of the module by dividing it by the module gain:

$$\Delta p_{\text{n@in},k} = \frac{p_{\text{noise},k+1}}{g_k} - \overline{k}T_0 B \tag{3.7a}$$

or

$$\Delta p_{\text{n@in},k} = \overline{k}BT_k. \tag{3.7b}$$

Here $\Delta p_{\text{n@in},k}$ is the additional noise in the source driving module k that would account for the observed noise. The contribution of the module to the noise factor is this power divided by the inherent source noise:

$$\Delta f_k = \frac{\Delta p_{\text{n@in},k}}{\overline{k} T_0 B}. \tag{3.8}$$

From Eqs. (3.7a) and (3.5) we see that this equals

$$\Delta f_k = \frac{p_{\text{noise},k+1}/g_k - \overline{k} T_0 B}{\overline{k} T_0 B} = f_k - 1, \tag{3.9a}$$

whereas, from Eqs. (3.7b) and (3.5), we see that it also equals

$$\Delta f_k = \frac{T_k}{T_0}. \tag{3.9b}$$

If the module is part of a cascade, its contribution to the cascade noise factor is reduced by the gain g_{pk} preceding the module (the product of the preceding module gains), since the cascade noise factor indicates the effective increase in the noise of the source for the whole cascade:

$$\Delta f_{\text{source},k} = \frac{f_k - 1}{g_{pk}} = \frac{f_k - 1}{\prod_{i=1}^{k-1} g_i} \tag{3.10a}$$

$$= \frac{T_k/T_0}{g_{pk}} = \frac{T_k}{T_0 \prod_{i=1}^{k-1} g_i}. \tag{3.10b}$$

While we have dropped the a subscript on the gain and the circumflex from f, all of the gains here are available power gains and f is still the theoretical noise factor $\hat{f}$.

The total equivalent noise from the source is

$$p_{\text{noise,equiv source}} = \overline{k} T_0 B + \sum_{k=1}^{n} \frac{\Delta p_{\text{n@in},k}}{g_{pk}}. \tag{3.11}$$

We divide Eq. (3.11) by the inherent available source noise power $\overline{k} T_0 B$ to get the total noise factor for the cascade:

$$f_{\text{cas}} = 1 + \sum_{k=1}^{N} \Delta f_{\text{source},k}. \tag{3.12a}$$

We can also divide Eq. (3.11) by $\overline{k}B$ to obtain the noise temperature for a system, source plus cascade:

$$T_{\text{sys}} = T_0 + T_{\text{cas}} = T_0 + \sum_{k=1}^{N} \frac{T_k}{g_{pk}}. \tag{3.12b}$$

By Eq. (3.10a), Eq. (3.12a) is

$$f_{\text{cas}} = 1 + \sum_{k=1}^{N} \frac{f_k - 1}{g_{pk}} = 1 + \sum_{k=1}^{N} \frac{f_k - 1}{\prod_{i=1}^{k-1} g_i}. \tag{3.13}$$

There is no gain preceding the first module so the denominator should be 1 for $k = 1$. This can be made clearer if the contribution from the first cascade element, $f_1 - 1$, is written separately. This also has the advantage of not requiring some unnecessary arithmetic.

$$f_{\text{cas}} = f_1 + \sum_{k=2}^{N} \frac{f_k - 1}{g_{pk}} = f_1 + \sum_{k=2}^{N} \frac{f_k - 1}{\prod_{i=1}^{k-1} g_i}. \tag{3.14}$$

This expression is somewhat awkward to compute because noise figure and gain (F and G) are usually given in dB and they must be converted from dB, using, for example,

$$f = 10^{F/(10\ \text{dB})}, \tag{3.15}$$

before they can be used in Eq. (3.14). Of course, ΣG can be computed before conversion to $\prod g$, but the summation in (3.14) cannot be done before all variables are converted from dB.

For two elements in cascade ($N = 2$), Eq. (3.14) simplifies to

$$f_{\text{cas}} = f_1 + (f_2 - 1)/g_1. \tag{3.16}$$

Example 3.1 Cascade Noise Figure Two modules in series each have a 3-dB noise figure and a 6-dB gain. What is the cascade noise figure?

From Eq. (3.14),

$$f_{\text{cas}} = 10^{3\ \text{dB}/10\ \text{dB}} + \frac{10^{3\ \text{dB}/10\ \text{dB}} - 1}{10^{6\ \text{dB}/10\ \text{dB}}} = 2 + 0.25 = 2.25 \Rightarrow F_2 = 3.52\ \text{dB}. \tag{3.17}$$

What will be the noise figure if another such stage is added to the cascade?

$$\begin{aligned} f_{\text{cas}} &= 10^{3\ \text{dB}/10\ \text{dB}} + \frac{10^{3\ \text{dB}/10\ \text{dB}} - 1}{10^{6\ \text{dB}/10\ \text{dB}}} + \frac{10^{3\ \text{dB}/10\ \text{dB}} - 1}{10^{12\ \text{dB}/10\ \text{dB}}} \\ &= 2 + 0.25 + 0.0625 = 2.31 \Rightarrow F_3 = 3.64\ \text{dB}. \end{aligned} \tag{3.18}$$

Here we can see that the noise factor has less effect further down the cascade where it is preceded by more gain.

All of this has been done for a source temperature of T_0 in accordance with the definition of noise figure. If the operational source temperature is T_s, Eq. (3.12b) can be modified to give a system noise temperature of

$$T_{\text{sys,op}} = T_s + \sum_{k=1}^{N} \frac{T_k}{g_{pk}}. \tag{3.19}$$

The source is often an antenna and the source temperature is then identified as $T_s = T_{\text{ant}}$.

The value of $T_{\text{sys,op}}$ determines how much noise occurs at the output of the system in its operational environment, where the source temperature is T_s, and this is the equation of importance in determining system performance. However, once the allowable value of the summation term T_{cas} has been determined, T_{sys} in Eq. (3.12b) can be computed with $T_s = T_0$ and, from that, f_{cas} can be obtained, permitting the required cascade noise factor or noise figure to be specified. These relationships are summarized in Table 3.1.

Example 3.2 Specifying Noise Figure to Meet System Requirement What noise figure is required for the cascade so the system noise temperature will be 400 K when the source temperature is 50 K (perhaps from an antenna looking at a cool sky)?

From Eq. (3.19), in the operating environment,

$$T_{\text{sys,op}} = 400 \text{ K} = 50 \text{ K} + T_{\text{cas}}, \tag{3.20}$$

leading to

$$T_{\text{cas}} = \sum_{k=1}^{N} \frac{T_k}{g_{pk}} = 350 \text{ K}. \tag{3.21}$$

Then Eq. (3.12b) gives, at the standard source temperature,

$$T_{\text{sys}} = T_0 + 350 \text{ K} = 640 \text{ K}. \tag{3.22}$$

Dividing by T_0, we obtain the allowed noise figure:

$$f_{\text{cas}} = \frac{T_{\text{cas}}}{T_0} + 1 = \frac{T_{\text{sys}}}{T_0} = 2.21 \Rightarrow 3.44 \text{ dB}. \tag{3.23}$$

TABLE 3.1 Summary of Noise Relationships

Source T		$f_k - 1 = T_k/T_0$ (1)	
T_0	Noise at output of **module** k having gain g_k	$(\bar{k}BT_0)f_k g_k$ (2F)	$(\bar{k}B)(T_0 + T_k)g_k$ (2T)
	Equivalent noise at **module** source	$(\bar{k}BT_0)f_k$ (3F)	$(\bar{k}B)(T_0 + T_k)$ (3T)
T_s	Noise at output of **module** k having gain g_k		$(\bar{k}B)(T_s + T_k)g_k$ (4T)
	Equivalent noise at **module** source		$(\bar{k}B)(T_s + T_k)$ (5T)
Any	Equivalent **module** source noise due to module k	$(\bar{k}BT_0)(f_k - 1)$ (6F)	$(\bar{k}B)T_k$ (6T)
	Equivalent **cascade** source noise due to module k, preceded by gain g_{pk}	$(\bar{k}BT_0)\dfrac{f_k - 1}{g_{pk}}$ (7F)	$(\bar{k}B)\dfrac{T_k}{g_{pk}}$ (7T)
	Equivalent **cascade** source noise due to all modules	$(\bar{k}BT_0)\left[f_1 - 1 + \dfrac{f_2 - 1}{g_{p2}} + \dfrac{f_3 - 1}{g_{p3}} + \cdots\right]$ (8F) $\equiv (\bar{k}BT_0)\left[f_1 - 1 + \sum_{k=2}^{n} \dfrac{f_k - 1}{g_{pk}}\right]$ (9F) $= (\bar{k}BT_0)(f_{cas} - 1)$ (10F) where $f_{cas} \equiv f_1 + \sum_{k=2}^{n} \dfrac{f_k - 1}{g_{pk}}$ (11F)	$(\bar{k}B)\left[T_1 + \dfrac{T_2}{g_{p2}} + \dfrac{T_3}{g_{p3}} + \cdots\right]$ (8T) $\equiv (\bar{k}B)\left[T_1 + \sum_{k=2}^{n} \dfrac{T_k}{g_{pk}}\right]$ (9T) $= (\bar{k}B)T_{cas}$ (10T) where $T_{cas} \equiv T_1 + \sum_{k=2}^{n} \dfrac{T_k}{g_{pk}}$ (11T)
T_0	Equivalent **system** source noise	$(\bar{k}BT_0)f_{cas}$ (12F)	$(\bar{k}B)T_{sys}$ (12T) where $T_{sys} = T_0 + T_{cas}$ (13T)
T_s			$(\bar{k}B)T_{sys,op}$ (14T) where $T_{sys,op} = T_s + T_{cas}$ (15T)

3.3 APPLICABLE GAINS AND NOISE FACTORS

For several practical reasons, noise factor is ordinarily measured using a standard source impedance. This is the theoretical noise factor only if the tested module is to be driven by that standard impedance in the cascade, a usual, but practically unattainable, goal.

While the gains in Eq. (3.13) are supposed to be available gains, Appendix N shows that the gains that we have used in Section 2.3 for our standard cascade are appropriate when using noise factors as they are usually measured, assuming unilateral modules ($Z_{12k} = 0$) with isolated noise sources. In other words, the theoretical relationship involving $\hat{f}$ and g_a also applies to f and g as defined for our standard cascade. We have represented the noise source in Fig. 3.1 as isolated, making its contribution independent of the driving source. While this is important to our analysis, we would expect to see some dependence of module noise on the impedance of the driving source. This will be considered in Section 3.8.

Figure 3.2 illustrates the usual method for determination of noise factor for a module and its contribution to the noise factor of a cascade. In both cases, the noise from an effective source that would produce the observed output noise is to be compared to the ideal source noise. Switch position 1 would be used to measure (actually or theoretically) these values. Unlike Fig. 3.1, the source in Fig. 3.2 has standard interface impedance R_0.

During module test, switch position 3 would be used to send the available source power through a cable (of standard interface impedance R_0) to the module.

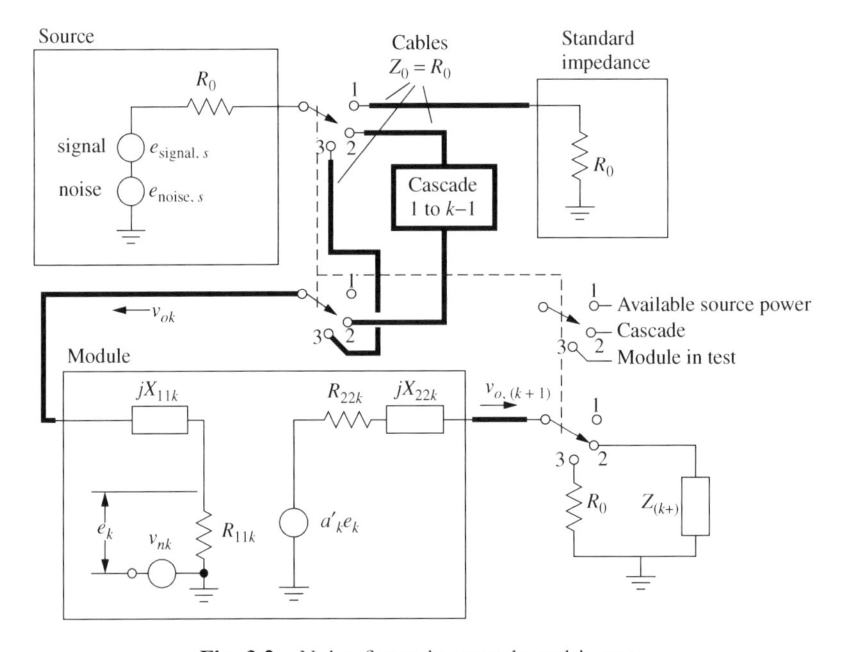

Fig. 3.2 Noise figure in cascade and in test.

Theoretically, if we could turn off the noise source in the module, we could then increase $e_{\text{noise},s}$ until the noise level at $v_{o,(k+1)}$ would be reestablished. Then we could move to switch position 1 and measure the increased noise level. The ratio of this level to the originally measured thermal noise would be the module noise factor. Since we cannot actually do this, we compute what would happen if we did.

In the cascade (switch position 2), the part of the cascade preceding the module would replace the cable from the source. If we could follow the same theoretical procedure that we have just described for the module, removing only the module noise, we could measure the module's contribution to the cascade noise figure. Again, we compute what we cannot measure directly.

The module test will establish the increase in the noise in the forward wave v_{ok} that is required to reproduce the observed module noise in a noiseless module. This will be the same whether the module is being tested or is in a cascade. Once this is established, the effective increase of the available noise in the source can be related to the noise in v_{ok} by the gain from the source to v_{ok} in the cascade. Because v_{ok} is the variable we have used in our standard-cascade calculations, the gains employed there also apply to noise figure calculations.

While R_0 is usually the same for all modules and the cascade, this is not necessary. There can be a change in the standard impedance along the cascade. Where this occurs, the input and output of some module (and their interconnects) would have different standard impedances. Each module would be tested with its standard input impedance (in switch positions 3 and 1), and the cascade would be tested with its standard input impedance (in switch positions 2 and 1).

We now show how the contribution from lossy interconnects is appropriately incorporated in our model.

3.4 NOISE FIGURE OF AN ATTENUATOR

The noise figure of a (ideal) passive attenuator at a temperature of T_0 (290 K) equals its attenuation. This is because the available noise at the output of the attenuator is the available noise from the Thevenin resistance of the attenuator, presumably the same as the standard impedance of the cables at that point in the cascade. This is the same as the available noise from the source, at the input to the attenuator, during characterization. Thus the noises in Eq. (3.1) cancel and f becomes the ratio of input signal power to output signal power, which equals the attenuation.

If we did a circuit-noise analysis of an attenuator, say a π or T network, we would get the same results (but less efficiently). We can do it either way (but must not add the two effects).

The combined noise figure of a module preceded by an attenuator at T_0 equals the module noise figure plus the attenuation. (The gain of the combination is, of course, lowered by the attenuation also.) To see this, write Eq. (3.14) for an attenuator followed by a module, using $1/g_1$ for the attenuation of the attenuator:

$$f = \frac{1}{g_1} + \frac{f_2 - 1}{g_1} = \frac{f_2}{g_1}. \tag{3.24}$$

In dB, this is

$$F = F_2 + (-G_1), \tag{3.25}$$

where $-G_1$ is the attenuation of the attenuator ($F > F_2$ because $G_1 < 1$). Here g_1 is available power gain, which suits well the definition of the attenuation.

If the attenuator is at a temperature T, the output noise that is not attributable to the source (which is at T_0 by definition) changes proportionally to T, giving a noise factor of (Pozar, 2001, p. 91)

$$f(T) = 1 + (1/g - 1)T/T_0, \tag{3.26}$$

which reduces to $1/g$ at $T = T_0$.

3.5 NOISE FIGURE OF AN INTERCONNECT

The transmission line interconnects, described in Section 2.3.2, will generally have some loss, but the gain we have ascribed to them also involves the effects of multiple reflections, so we might suspect that they do not act like simple attenuators. A lengthy analysis in Appendix N, Section N.6, shows that the proper noise figure for an interconnect in a standard cascade at $T = T_0$ is

$$f_{\text{cbl}} = 1/g_2 + |\rho_1|^2(1 - g_2), \tag{3.27}$$

where $1/g_2$ is the attenuation of the properly terminated interconnect and ρ_1 is the reflection coefficient looking into the output of the preceding module. This can also be expressed as

$$f_{\text{cbl}}(\text{SWR}) = \frac{1}{g_2} + \left[\frac{\text{SWR}_1 - 1}{\text{SWR}_1 + 1}\right]^2 (1 - g_2). \tag{3.28}$$

If the cable is at a temperature other than T_0, f_{cbl} will be modified in a manner similar to the change in f for a simple attenuator [Eq. (3.26)]:

$$f_{\text{cbl}}(T, \text{SWR}) = 1 + [f_{\text{cbl}}(\text{SWR}) - 1]T/T_0. \tag{3.29}$$

This general expression includes Eqs. (3.26) and (3.28) as particular cases.

3.6 CASCADE NOISE FIGURE

Example 3.3 Cascade Noise Figure Figure 3.3 shows the spreadsheet used in the previous analysis with added noise figure information. We compute the cascade noise figure for several combinations of values of noise figures and gains. Cells G4–H10 give mean and maximum noise figures defined for the modules. The interconnect noise figures, in cells G to L, 15, 17, and 19, are obtained

	A	B	C	D	E	F	G	H	I	J	K	L
2		Gain	Gain	SWR			NF			Temperature		
3		nom	+/–	at out	$\lvert a_{RT}\rvert$		mean	max		290 K		
4	Module 1	12.0 dB	1.0 dB	1.5			2.0 dB	2.6 dB				
5	Cable 1	–1.5 dB		1.5	0.0283							
6	Module 2	10.0 dB	2.0 dB	2			4.0 dB	5.0 dB				
7	Attenuator	–8.0 dB	0.5 dB	1.5	0.0106							
8	Module 3	7.0 dB	2.0 dB	2.8			3.0 dB	3.7 dB				
9	Cable 2	–0.8 dB		3.2	0.2064							
10	Module 4	30.0 dB	2.0 dB				5.0 dB	5.5 dB				
11	DERIVED											
12				Gain			NF using mean NFs (see Note *) at			NF using max NFs (see Note *) at		
13		mean	max	min	±	σ	mean G	max G	min G	mean G	max G	min G
14	Module 1	12.00 dB	13.00 dB	11.00 dB	1.00 dB	0.50 dB	2.00 dB	2.00 dB	2.00 dB	2.60 dB	2.60 dB	2.60 dB
15	Cable 1	–1.50 dB	–1.25 dB	–1.74 dB	0.25 dB	0.17 dB	1.54 dB	1.54 dB	1.54 dB	1.54 dB	1.54 dB	1.54 dB
16	Module 2	10.00 dB	12.00 dB	8.00 dB	2.00 dB	1.00 dB	4.00 dB	4.00 dB	4.00 dB	5.00 dB	5.00 dB	5.00 dB
17	Attenuator	–8.00 dB	–7.41 dB	–8.59 dB	0.59 dB	0.41 dB	8.06 dB	7.57 dB	8.56 dB	8.06 dB	7.57 dB	8.56 dB
18	Module 3	7.00 dB	9.00 dB	5.00 dB	2.00 dB	0.80 dB	3.00 dB	3.00 dB	3.00 dB	3.70 dB	3.70 dB	3.70 dB
19	Cable 2	–0.61 dB	1.21 dB	–2.43 dB	1.82 dB	1.27 dB	0.93 dB	0.93 dB	0.93 dB	0.93 dB	0.93 dB	0.93 dB
20	Module 4	30.00 dB	32.00 dB	28.00 dB	2.00 dB	1.30 dB	5.00 dB	5.00 dB	5.00 dB	5.50 dB	5.50 dB	5.50 dB
21	CUMULATIVE											
22				Gain			NF using mean NFs at			NF using max NFs at		
23	at output of	mean	max	min	±	σ	mean G	max G	min G	mean G	max G	min G
24	Module 1	12.00 dB	13.00 dB	11.00 dB	1.00 dB	0.50 dB	2.00 dB	2.00 dB	2.00 dB	2.60 dB	2.60 dB	2.60 dB
25	Cable 1	10.50 dB	11.75 dB	9.26 dB	1.25 dB	0.53 dB	2.07 dB	2.06 dB	2.09 dB	2.66 dB	2.65 dB	2.68 dB
26	Module 2	20.50 dB	23.75 dB	17.26 dB	3.25 dB	1.13 dB	2.42 dB	2.32 dB	2.55 dB	3.09 dB	2.98 dB	3.24 dB
27	Attenuator	12.50 dB	16.34 dB	8.67 dB	3.84 dB	1.20 dB	2.54 dB	2.37 dB	2.82 dB	3.20 dB	3.02 dB	3.48 dB
28	Module 3	19.50 dB	25.34 dB	13.67 dB	5.84 dB	1.45 dB	2.67 dB	2.43 dB	3.12 dB	3.35 dB	3.09 dB	3.82 dB
29	Cable 2	18.89 dB	26.55 dB	11.24 dB	7.66 dB	1.93 dB	2.68 dB	2.43 dB	3.14 dB	3.36 dB	3.09 dB	3.84 dB
30	Module 4	48.89 dB	58.55 dB	39.24 dB	9.66 dB	2.32 dB	2.74 dB	2.44 dB	3.47 dB	3.42 dB	3.10 dB	4.17 dB
31		* Cable NF is based on SWRs, which are taken as fixed for analysis.										

Fig. 3.3 Spreadsheet with noise figures.

using Eqs. (3.28) and (3.29). The temperature is entered in cell J3. SWRs are assumed to be fixed at the values given in cells D4–D9 so f_{cbl} varies only if its attenuation (cells B5, B7, and B9) has a specified variation (cells C5, C7, and C9). In this example, a variation is given for the attenuator (line 7) but not for the other interconnects.

Cumulative noise figure (cells G24–L30) through stage j is computed according to Eq. (3.16), where the subscript 1 refers to the cascade preceding stage j and 2 refers to stage j. If all modules and interconnects were treated separately, using Eq. (3.14), the results would be the same but the formulas would be longer.

3.7 EXPECTED VALUE AND VARIANCE OF NOISE FIGURE

Figure 3.3 gives the noise figure when all gains are mean, but not the mean, or expected, noise figure. As can be seen from a plot of the computed values (Fig. 3.4), the mean noise figure should be expected to be higher than the noise figure at the mean gain since it increases more at low gains than it decreases with the same deviation on the high side. A Monte Carlo analysis would give us a distribution from which we could obtain mean gain and standard deviation or variance. Short of that, we might estimate the mean value as being on the high side of the value obtained with mean gains (e.g., 2.9 or 3 dB with mean noise figures in Fig. 3.4).

For small variances we can use a sensitivity analysis to determine the variance of the noise figure of a cascade from the variances of individual element parameters according to (see Appendix V)

$$\sigma_{F_{\text{cas}}}^2 = \sum_i [\hat{S}_{fi}^2 \sigma_{fi}^2 + \hat{S}_{gi}^2 \sigma_{gi}^2 + \hat{S}_{\text{SWR}i}^2 \sigma_{\text{SWR}i}^2]. \tag{3.30}$$

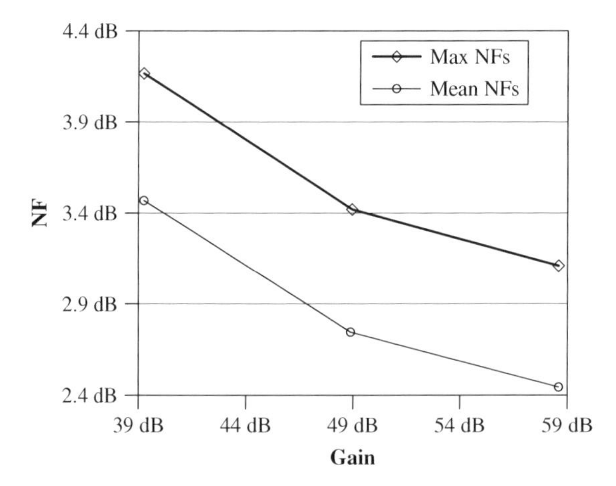

Fig. 3.4 Cascade noise figure from Fig. 3.3.

The sensitivities $\hat{S}_{xi}$ can be determined by making small changes in the variables and observing their effects on F_{cas}. Except for the variables involved, this is similar to what was done in Example 2.6 (see Fig. 2.26), and the spreadsheet can be used to aid in computing $\hat{S}_{xi}$, as is done there, and in giving the variance according to Eq. (3.30) once the sensitivities have been determined.

Unfortunately, this process is somewhat time consuming and has to be done anew whenever the system is modified so we would like to obtain Eq. (3.30) in closed form. This can be rather complex but is done in Appendix V for the simplified case where only the module noise figures vary (i.e., with fixed gains and fixed SWRs). In this case, we can write the resulting variance of the cascade noise figure $F_{cas,n}$ at stage n in terms of the noise figure $F_{cas,(n-1)}$ one stage earlier as

$$\sigma^2_{F_{cas,n}} = 10^{-F_{cas,n}/5\ dB}\{10^{F_{cas(n-1)}/5\ dB}\sigma^2_{F_{cas(n-1)}} + 10^{(F_{cas,n}-G_{cas(n-1)})/5\ dB}\sigma^2_{F_n}\}, \quad (3.31)$$

where $G_{cas(n-1)}$ is the cascade gain through the previous stage and F_n is the noise figure of the nth stage. This restriction of variances to module noise figures is consistent with our spreadsheet where the SWRs are fixed and where computations are made for several sets of fixed gains.

In Fig. 3.5 some cells not of current interest have been removed from Fig. 3.3, and two columns of cumulative estimated noise figure standard deviations have been added at cells I25–J31. Equation (3.31) has been implemented in these cells. The cells from which data is drawn for cell J29 (its precedents) are indicated by arrows, with circles at their origins (under Excel 98's menu item, Tools; Auditing; Trace Precedents).

Cell I31 gives $\sigma_{F_{cas}}$ when all elements have mean gains and cell J31 gives it for minimum gains, in which case F_{cas} (cell H31) is maximum. Note that, in this example, the variance of F_{cas} decreases as elements are added. This is a variance of noise figure in dB and therefore represents a larger absolute variance as the value of F_{cas} to which it applies increases. Let us now consider a potential source of variations in the module noise factors.

3.8 IMPEDANCE-DEPENDENT NOISE FACTORS

We have represented the noise contribution of a module by an equivalent noise source at the input to the cascade. This can be multiplied by the transducer gain to the module output to obtain the noise delivered to a standard impedance at the output of the module. It can also be multiplied by the transducer gain to the module's input to determine the equivalent noise that would be delivered to a standard impedance there, or it can be multiplied by available gain to obtain the noise that would be delivered to a matched load.

If the module noise source is isolated, the equivalent cascade source can be computed using a module noise factor that was measured in a standard-impedance environment. Since this determines the noise power that would be delivered to a

	A	B	C	D	E	F	G	H	I	J
2		Gain	Gain	SWR		Temp.	NF			
3		nom	+/−	at out	$\lvert a_{RT} \rvert$	290 K	mean	max	σ	
4	Module 1	12.0 dB	1.0 dB	1.5			2.0 dB	2.6 dB	0.3 dB	
5	Cable 1	−1.5 dB		1.5	0.0283					
6	Module 2	10.0 dB	2.0 dB	2			4.0 dB	5.0 dB	0.6 dB	
7	Attenuator	−8.0 dB	0.5 dB	1.5	0.0106					
8	Module 3	7.0 dB	2.0 dB	2.8			3.0 dB	3.7 dB	0.4 dB	
9	Cable 2	−0.8 dB		3.2	0.2064					
10	Module 4	30.0 dB	2.0 dB				5.0 dB	5.5 dB	0.3 dB	
11	DERIVED									
12		Gain					NF using mean NFs at			
13		mean	max	min	±	σ	mean G	min G		
14	Module 1	12.00 dB	13.00 dB	11.00 dB	1.00 dB	0.50 dB	2.00 dB	2.00 dB		
15	Cable 1	−1.50 dB	−1.25 dB	−1.74 dB	0.25 dB	0.17 dB	1.54 dB	1.54 dB		
16	Module 2	10.00 dB	12.00 dB	8.00 dB	2.00 dB	1.00 dB	4.00 dB	4.00 dB		
17	Attenuator	−8.00 dB	−7.41 dB	−8.59 dB	0.59 dB	0.41 dB	8.06 dB	8.56 dB		
18	Module 3	7.00 dB	9.00 dB	5.00 dB	2.00 dB	0.80 dB	3.00 dB	3.00 dB		
19	Cable 2	−0.61 dB	1.21 dB	−2.43 dB	1.82 dB	1.27 dB	0.93 dB	0.93 dB		
20	Module 4	30.00 dB	32.00 dB	28.00 dB	2.00 dB	1.30 dB	5.00 dB	5.00 dB		
21	CUMULATIVE									
22							cum. NF using		cum. NFσ using	
23		Gain					mean NFs at		mean NF at	
24	at output of	mean	max	min	±	σ	mean G	min G	mean G	min G
25	Module 1	12.00 dB	13.00 dB	11.00 dB	1.00 dB	0.50 dB	2.00 dB	2.00 dB	0.30 dB	0.30 dB
26	Cable 1	10.50 dB	11.75 dB	9.26 dB	1.25 dB	0.53 dB	2.07 dB	2.09 dB	0.30 dB	0.29 dB
27	Module 2	20.50 dB	23.75 dB	17.26 dB	3.25 dB	1.13 dB	2.42 dB	2.55 dB	0.28 dB	0.27 dB
28	Attenuator	12.50 dB	16.34 dB	8.67 dB	3.84 dB	1.20 dB	2.54 dB	2.82 dB	0.27 dB	0.26 dB
29	Module 3	19.50 dB	25.34 dB	13.67 dB	5.84 dB	1.45 dB	2.67 dB	3.12 dB	0.26 dB	0.25 dB
30	Cable 2	18.89 dB	26.55 dB	11.24 dB	7.66 dB	1.93 dB	2.68 dB	3.14 dB	0.26 dB	0.25 dB
31	Module 4	48.89 dB	58.55 dB	39.24 dB	9.66 dB	2.32 dB	2.74 dB	3.47 dB	0.26 dB	0.23 dB

Fig. 3.5 Spreadsheet with noise figure variances and showing data sources for cell J29.

standard impedance, we can find the equivalent cascade source noise power by dividing by transducer gain.

However, if the module noise source is not isolated, if its value depends on the source impedance, accurate determination of the module noise factor requires that it be measured using the same source impedance that the module sees in the cascade. That measurement determines the equivalent noise power that would be delivered by the driving source to a matched load at the module input so the equivalent cascade noise source is obtained by dividing that power by available gain (i.e., the gain into a matched load) from the cascade input to the module input. Multiplying the equivalent cascade noise source, so obtained, by transducer gain still determines how much noise is delivered to a standard impedance, but we cannot, without loss of accuracy, use a noise factor that was measured in a standard-impedance environment to find the value of the equivalent cascade noise source.

3.8.1 Representation

The dependence of noise factor on input impedance has been represented as shown in Fig. 3.6 (Haus et al., 1960b). Here a noisy module (1–2) consists of

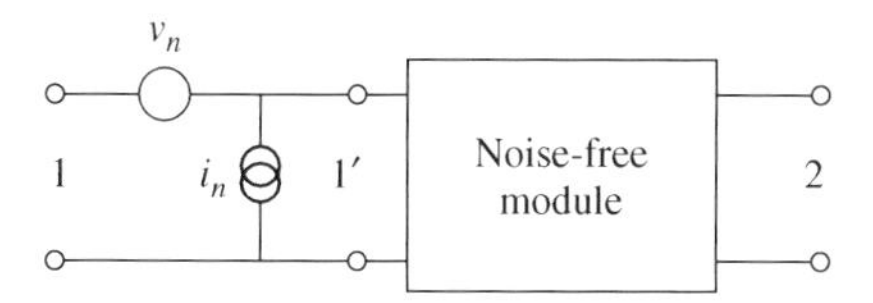

Fig. 3.6 Module with input noise sources.

a noise-free module (1′–2) proceeded by a pair of noise sources. (The noise sources, voltage v_n and current i_n, are often specified for op amps, for example.) These two sources are, in general, partly correlated and this must be taken into account. All of the noise in the module can be represented by i_n and v_n, and these can be used to determine the dependence of noise figure on source impedance.

For completeness, it might seem that another pair of sources would be required at the output to represent the dependence of noise figure on load impedance. However, there is no such dependency. Whereas the noise sources in Fig. 3.6 can be absorbed into the driving source when noise factor is determined, the load identically converts all preceding sources, signal or noise, to output power. Therefore, the ratio of signal to noise does not depend on load impedance. If we should redefine port 1 as the output, we could then show that noise appearing in the *source* depends on the *load* impedance, so there is a symmetry.

The source-dependent noise factor can be expressed as

$$\hat{f} = f_0 + \frac{R_n}{G_s}[(G_s - G_0)^2 + (B_s - B_0)^2] \tag{3.32}$$

$$= f_0 + \frac{R_n}{G_s}|Y_s - Y_0|^2. \tag{3.33}$$

Here $Y_s = G_s + jB_s$ is the source admittance connected to port 1 and Y_0 is the optimum value of that source admittance, for which $\hat{f}$ has its minimum value, f_0. Part of Y_0 represents the correlation between the two sources; R_n is a constant, called the equivalent noise resistance. We mark $\hat{f}$ as a theoretical noise factor because Fig. 3.1 represents its test procedure wherein

$$Y_s = \frac{1}{R_{22(k-)} + jX_{22(k-)}}. \tag{3.34}$$

3.8.2 Constant-Noise Circles

For given values of $\hat{f}$ and f_0, Eq. (3.33) describes a circle on the Smith chart (Gonzalez, 1984, pp. 142–145; Pozar, 2001, pp. 214–216; Section F.5). Figure 3.7 shows two such circles. The one for $\hat{f} = \hat{f}_2$ passes through the point that represents a particular source admittance Y_s', indicating that, with that source admittance, the module has noise factor $\hat{f}_2$.

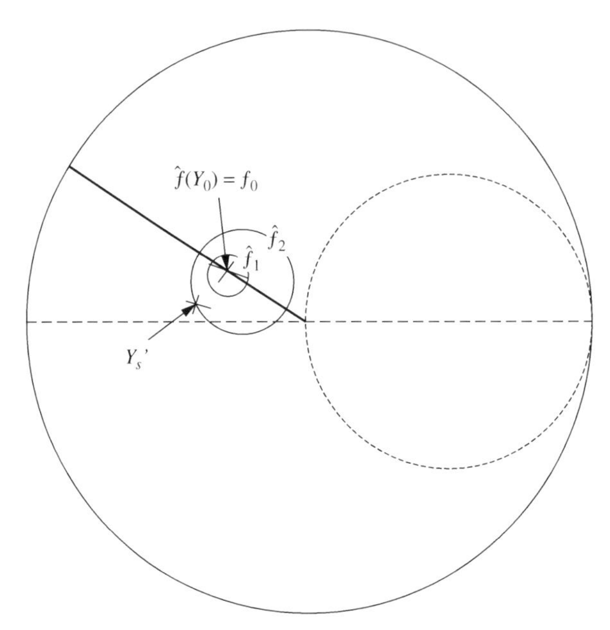

Fig. 3.7 Constant $\hat{f}$ curves on Smith chart. These are theoretical noise factors $\hat{f}$ rather than standard noise factors f.

If the source impedance seen by the module changes while the reflection coefficient (SWR) remains constant, as when the length of a lossless interconnect changes or the phase of the reflection, but not its magnitude, changes, the impedance (and admittance) seen by the module will be represented by a circle, as shown in Fig. 3.8. Here additional constant-$\hat{f}$ curves have been drawn. We see that the noise figure varies between $\hat{f}_1$ and $\hat{f}_4$ as the phase goes through all values. This shows us the range of noise factors corresponding to a given SWR. Ideally, the SWR will be small so $\hat{f}$ will not change much. It also helps if the optimum f_0 occurs at the standard impedance value R_0, in the center of the Smith chart.

3.8.3 Relation to Standard Noise Factor

In the center of the chart, $\hat{f} = f$ since the standard noise factor occurs when the source impedance is the standard impedance R_0. Elsewhere on the chart the theoretical noise factor $\hat{f}$ for the given source impedance (Fig. 3.1) is shown. Our standard noise factor, referred to a cascade input as described in Section 3.3, accurately indicates the cascade noise figure if the noise source is isolated (Figs. 3.1 and 3.2). Even this isolated noise source produces theoretical noise factors that are represented as shown in Fig. 3.8 (see Appendix N). Therefore, a noise figure that is described by constant-noise-figure circles on a Smith chart does not imply that our standard treatment is inaccurate.

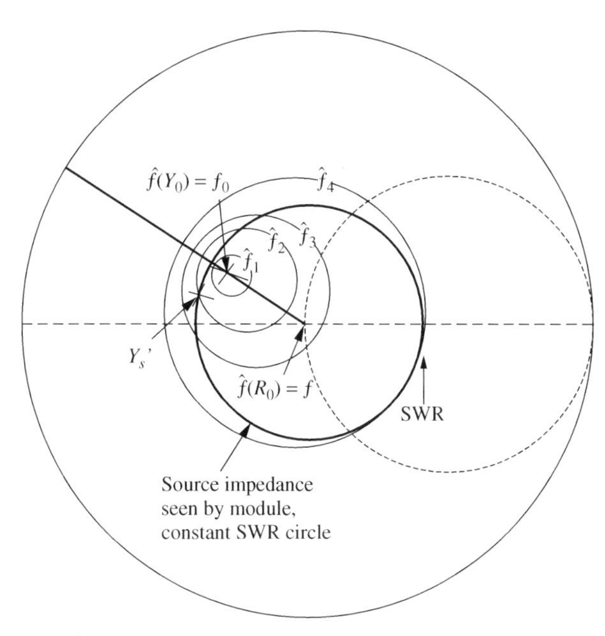

Fig. 3.8 Locus of $\hat{f}$ with changing line length. In the center, the theoretical noise factor $\hat{f}$ is the same as standard noise factor f.

We can check on the accuracy of our treatment that uses an isolated noise source by comparing $\hat{f}$, given by constant-noise-figure circles for a particular module, to $\hat{f}$ calculated (as shown in the next paragraph) for our isolated-source model. We can make the comparison along a circle representing the SWR seen at the output of the cable that drives the module whose noise figure is under consideration. If Fig. 3.8 represents $\hat{f}$ for a module and the constant-SWR circle represents the impedance at the output of the cable, we can compare $\hat{f}$ computed for an isolated noise source to that indicated by the constant-noise circles. If the value of $\hat{f}$ is the same in both cases, the noise source is isolated, as assumed. Otherwise, the ratio of the two noise factors will indicate how much correction is required to f. Essentially, we could consider f to be a function of the source impedance as we move along the constant-SWR circle.

The value of $\hat{f}_k$, for module k having an *isolated noise source*, can be computed at a point **P** on the Smith chart, from

$$\frac{\hat{f}_k - 1}{f_k - 1} = \frac{|Z_{11k} + Z_{22(k-)}|^2/R_{22(k-)}}{|Z_{11k} + R_0|^2/R_0}, \tag{3.35}$$

where Z_{22k-} is the impedance at **P**, f_k is the noise factor in the center of the chart, and Z_{11k} is the impedance looking into the input of module k. Equation (3.35) is developed in Section N.2. It is reasonable to expect that Z_{11k} will be known if $\hat{f}_k$ is known in such detail.

3.8.4 Using the Theoretical Noise Factor

The SWR at the cable output can be obtained from the SWR specified for the preceding module output by converting SWR to reflection coefficient ρ, reducing ρ, by the round-trip loss in the cable, and reconverting to SWR (see Section F.2).

As we move around the circle that represents maximum SWR, if $\hat{f}_k(Z_{22k-})$ deviates from the value given by Eq. (3.35), we might use that deviation in establishing the tolerance for f_k. We have given up some information, though, because the gain that references $(f_k - 1)$ to the preceding module also depends on the variation in output impedance around the constant-SWR circle. Thus we might, for example, use maximum noise factor with minimum gain even though they do not occur at the same point on the circle.

We can retain more information by using $\hat{f}_k$, rather than f_k, for a particular module for which it is known, but we must then reference the added noise to the cascade input using *available gain*. Available gain is higher than the transducer gain into R_0 by a factor,

$$\frac{g_a}{g_t} = \frac{1}{1 - |\rho|^2}, \tag{3.36}$$

where ρ corresponds to the SWR for the circle [see Appendix N, Eq. (36)]. The gain to the output of the previous module in a standard cascade is the transducer gain $g_{tp,k-1}$ for that part of the cascade (Fig. 3.9). To obtain the available gain g_{apk} at the module input, decrease $g_{tp,k-1}$ by the one-way loss of the cable, $1/|\tau|^2$, and then divide by $(1 - |\rho|^2)$. Thus Eq. (3.10a) becomes

$$\Delta f_{\text{source},k} = \frac{1 - |\rho|^2}{g_{tp,k-1}|\tau|^2}(\hat{f}_k - 1). \tag{3.37}$$

The contribution to the cascade noise factor, $(\hat{f}_k - 1)$, is thereby divided by g_{apk} to reference it to the input.

By this procedure, we refer a varying noise factor $\hat{f}_k$ to the cascade input using a gain g_a that is independent of the reflections in the preceding cable. In the standard procedure, the gain varies due to varying phases but f is fixed. The results are the same for an isolated noise source (see proof in Section N.4).

If we know $Z_{22(k-)}$ (i.e., the location on the SWR circle), we can obtain $\Delta f_{\text{source},k}$ exactly. Otherwise, we obtain a *range* of values for $\Delta f_{\text{source},k}$. While

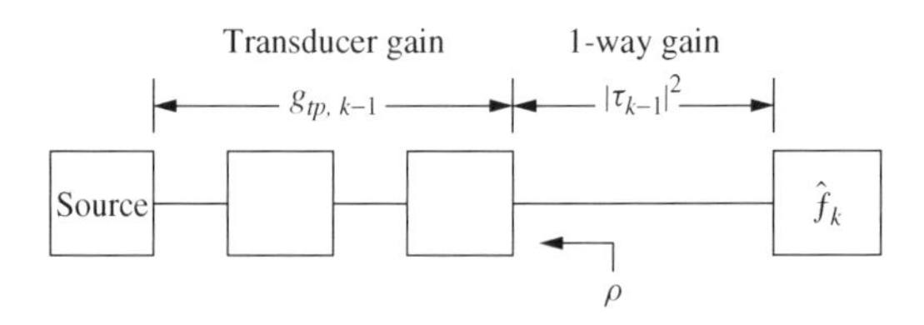

Fig. 3.9 Power gains for referencing theoretical noise factor to source.

the process that we have established for summing the effects of noise contributions and variations in the standard cascade will be modified when one or more modules are to be treated differently, all of the contributions at the source $\Delta f_{\text{source},k}$ must be summed [Eq. (3.12)], no matter how obtained.

Perhaps the most likely module to be treated in a special manner is the first amplifier in a system since it is not preceded by gain and is therefore very influential in establishing noise figure. For this case, $g_{tp,k-1}$ in Eq. (3.37) would be 1. However, rather than taking the source (perhaps an antenna) as characterized by a SWR in a standard-impedance system, more information could be obtained if the actual impedance of the source were used, plotting it on the same Smith chart with the constant-noise circles. Then the system signal and noise levels at the output of the amplifier could be established by using that noise factor and the gain of the amplifier when driven by the actual source.

3.8.5 Summary

- The effect of an isolated noise source is simply represented in the standard cascade.
- If a plot of constant-noise circles is available for a module, it may be used to verify that the noise source is isolated or to determine the deviation of the noise factor from that case.
- If there is a deviation from the isolated case, that deviation may be taken into account in determining the expected variations in the noise factor.
- It is possible (if complicated) to use the noise circles, and the noise factors that they imply along the constant-SWR circle, together with the available gain to the module input, to determine more exactly the contribution to the cascade noise factor.

3.9 IMAGE NOISE, MIXERS

When a mixer, used for frequency conversion, appears in a cascade, there is usually an opportunity for additional noise to enter. This is because the mixer translates two frequency bands into the intended output frequency band. While only one of them normally carries a signal, both the intended input band and the other, image, band carry noise. Frequency conversions will be discussed in detail in Chapter 7; here we treat the mixer as a component in the cascade whose effective noise figure must be determined, based on the image noise that enters through it. Additional increases in mixer noise factor due to LO noise will be discussed in Section 8.4.

In the less common case where the mixer is designed to reject the image band, either due to an internal filter or an image rejection configuration in which the image response is canceled, the mixer can be treated like any other module, characterized by a gain and noise figure. However, that is not the case being treated here.

If the mixer is preceded directly by an image-rejecting (image) filter that presents a match, supplying only thermal noise ($\overline{k}TB$) at the image frequency, the mixer's effective noise figure will be its measured (specified) single-sideband noise figure. Otherwise the mixer will convert two bands of noise to its output [intermediate frequency (IF)]. Assuming there is to be a signal in only one of these bands, so that the theoretical source noise is considered to be only the noise in that one band, the noise factor, defined by Eq. (3.1), will be increased due to the insertion of this additional noise. If the circuitry preceding the mixer is high-gain broadband (same gain at all frequencies of importance), the cascade noise figure can increase as much as 3 dB. If a filter appears at some intermediate point, after the front end of that cascade but not immediately before the mixer, the increase in *cascade noise figure* will be somewhere between 0 and 3 dB. The increase in the *effective noise figure of the mixer* will be much greater. We will determine exactly what the increases will be for this general case.

3.9.1 Effective Noise Figure of the Mixer

The single-sideband gain of a mixer is measured by inputting a signal at frequency f_R and measuring the output at frequency f_I, where

$$f_I = f_{I+} \triangleq f_L + f_R \tag{3.38}$$

or

$$f_I = f_{I-} \triangleq |f_L - f_R|, \tag{3.39}$$

and f_L is the local oscillator (LO) frequency. The part of the cascade preceding the mixer operates in the vicinity of f_R and the part after the mixer operates near f_I.

Both output frequencies (f_{I+} and f_{I-}) occur, but only one is used to determine single-sideband gain. Likewise, the signal at only one of these output frequencies, and the noise in its vicinity, are used to measure single-sideband noise figure. Broadband terminations are commonly used on all three ports for these measurements.

The fact that two IF signals are created by each RF signal implies that each IF can be created by two different RFs (Fig. 3.10);

$$f_{R+} \triangleq f_L + f_I \tag{3.40}$$

and

$$f_{R-} \triangleq |f_L - f_I|. \tag{3.41}$$

A signal exists at only one of these frequencies — the other is termed the image frequency — in most applications, but noise is converted to the IF from both.

Figure 3.11 shows a generic cascade, beginning with a matched source impedance, followed by an amplifier, an image rejection filter, another amplifier, the

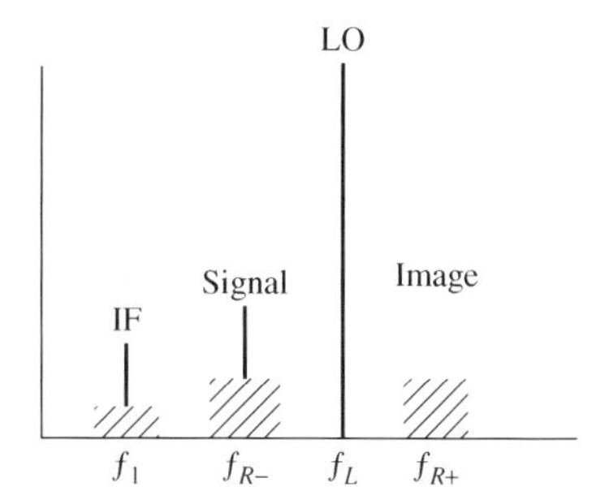

Fig. 3.10 Conversion frequencies. The noise bands shown are those that eventually appear in the IF.

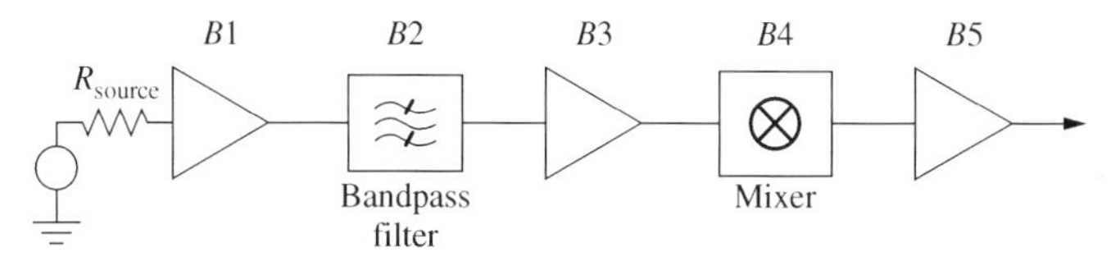

Fig. 3.11 Cascade with mixer. The "Amplifier" blocks ($B1$, $B3$, $B5$) can each represent cascades of other elements.

mixer, and a final amplifier. Each module, or block, is unique because of its location relative to the mixer or filter, and each may represent a cascade of other modules. Block B_j has gain g_j and noise factor f_j. The filter should ideally be a triplexer, allowing the cascade to see the environment encountered during characterization, or at least a diplexer, presenting a matching impedance at the image frequency.[2] This is especially important in the degenerate case in which $B3$ disappears. It is also important for any filter at the IF output (see Section 7.2.2). Equation (3.14) written explicitly for this arrangement is

$$f_{\text{cas}} = f_{B1} + \frac{f_{B2} - 1}{g_{B1}} + \frac{f_{B3} - 1}{g_{B1} g_{B2}} + \frac{f_{B4} - 1}{g_{B1} g_{B2} g_{B3}} + \frac{f_{B5} - 1}{g_{B1} g_{B2} g_{B3} g_{B4}}. \tag{3.42}$$

The image noise, which appears at the input to the mixer, is available thermal noise N_T times $f'_{B3} g'_{B3}$, where primes are used in case parameters are different at the image frequency than they are at the desired signal frequency. Again, these may represent the composite parameters for a cascade that is represented here by block $B3$. The difference between this image noise and the noise that was present when the mixer was characterized is $N_T(f'_{B3} g'_{B3} - 1)$. The change appears at the mixer output multiplied by the mixer gain at the image frequency g'_{B4}. The input noise in the signal band that would produce the same output is obtained by dividing this by the mixer gain at the signal frequency g_{B4}. Thus the effective change in input noise is $N_T \Delta f_{B4}$, where Δf_{B4} is the effective change in the mixer noise figure due to the image noise:

$$\Delta f_{B4} = (f'_{B3} g'_{B3} - 1) \frac{g'_{B4}}{g_{B4}}. \tag{3.43}$$

The system noise with image noise is then

$$f_{\text{cas}} = f_{B1} + \frac{f_{B2} - 1}{g_{B1}} + \frac{f_{B3} - 1}{g_{B1} g_{B2}} + \frac{(f'_{B3} g'_{B3} - 1)(g'_{B4}/g_{B4}) + f_{B4} - 1}{g_{B1} g_{B2} g_{B3}} + \frac{f_{B5} - 1}{g_{B1} g_{B2} g_{B3} g_{B4}}. \tag{3.44}$$

From this, we can write, for the fourth module

$$\Delta f_{B4} = \frac{(f'_{B3} g'_{B3} - 1)(g'_{B4}/g_{B4}) + f_{B4} - 1}{g_{B1} g_{B2} g_{B3}}, \tag{3.45}$$

or we can use Eq. (3.42) but substitute f_{e4}, the effective noise factor of the mixer with image noise, for the measured noise factor f_{B4}:

$$f_{e4} = f_{B4} + (f'_{B3} g'_{B3} - 1) \frac{g'_{B4}}{g_{B4}}. \tag{3.46}$$

When we use the same mixer gain for the signal and the image, Eq. (3.45) becomes

$$\Delta f_{B4}|_{g'_{B4} = g_{B4}} = \frac{f'_{B3} g'_{B3} + f_{B4} - 2}{g_{B1} g_{B2} g_{B3}}. \tag{3.47}$$

If the filter is not a triplexer or diplexer but is reactive at the image frequency, the value of $f'_{B3} g'_{B3}$ may have to be modified to give the correct noise output at the image frequency under that condition.

If the cascade begins with the filter $B2$, we set $g_{B1} = f_{B1} = 1$ (as if $B1$ were a short cable). If also there is no filter, we also set $g_{B2} = f_{B2} = 1$ and the cascade effectively begins with thermal noise at the input to $B3$. In this latter case, Eq. (3.44) would become

$$\begin{aligned} f_{\text{cas}} &= f_{B3} + \frac{(f'_{B3} g'_{B3} - 1)(g'_{B4}/g_{B4}) + f_{B4} - 1}{g_{B3}} + \frac{f_{B5} - 1}{g_{B3} g_{B4}} \\ &= f_{B3} + \frac{f_{e4} - 1}{g_{B3}} + \frac{f_{B5} - 1}{g_{B3} g_{B4}}. \end{aligned} \tag{3.48}$$

As an alternative, we could represent by $B3$ the whole cascade preceding the mixer (see Example 3.6). In that case, Eq. (3.48) would be used and the effect of the filter would be represented by its great attenuation at the image frequency rather than by complete elimination of the image. This could sometimes be awkward, requiring us to designate parameters at the image frequency for many modules preceding the filter, even when their contribution to the effective noise factor of the mixer is negligible.

3.9.2 Verification for Simple Cases

Other presentations of this theory have come up with results that are close, but not quite identical, to this; so we should check some simple cases to see if it makes sense.

A simple case that fails in some other representations is that where the system consists of the mixer alone. Assume that g_{B1} through g_{B3} and g_{B5} represents short pieces of matched cable. Then, for those four modules, $g = 1$ and $f = 1$ and (3.44) is

$$f_{\text{cas}} = 1 + \frac{1-1}{1} + \frac{1-1}{1} + \frac{(1-1)(g'_{B4}/g_{B4}) + f_{B4} - 1}{1} + \frac{1-1}{g_{B4}} = f_{B4} \tag{3.49}$$

as it should be.

For another test, replace $B3$ with a short cable so the mixer sees, at the image frequency, only a termination. Then

$$f_{\text{cas}} = f_{B1} + \frac{f_{B2}-1}{g_{B1}} + \frac{1-1}{g_{B1}g_{B2}} + \frac{(1-1)(g'_{B4}/g_{B4}) + f_{B4} - 1}{g_{B1}g_{B2}} + \frac{f_{B5}-1}{g_{B1}g_{B2}g_{B4}} \tag{3.50}$$

$$= f_{B1} + \frac{f_{B2}-1}{g_{B1}} + \frac{f_{B4}-1}{g_{B1}g_{B2}} + \frac{f_{B5}-1}{g_{B1}g_{B2}g_{B4}}, \tag{3.51}$$

which is a normal representation without image noise.

3.9.3 Examples of Image Noise

Example 3.4 Effect in a Simple Front End A simple RF front end is illustrated in Fig. 3.12 ($f_{B1} = f_{B2} = g_{B1} = g_{B2} = 1$, $f'_{B3} = f_{B3}$, $f'_{B4} = f_{B4}$, $g'_{B3} = g_{B3}$ and $g'_{B4} = g_{B4}$ in Fig. 3.11) and its noise figure is plotted in Fig. 3.13 as a function of the preamplifier ($B3$) gain. Curve 1 shows the noise figure when

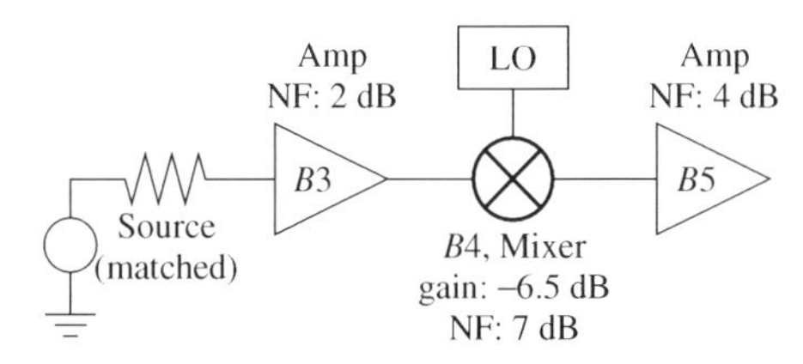

Fig. 3.12 Simple RF front end. Components are assumed to be broadband and all ports are matched.

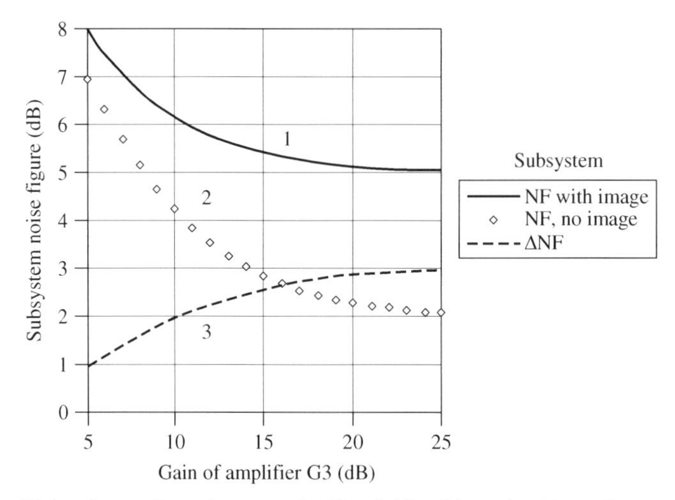

Fig. 3.13 Noise figure for subsystem in Fig. 3.12 with and without image noise and difference between the two.

image noise is accounted for [Eq. (3.44)]; curve 2 shows the noise figure with no image noise [Eq. (3.42)]; and curve 3 shows the difference. This difference could represent an error in the system performance estimate, if existing image noise is not taken into account. It could also represent a loss in performance because image noise was not properly filtered out.

Example 3.5 Spreadsheet with Image Noise, Broadband System Figure 3.14 is a spreadsheet with gain and noise figure given for seven modules (cells C4–D10) plus cumulative gain and noise figure (cells C14–D20) computed as before, but using cells E4–E10 for derived noise figure. The latter differ from the values in the column to their left only where a module is identified in cells B4–B10 as being a mixer. Then the effective noise figure of the mixer is used [Eq. (3.46)]. Here we have assumed broad bandwidth, that is, that the gain and noise figures in the image band are the same as in the desired signal band ($f = f'$, $g = g'$), except, of course, in the filter, which is assumed to reject the image completely.

The "mixer" and "filter" designations in cells B4–B10 can be moved so the effect of their placement on total noise figure (cell D20) can be observed. These words must not be moved using a cut operation or by dragging because the spreadsheet will then outsmart itself by moving all references to the cells that contain these words, following the words. This will defeat any change as a result of the movement and will corrupt the spreadsheet for further use. Move the words by retyping or by first copying and then erasing their former locations.

Cells F5 and G5 contain the cumulative gain and noise figure, respectively, at the filter position. They are copied from the corresponding cells in C14–D20 (i.e., F5 = C15, etc.). Columns F4–F10 and G4–G10 are summed to find the values in these two cells, whichever row they are in, since no other cells in these ranges contain values. These two values are then used in the cell on the same

	A	B	C	D	E	F	G
2		enter	Gain	NF		cumulative at filter	
3		below	expected	expected	derived	gain	NF
4	Module 1		12.00 dB	2.00 dB	2.00 dB		
5	Module 2	filter	-4.00 dB	4.00 dB	4.00 dB	8	2.2538
6	Module 3		6.00 dB	2.50 dB	2.50 dB		
7	Module 4		-2.00 dB	2.00 dB	2.00 dB		
8	Module 5		8.00 dB	3.00 dB	3.00 dB		
9	Module 6	mixer	-7.50 dB	8.00 dB	16.24 dB		
10	Module 7		20.00 dB	3.00 dB	3.00 dB		
11							
12			CUMULATIVE				
13	at output of		Gain	NF			
14	Module 1		12.00 dB	2.00 dB			
15	Module 2		8.00 dB	2.25 dB			
16	Module 3		14.00 dB	2.56 dB			
17	Module 4		12.00 dB	2.62 dB			
18	Module 5		20.00 dB	2.76 dB			
19	Module 6		12.50 dB	3.62 dB			
20	Module 7		32.50 dB	3.72 dB			

Fig. 3.14 Spreadsheet with image noise.

line as "mixer" in E4–E10 to give effective noise figure according to Eq. (3.46). The following development will show how Eq. (3.46) is reorganized in terms of individual component modules (e.g., "Module 1," rather than effective modules consisting of multiple component modules, like "$B1$") to enable its computation from the spreadsheet. However, it may be simpler just to study the spreadsheet.

The value of f_{B3}, for the cascade from the module just after the filter through the module just before the mixer (composite module $B3$ in Fig. 3.11), is obtained from Eq. (3.14) as

$$f_{B3} = f_{\text{cas}}\Big|_{k(\text{F})+1}^{k(\text{M})-1} = 1 + \sum_{j=k(\text{F})+1}^{k(\text{M})-1} \frac{f_j - 1}{\prod\limits_{i=k(\text{F})+1}^{j-1} g_i}, \tag{3.52}$$

where $k(\text{M})$ is the index of the mixer and $k(\text{F})$ is the index of the filter, and $x\big|_{n1}^{n2}$ represents parameter x of the cascade starting with element $n1$ and ending with element $n2$. [Similarly to Eq. (3.13), the denominator is one when $j = k(\text{F}) + 1$.] We can write this in terms of the noise factor preceding the mixer and the noise factor preceding and including the filter:

$$f_{B3} = 1 + \sum_{j=1}^{k(\text{M})-1} \frac{f_j - 1}{\prod\limits_{i=k(\text{F})+1}^{j-1} g_i} - \sum_{j=1}^{k(\text{F})} \frac{f_j - 1}{\prod\limits_{i=k(\text{F})+1}^{j-1} g_i} \tag{3.53}$$

$$= 1 + \left[\sum_{j=1}^{k(\mathrm{M})-1} \frac{f_j - 1}{\prod_{i=1}^{j-1} g_i} - \sum_{j=1}^{k(\mathrm{F})} \frac{f_j - 1}{\prod_{i=1}^{j-1} g_i} \right] \prod_{i=1}^{k(\mathrm{F})} g_i \tag{3.54}$$

$$= 1 + [f_{\mathrm{cas},k(\mathrm{M})-1} - f_{\mathrm{cas},k(\mathrm{F})}] g_{\mathrm{cas},k(\mathrm{F})}, \tag{3.55}$$

where $f_{\mathrm{cas},j}$ is the noise factor for the cascade of modules from 1 to j.

The gain of block $B3$ can be written

$$g_{B3} = g_{\mathrm{cas}}\Big|_{k(\mathrm{F})+1}^{k(\mathrm{M})-1} = \frac{g_{\mathrm{cas},k(\mathrm{M})-1}}{g_{\mathrm{cas},k(\mathrm{F})}}; \tag{3.56}$$

so the product of the noise factor and the gain is

$$f_{B3} g_{B3} = \{1 + [f_{\mathrm{cas},k(\mathrm{M})-1} - f_{\mathrm{cas},k(\mathrm{F})}] g_{\mathrm{cas},k(\mathrm{F})}\} \frac{g_{\mathrm{cas},k(\mathrm{M})-1}}{g_{\mathrm{cas},k(\mathrm{F})}}. \tag{3.57}$$

Similarly, at the image frequency,

$$f'_{B3} g'_{B3} = \{1 + [f'_{\mathrm{cas},k(\mathrm{M})-1} - f'_{\mathrm{cas},k(\mathrm{F})}] g'_{\mathrm{cas},k(\mathrm{F})}\} \frac{g'_{\mathrm{cas},k(\mathrm{M})-1}}{g'_{\mathrm{cas},k(\mathrm{F})}}. \tag{3.58}$$

When a cell in B5–B10 contains "mixer," the corresponding line in cells E5–E10 uses Eq. (3.46), where $f'_3 g'_3 = f_3 g_3$ is obtained from Eq. (3.58). In that equation, $g'_{\mathrm{cas},k(F)}$ and $f'_{\mathrm{cas},k(F)}$ come from the nonblank cell in F4–F10 or G4–G10, respectively, while $g'_{\mathrm{cas},k(M)-1}$ and $f'_{\mathrm{cas},k(M)-1}$ come from the appropriate cell in C14–C20 or D14–D20, respectively. The appropriate cells are in the line for the module before the one marked "mixer" in cells B4–B10.

Example 3.6 Parameters Differing at Image Frequency Figure 3.15*a* is similar to Fig. 3.14 but allows for different values of g and f at the image frequency (columns F and G). The conversion from Fig. 3.14 is straightforward (although the ratio g'_{B4}/g_{B4} must now be included). This allows also for an alternative, simpler, realization of the spreadsheet since the filter can now be represented as part of module $B3$ in Fig. 3.11, an individual module that is characterized as having much more loss at the image frequency than at the desired frequency. This is done in cells F5 and G5 in Fig. 3.15*b*. Columns H and I of Fig. 3.15*a* are gone. There is no need to determine f and g for modules $B1$ and $B2$ in Fig. 3.11. They have now disappeared ($f_{B1} = g_{B1} = f_{B2} = g_{B2} = 1$), as in Fig. 3.12, and the filter has become part of module $B3$. The noise figure at the mixer (cell E9) uses Eq. (3.45) directly, obtaining f'_3 and g'_3 from the corresponding cells in F14–G20. This can be more accurate because it allows the filter to be given a finite attenuation at the image frequency, whereas the attenuation of

	A	B	C	D	E	F	G	H	I
2		enter	Gain	NF		expected at image		cumulative at filter	
3		below	expected	expected	derived	Gain	NF	gain	NF
4	Module 1		12.00 dB	2.00 dB	2.00 dB	11.00 dB	2.20 dB		
5	Module 2	filter	−4.00 dB	4.00 dB	4.00 dB	−20.00 dB	20.00 dB	−9	9.788
6	Module 3		6.00 dB	2.50 dB	2.50 dB	5.00 dB	2.50 dB		
7	Module 4		−2.00 dB	2.00 dB	2.00 dB	−2.30 dB	2.30 dB		
8	Module 5		8.00 dB	3.00 dB	3.00 dB	8.00 dB	3.00 dB		
9	Module 6	mixer	−7.50 dB	8.10 dB	15.06 dB	−8.00 dB	8.60 dB		
10	Module 7		20.00 dB	3.00 dB	3.00 dB	17.00 dB	3.00 dB		
11									
12			CUMULATIVE			CUMULATIVE at image			
13	at output of		Gain	NF		Gain	NF		
14	Module 1		12.00 dB	2.00 dB		11.00 dB	2.20 dB		
15	Module 2		8.00 dB	2.25 dB		−9.00 dB	9.79 dB		
16	Module 3		14.00 dB	2.56 dB		−4.00 dB	11.96 dB		
17	Module 4		12.00 dB	2.62 dB		−6.30 dB	12.42 dB		
18	Module 5		20.00 dB	2.76 dB		1.70 dB	13.37 dB		
19	Module 6		12.50 dB	3.43 dB		−6.30 dB	14.14 dB		
20	Module 7		32.50 dB	3.53 dB		10.70 dB	14.80 dB		

(*a*)

	A	B	C	D	E	F	G
2		enter	Gain	NF		expected at image	
3		below	expected	expected	derived	Gain	NF
4	Module 1		12.00 dB	2.00 dB	2.00 dB	11.00 dB	2.20 dB
5	Module 2		−4.00 dB	4.00 dB	4.00 dB	−20.00 dB	20.00 dB
6	Module 3		6.00 dB	2.50 dB	2.50 dB	5.00 dB	2.50 dB
7	Module 4		−2.00 dB	2.00 dB	2.00 dB	−2.30 dB	2.30 dB
8	Module 5		8.00 dB	3.00 dB	3.00 dB	8.00 dB	3.00 dB
9	Module 6	mixer	−7.50 dB	8.10 dB	15.34 dB	−8.00 dB	8.60 dB
10	Module 7		20.00 dB	3.00 dB	3.00 dB	17.00 dB	3.00 dB
11							
12			CUMULATIVE			CUMULATIVE at image	
13	at output of		Gain	NF		Gain	NF
14	Module 1		12.00 dB	2.00 dB		11.00 dB	2.20 dB
15	Module 2		8.00 dB	2.25 dB		−9.00 dB	9.79 dB
16	Module 3		14.00 dB	2.56 dB		−4.00 dB	11.96 dB
17	Module 4		12.00 dB	2.62 dB		−6.30 dB	12.42 dB
18	Module 5		20.00 dB	2.76 dB		1.70 dB	13.37 dB
19	Module 6		12.50 dB	3.47 dB		−6.30 dB	14.14 dB
20	Module 7		32.50 dB	3.57 dB		10.70 dB	14.80 dB

(*b*)

Fig. 3.15 Spreadsheets with parameters differing at image frequency. The filter eliminates the image at (*a*), as in Fig. 3.14. At (*b*) the filter presents a high, but finite, attenuation of the image.

image noise is infinite in the other representation. (The image frequency parameters given for the filter and preceding modules in Fig. 3.15*a* ultimately have no effect on the derived mixer noise figure.) However, accounting for the image response of modules preceding the mixer can be a nuisance if there are many of them, especially if their effect at the filter output is small. The representations of Fig. 3.15*a* and 3.15*b* are equivalent in the limit where the filter has infinite attenuation at the image frequency. That attenuation has been purposefully set rather low in Fig. 3.15 in order that there be some difference between the values in cells D20 in the two figures. One might increase it to see how large it must be for the overall noise figures in the two representations to be equal within some tolerance.

Example 3.7 Combined with Interconnects in a Standard Cascade
Figure 3.16 is similar to Fig. 3.5, showing the effects of mismatches at interfaces, except that only noise figures for mean gain and mean individual noise figures have been retained (for simplicity) and the equations for noise figure in cells I16, I18, and I20 use the conditional formulas for effective noise figure with image noise that were used in Fig. 3.14. Cells B14 and B20 designate the corresponding modules as filter and mixer, respectively. This illustrates how image noise and mismatches can be included in the same analysis. Of course, this can also be done with combinations of gain and noise figure extremes as used in Fig. 3.3, and we could use the technique in Fig. 3.15*b* of listing separate parameters at the desired and image frequencies.

However, the mixer is not particularly well represented as a unilateral module, as is assumed in our standard cascade analysis. Unbalanced mixers provide little RF-to-IF (the signal path) isolation. Fortunately, doubly balanced mixers are commonly used and they do provide some isolation. RF-to-IF isolation, which indicates how much of the RF signal is seen in the IF, is often greater than 20 dB, sometimes much greater, providing significant round-trip loss. In that case mismatches at the mixer output have little effect on the signal at its input. However, the two-way conversion loss provides another path, from RF-to-IF-to-RF, and the conversion loss usually ranges from 5 to 10 dB, providing as little as 10-dB two-way loss. On the other hand, good design practice promotes care in providing the specified termination for a mixer. The SWRs obtained in characterization will, in that case, also occur in the cascade, and reflections at the output will be minimized, reducing the impact of the reverse transmission on the analysis.

3.10 EXTREME MISMATCH, VOLTAGE AMPLIFIERS

In some cases, particularly at lower frequencies, amplifiers that are characterized by high input impedances (and often low output impedances) may be used in cascade. The amplifier stages often consist of elementary amplifiers and associated input and feedback impedances (Egan, 1998, pp. 49–54). Often the voltage gain and equivalent input noise generators are specified for the elementary amplifier circuit, the extreme mismatch at interfaces is a very bad approximation to a standard interface, and it is difficult to analyze these cascades except in terms

	A	B	C	D	E	F	G	H	I	J	K
2			Gain	Gain	SWR			specified NF		Temperature	
3			nom	+/−	at out	$\lvert a_{RT} \rvert$		mean		290 K	
4	Module 1		12.0 dB	1.0 dB	1.5			2.0 dB			
5	Cable 1		−1.5 dB		1.5	0.0283					
6	Module 2		10.0 dB	2.0 dB	2			4.0 dB			
7	Attenuator		−8.0 dB	0.5 dB	1.5	0.0106					
8	Module 3		7.0 dB	2.0 dB	2.8			3.0 dB			
9	Cable 2		−0.8 dB		3.2	0.2064					
10	Module 4		30.0 dB	2.0 dB				5.0 dB			
11					DERIVED						
12			Gain	Gain	Gain	Gain	Gain		mean NF	gain, cum	NF, cum
13			mean	max	min	±	σ		at mean gain	at filter	at filter
14	Module 1	filter	12.00 dB	13.00 dB	11.00 dB	1.00 dB	0.50 dB		2.00 dB	12	2
15	Cable 1 (no mixer here)		−1.50 dB	−1.25 dB	−1.74 dB	0.25 dB	0.17 dB		1.54 dB	0	0
16	Module 2		10.00 dB	12.00 dB	8.00 dB	2.00 dB	1.00 dB		4.00 dB	0	0
17	Attenuator (no mixer)		−8.00 dB	−7.41 dB	−8.59 dB	0.59 dB	0.41 dB		8.06 dB	0	0
18	Module 3		7.00 dB	9.00 dB	5.00 dB	2.00 dB	0.80 dB		3.00 dB	0	0
19	Cable 2 (no mixer here)		−0.61 dB	1.21 dB	−2.43 dB	1.82 dB	1.27 dB		0.93 dB	0	0
20	Module 4	mixer	30.00 dB	32.00 dB	28.00 dB	2.00 dB	1.30 dB		14.45 dB	0	0
21					CUMULATIVE						
22					Gain				NF using mean NFs at		
23	at output of		mean	max	min	±	σ		mean G		
24	Module 1		12.00 dB	13.00 dB	11.00 dB	1.00 dB	0.50 dB		2.00 dB		
25	Cable 1		10.50 dB	11.75 dB	9.26 dB	1.25 dB	0.53 dB		2.07 dB		
26	Module 2		20.50 dB	23.75 dB	17.26 dB	3.25 dB	1.13 dB		2.42 dB		
27	Attenuator		12.50 dB	16.34 dB	8.67 dB	3.84 dB	1.20 dB		2.54 dB		
28	Module 3		19.50 dB	25.34 dB	13.67 dB	5.84 dB	1.45 dB		2.67 dB		
29	Cable 2		18.89 dB	26.55 dB	11.24 dB	7.66 dB	1.93 dB		2.68 dB		
30	Module 4		48.89 dB	58.55 dB	39.24 dB	9.66 dB	2.32 dB		3.42 dB		

Fig. 3.16 Spreadsheet with mismatch and image.

of terminal voltages. We will term such amplifiers and cascades "hi-Z" and will see how to determine the noise figure for a hi-Z cascade so it can be treated as a module driven by the standard impedance R_0 that precedes it.

3.10.1 Module Noise Factor

Refer to Fig. 3.17, which is the same as Fig. 3.1 except some variables have been added and some deleted and zero reverse transmission is assumed. Equation (3.1) can be written in terms of open-circuit voltage sources, e, as

$$\hat{f}_k = \frac{\dfrac{|e_{\text{signal},s}/2|^2}{R_{22(k-)}} \Big/ \dfrac{|e_{\text{noise},s}/2|^2}{R_{22(k-)}}}{\dfrac{|e_{\text{signal,out},k}/2|^2}{R_{22k}} \Big/ \dfrac{|e_{\text{noise,out},k}/2|^2}{R_{22k}}} \tag{3.59}$$

$$= \frac{|e_{\text{signal},s}/2|^2/|e_{\text{noise},s}/2|^2}{|e_{\text{signal,out},k}/2|^2/|e_{\text{noise,out},k}/2|^2} \tag{3.60}$$

$$= \frac{|e_{\text{noise,out},k}/2|^2}{\bar{k}T_0BR_{22(k-)}} \Big/ \frac{|e_{\text{signal,out},k}/2|^2}{|e_{\text{signal},s}/2|^2}, \tag{3.61}$$

where $R_{22(k-)}$ is the resistance looking into the part of the cascade preceding module k (equal to $R_{22(k-1)}$ if module $k-1$ is unilateral). The ratio of the module's output open-circuit voltage to the source's open-circuit voltage is

$$\frac{e_{\text{signal,out},k}}{e_{\text{signal},s}} = c_k a_k, \tag{3.62}$$

where

$$c_k \stackrel{\Delta}{=} \frac{v_{\text{signal},k}}{e_{\text{signal},s}} = \frac{Z_{11k}}{Z_{11k} + Z_{22(k-)}} \tag{3.63}$$

is the ratio of the interface voltage to the source voltage that produces it,

$$a_k = \frac{e_{\text{signal,out},k}}{v_{\text{signal},k}} \tag{3.64}$$

is the open-circuit (no module load) voltage gain of module k, and

$$v_k = v_{\text{signal},k} + v_{\text{noise},k}. \tag{3.65}$$

Combining Eq. (3.62) with Eq. (3.61), we obtain the noise factor for module k:

$$\hat{f}_k = \frac{|e_{\text{noise,out},k}/2|^2}{\bar{k}T_0BR_{22(k-)}|c_k a_k|^2}. \tag{3.66}$$

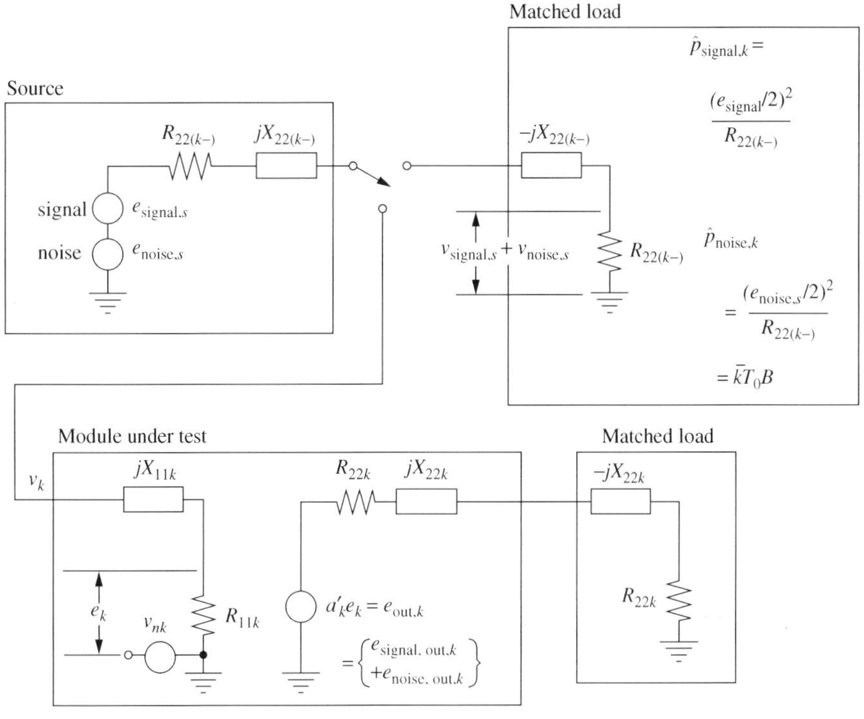

Fig. 3.17 Noise figure test, theoretical. This is the same as Fig. 3.1 with some other variables shown.

If the module were noiseless, $|e^2_{\text{noise,out},k}/2|^2$ would equal the denominator of Eq. (3.66), giving $f_k = 1$. Thus the noise contributed by the module is equivalent to an additional effective noise source, in the Source, with an rms value

$$\tilde{\Delta} v_{nk} = 2\sqrt{\bar{k}T_0BR_{22(k-)}(\hat{f}_k - 1)}, \tag{3.67}$$

which would produce

$$\Delta p_{nk} = \bar{k}T_0B(\hat{f}_k - 1) \tag{3.68}$$

into $R_{22(k-)}$. Note, however, that this voltage would produce

$$\Delta p_{nk} = \bar{k}T_0B(\hat{f}_k - 1)R_{22(k-)}/R_0 \tag{3.69}$$

into a matched load if it were in series with the cascade source impedance R_0 (Fig. 3.2, switch position 1). (Here we are neglecting any reactances, which would have to be canceled by their conjugates.) The ratio, $R_{22(k-)}/R_0$, had not appeared in our standard cascade because we employed power gains there whereas, here, we are using voltage gains.

3.10.2 Cascade Noise Factor

We assume that each hi-Z module will be measured with the same driving impedance $Z_{22(k-)}$ that it sees in the cascade or that the noise factor will be calculated (Appendix A, Section A.3) for such a driving impedance. Calculations can be facilitated by information giving equivalent input noise voltage and noise current generators, which is often provided for op amps (Steffes, 1998; Baier, 1996) (see also Section 3.8).

In a cascade, the effective cascade Source noise voltage that is equivalent to the noise in module k, is reduced by the gain of the other modules between the source and the noise:

$$\frac{e_{\text{signal,out},(k-1)}}{e_{\text{signal},s}} = \frac{e_{\text{signal,out},1}}{e_{\text{signal},s}} \frac{e_{\text{signal,out},2}}{e_{\text{signal,out},1}} \cdots \frac{e_{\text{signal,out},(k-1)}}{e_{\text{signal,out},(k-2)}} = \prod_{1}^{k-1} c_j a_j. \tag{3.70}$$

Division by this gain places the equivalent noise source in series with the cascade Source impedance R_0. Therefore, the available power from the total equivalent added noise voltage at the cascade source is the sum of the noise powers given by Eq. (3.69), each divided by the preceding gain:

$$\Delta p_n = \overline{k} T_0 B \sum_{k=1}^{N} \frac{(\hat{f}_k - 1)}{\displaystyle\prod_{i=1(k\neq 1)}^{k-1} |c_i a_i|^2} \frac{R_{22(k-)}}{R_0}, \tag{3.71}$$

and the total noise factor is

$$f_{\text{total}} = 1 + \frac{\Delta p_n}{\overline{k} T_0 B} \tag{3.72}$$

$$= 1 + \sum_{k=1}^{N} \frac{(\hat{f}_k - 1)}{\displaystyle\prod_{i=1(k\neq 1)}^{k-1} |c_i a_i|^2} \frac{R_{22(k-)}}{R_0} \tag{3.73}$$

$$= \hat{f}_1 + \sum_{k=2}^{N} \frac{(\hat{f}_k - 1)}{\displaystyle\prod_{i=1}^{k-1} |c_i a_i|^2} \frac{R_{22(k-)}}{R_0}. \tag{3.74}$$

Here we have used $R_{22(1-)} = R_0$. That is, the first module in the hi-Z cascade is driven from a source, the real part of which is R_0. If R_0 is the standard impedance at the input interface to the hi-Z cascade, the hi-Z cascade can be treated like any module in a standard cascade as can its noise figure. In other words, if the standard impedance at the input to the hi-Z cascade is R_0, Eq. (3.73) gives the noise factor to be used for the hi-Z cascade as if it were a module in a standard cascade. (The gain used for this equivalent module would be its transducer gain, as for any other module.)

3.10.3 Combined with Unilateral Modules

A cascade of voltage amplifiers can be considered an equivalent standard module, driven by the standard impedance at the output of the preceding cascade, as in Fig. 3.2, switch position 2. R_0 might represent the well-controlled output impedance from the preceding part of a cascade or it might be the standard interface impedance of a cable connecting the cascade of voltage amplifiers to preceding standard-impedance stages. Recall that the noise factor used in Section 3.3 was also measured with a standard interface impedance.

If the input to the voltage-amplifier section is not well matched to R_0, it will be important that the output of the last module in the preceding section be well matched to the cable impedance to prevent excessive variations in cable gain at the interface.

3.10.4 Equivalent Noise Factor

We may want to use a noise factor program or spreadsheet that is built for the standard cascade relationships, Eq. (3.13) or its equivalent Eq. (3.14). To enable us to do so, we can define parameters that can be put into that equation for gain and noise factor but will give us results according to Eq. (3.73). To this end, we define

$$\breve{f}_k = 1 + (\hat{f}_k - 1)R_{22(k-)}/R_0 \tag{3.75}$$

and

$$\breve{g}_k = |c_k a_k|^2. \tag{3.76}$$

Replacing f_k and g_k with these variables in Eq. (3.13) [or in a program that realizes Eq. (3.13)] will cause f to be computed according to Eq. (3.73).

3.11 USING NOISE FIGURE SENSITIVITIES

Sensitivities of cascade noise figure to module parameters can be especially useful in identifying critical modules in a cascade. We can write

$$dF_{\text{cas}} = \sum_k (\hat{S}_{fk} df_k + \hat{S}_{gk} dg_k + \hat{S}_{\text{SWR}k} d\text{SWR}_k), \tag{3.77}$$

where

$$\hat{S}_{xk} = \frac{\partial F_{\text{cas}}}{\partial x_k} \tag{3.78}$$

is the sensitivity of F_{cas} to the parameter x_i. This is based on the Taylor series [(see Eq. (2) in Appendix V]. Equation (3.77) is further developed in Appendix V for the case where gains and SWRs are fixed and only the module noise figures vary, leading to

$$dF_{\text{cas}}(dF_j) = 10^{-F_{\text{cas}}/10\text{ dB}}\{10^{F_1/10\text{ dB}} dF_1 + 10^{(F_3 - G_1 - G_2)/10\text{ dB}} dF_3 + \cdots\}, \tag{3.79}$$

where F_j is not shown for j odd based on the assumption that those elements are interconnects. An alternative is to determine sensitivities from the spreadsheet, as we did for gain in Example 2.4. An example of the use of this process for determining sensitivities of noise figure to module parameters is given in Section 3.12.3.

3.12 MIXED CASCADE EXAMPLE

Example 3.8 Figure 3.18 shows a cascade that begins as a standard cascade, unilateral modules interconnected by cables of standard impedance, and ends with a cascade of voltage amplifiers. The latter consists of Op Amps 1–3. Intermediate modules are treated as a simple cascade, appropriate for good impedance matches. Parameters are given in Fig. 3.19, rows 4–15. The emitter follower in the Transistor Amplifier has sufficient current gain to provide an effective transformation from 50 to 125 Ω. An impedance transformation from 125 Ω to 2 kΩ occurs in the Transformer (1-to-4 voltage ratio, 16-to-1 impedance ratio). The Filter is designed for 2-kΩ interfaces, which it sees at both ports. Op Amp 1 has high input impedance, so only the shunt 2 kΩ is seen, and the Filter provides a 2-kΩ source for the cascade of voltage amplifiers. The last two op amp circuits are inverting and have voltage gains of 1 and 10, respectively. We use 20-Ω effective output resistances for the three op amps in closed loop. These are the result of higher open-loop output resistances, which are reduced by the feedback. As a result, this value will change with frequency as the open-loop gains of the op amps change.

The reference resistance for the voltage-amplifier cascade is the 2-kΩ driving resistance. Power gains are used to the left of that point and transform the equivalent 2-kΩ source noise to equivalent noise at the overall source on the far left. No interconnect is assumed after cable 3, although we could have used effective cables to account for mismatches. However, good matches are likely at the Transistor-Amplifier output and Op Amp 1 input; so interconnect resonances would be killed there anyway.

Effective gains, according to Eq. (3.76), are computed in cells B13–B15 and effective noise factors, according to Eq. (3.75), are computed in cells F28–F30 (they are copied to the right since no gain variation is indicated for these amplifiers). Rows 34–45 contain cumulative values computed as before.

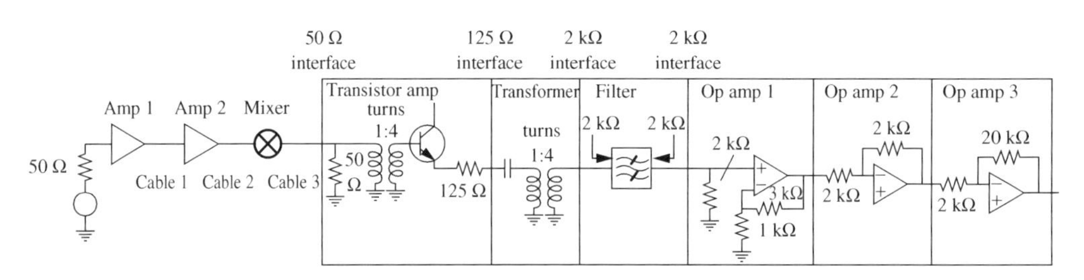

Fig. 3.18 Standard cascade feeding voltage amplifiers.

	A	B	C	D	E	F	G	H
2		Gain	Gain	SWR		NF		Temperature
3		nom	+/–	at out	$\lvert a_{RT} \rvert$	mean		290 K
4	Amp1	12.0 dB	1.0 dB	1.5		2.0 dB		
5	Cable 1	–1.5 dB		1.5	*0.028*			
6	Amp 2	12.0 dB	2.0 dB	2.5		4.0 dB		
7	Cable 2	–1.0 dB		3	*0.17*			
8	Mixer	–8.0 dB	2.0 dB	3		$1/g$ + 0.55 dB		
9	Cable 3	–0.2 dB		1	*0*			
10	Transistor Amp	1.4 dB	0.2 dB			5.0 dB		
11	Transformer	–0.4 dB	0.1 dB			$1/g$		
12	Filter	–7.0 dB	0.3 dB	R_0	R_{22k-}	$1/g$	c_k	a_k
13	Op Amp 1	*12.0 dB*		2000 Ω	2000 Ω	6.5000 dB	1	4
14	Op Amp 2	*–0.1 dB*		2000 Ω	20 Ω	27.8674 dB	0.990099	1
15	Op Amp 3	*19.9 dB*		2000 Ω	20 Ω	25.7646 dB	0.990099	10
16				DERIVED (B13-B15 are derived also.)				
17		Gain				NF using mean NFs (see Note*) at		
18		mean	max	min	±	mean G	max G	min G
19	Amp1	12.00 dB	13.00 dB	11.00 dB	1.00 dB	2.00 dB	2.00 dB	2.00 dB
20	Cable 1	–1.50 dB	–1.25 dB	–1.74 dB	0.25 dB	1.54 dB	1.54 dB	1.54 dB
21	Amp 2	12.00 dB	14.00 dB	10.00 dB	2.00 dB	4.00 dB	4.00 dB	4.00 dB
22	Cable 2	–0.87 dB	0.62 dB	–2.37 dB	1.49 dB	1.13 dB	1.13 dB	1.13 dB
23	Mixer	–8.00 dB	–6.00 dB	–10.00 dB	2.00 dB	8.55 dB	6.55 dB	10.55 dB
24	Cable 3	–0.20 dB	–0.20 dB	–0.20 dB	0.00 dB	0.25 dB	0.25 dB	0.25 dB
25	Transistor Amp	1.40 dB	1.60 dB	1.20 dB	0.20 dB	5.00 dB	5.00 dB	5.00 dB
26	Transformer	–0.40 dB	–0.30 dB	–0.50 dB	0.10 dB	0.40 dB	0.30 dB	0.50 dB
27	Filter	–7.00 dB	–6.70 dB	–7.30 dB	0.30 dB	7.00 dB	6.70 dB	7.30 dB
28	Op Amp 1	12.04 dB	12.04 dB	12.04 dB	0.00 dB	6.50 dB	6.50 dB	6.5000 dB
29	Op Amp 2	–0.09 dB	–0.09 dB	–0.09 dB	0.00 dB	8.52 dB	8.52 dB	8.5186 dB
30	Op Amp 3	19.91 dB	19.91 dB	19.91 dB	0.00 dB	6.78 dB	6.78 dB	6.7770 dB
31					CUMULATIVE			
32		Gain				NF using mean NFs at		
33	at output of	mean	max	min	±	mean G	max G	min G
34	Amp1	12.00 dB	13.00 dB	11.00 dB	1.00 dB	2.00 dB	2.00 dB	2.0000 dB
35	Cable 1	10.50 dB	11.75 dB	9.26 dB	1.25 dB	2.07 dB	2.06 dB	2.0914 dB
36	Amp 2	22.50 dB	25.75 dB	19.26 dB	3.25 dB	2.42 dB	2.32 dB	2.5478 dB
37	Cable 2	21.63 dB	26.37 dB	16.89 dB	4.74 dB	2.43 dB	2.32 dB	2.5563 dB
38	Mixer	13.63 dB	20.37 dB	6.89 dB	6.74 dB	2.53 dB	2.35 dB	3.0389 dB
39	Cable 3	13.43 dB	20.17 dB	6.69 dB	6.74 dB	2.54 dB	2.35 dB	3.0645 dB
40	Transistor Amp	14.83 dB	21.77 dB	7.89 dB	6.94 dB	2.77 dB	2.40 dB	3.9590 dB
41	Transformer	14.43 dB	21.47 dB	7.39 dB	7.04 dB	2.77 dB	2.40 dB	3.9934 dB
42	Filter	7.43 dB	14.77 dB	0.09 dB	7.34 dB	3.09 dB	2.47 dB	5.1914 dB
43	Op Amp 1	19.47 dB	26.81 dB	12.13 dB	7.34 dB	4.26 dB	2.74 dB	8.2600 dB
44	Op Amp 2	19.39 dB	26.72 dB	12.05 dB	7.34 dB	4.37 dB	2.77 dB	8.4958 dB
45	Op Amp 3	39.30 dB	46.64 dB	31.96 dB	7.34 dB	4.44 dB	2.79 dB	8.6377 dB
46			*Note: Cable NF depends on SWR, which is assumed to be fixed.					

Fig. 3.19 Spreadsheet for Fig. 3.18.

3.12.1 Effects of Some Resistor Changes

As should be expected, the overall noise factor is not changed if we redraw the boundaries between op amps to include part of the input resistor of op amp 2 or 3 as part of the previous stage. This is verified in Appendix A, Section A.1.

We have used 20 Ω as the output resistance of the op amps. The correct value may be difficult to ascertain and will not be constant, as we have assumed, since it depends on the closed-loop gain of the op amp. Section A.2 shows that, while doubling this assumed resistance changes the noise factor of the individual op amp stages significantly, it has little effect on the overall noise factor. This is only partly due to the magnitude of the preceding gain.

We might also be concerned with the effect of a change in the source resistance for the voltage-amplifier cascade, R_0 in Eq. (3.71), especially since the output impedance of the filter is likely to vary some. However, Section A.2 again shows that the overall noise figure is little affected in this example.

3.12.2 Accounting for Other Reflections

How might we discover the range of variations in cascade noise factor and gain that occur due to a mismatch at the filter input? We could treat the Transformer as part of the Transistor Amp, taking its losses into account in computing the latter's noise figure and gain and giving the new module the SWR of the transformer (which is well terminated at the Transistor Amp output). We should be able to treat the Filter as a unilateral module because it has a good termination at the input to Op Amp 1, the same termination with which it was presumably tested. Therefore there will be no reflections through the filter to contend with except those that are included in the measured input SWR. In addition, a round trip attenuation of 14 dB helps to isolate the input SWR from effects at the Filter output. Now that we would have two effectively unilateral modules, we could interconnect them with a zero-length 2-kΩ interconnect and use the equations for a standard cascade to include the range of variations to be expected due to this interface.

3.12.3 Using Sensitivities

Sensitivities of cascade noise figure to module gains and noise figures are shown in Fig. 3.20, cells I34–J45, for minimum gain.

To obtain these values we begin with the equation in cell I45, which gives the difference between the noise figure in cell H45 and the value in the same cell of

	A	B	C	D	E	F	G	H	I	J
31		CUMULATIVE							for min G	
32		Gain				NF using mean NFs at			Sensitivity, NF Change	
33	at output of	mean	max	min	±	mean G	max G	min G	per dB Gain	per dB NF
34	Amp 1	12.00 dB	13.00 dB	11.00 dB	1.00 dB	2.00 dB	2.00 dB	2.0000 dB	–0.781 dB	0.219 dB
35	Cable 1	10.50 dB	11.75 dB	9.26 dB	1.25 dB	2.07 dB	2.06 dB	2.0914 dB	–0.749 dB	
36	Amp 2	22.50 dB	25.75 dB	19.26 dB	3.25 dB	2.42 dB	2.32 dB	2.5478 dB	–0.752 dB	0.041 dB
37	Cable 2	21.63 dB	26.37 dB	16.89 dB	4.74 dB	2.43 dB	2.32 dB	2.5563 dB	0.540 dB	
38	Mixer	13.63 dB	20.37 dB	6.89 dB	6.74 dB	2.53 dB	2.35 dB	3.0389 dB	–0.754 dB	0.032 dB
39	Cable 3	13.43 dB	20.17 dB	6.69 dB	6.74 dB	2.54 dB	2.35 dB	3.0645 dB	–0.757 dB	
40	Transistor Amp	14.83 dB	21.77 dB	7.89 dB	6.94 dB	2.77 dB	2.40 dB	3.9590 dB	–0.657 dB	0.094 dB
41	Transformer	14.43 dB	21.47 dB	7.39 dB	7.04 dB	2.77 dB	2.40 dB	3.9934 dB	–0.679 dB	
42	Filter	7.43 dB	14.77 dB	0.09 dB	7.34 dB	3.09 dB	2.47 dB	5.1914 dB	–0.679 dB	
43	Op Amp 1	19.47 dB	26.81 dB	12.13 dB	7.34 dB	4.26 dB	2.74 dB	8.2600 dB	–0.082 dB	0.601 dB
44	Op Amp 2	19.39 dB	26.72 dB	12.05 dB	7.34 dB	4.37 dB	2.77 dB	8.4958 dB	–0.032 dB	0.052 dB
45	Op Amp 3	39.30 dB	46.64 dB	31.96 dB	7.34 dB	4.44 dB	2.79 dB	8.6377 dB	0.000 dB	0.033 dB
46			*Note: Cable NF depends on SWR, which is assumed to be fixed.							

Fig. 3.20 Sensitivities of cascade NF to module gain and NF for Fig. 3.18 at minimum gain. Missing cells are as in Fig. 3.19.

Fig. 3.19 (our reference value). Initially the value in cell I45 is zero, but, if we modify a module parameter, it will indicate the change in module noise figure due to the change in the module parameter. To make the sensitivity approximate a derivative [Eq. (3.78)], we will use small changes in module parameters, 0.1 dB, so we include a factor of 10 to the formula in I45 in order to get sensitivity in units of dB/dB. Then we copy that equation (cell I45) to all the cells in I34–J45 [maintaining its reference to cells H45 (one in Fig. 3.19 and one in Fig. 3.20) by designating them H45 before copying]. When we change the gain of Amp 1 (in cell B4) by 0.1 dB, all of the cells in I34–J45 will show the resulting change in cascade noise figure (times 10). We then copy cell I34 and paste it "by value" in place, replacing the formula by its numerical value as we do so. When we return cell B4 to its original value, all of the cells in I34–J45 return to zero value (indicating we have accurately restored the original value) except for cell I34, which retains the pasted value. We do this for each gain and each noise figure that is specified and that is not simply the negative of the gain (in dB). In the latter cases we blank the corresponding sensitivity cell. When we have completed this process, each cell in the range (except possibly I45) contains a number, rather than a formula.

Analyzing the results, we note that all of the gains up to Op Amp 1 are fairly significant. This is consistent with the fact that the cumulative gain just before Op Amp 1 is close to zero, dropping the signal into thermal noise. (We would expect these sensitivities to be considerably smaller if we were analyzing the cascade with mean gains rather than minimum values.)

In column J, we see a significant sensitivity to Op-Amp-1 noise figure. This might lead us to attempt to improve its noise figure (12 dB, $f = 16$). The matching resistor across its input (which we need there) automatically contributes 1 to its noise factor and the 1-kΩ and 3-kΩ resistors together contribute 1.5. We might reduce the latter some but would probably look for a lower-noise op amp to improve performance significantly.

The transformer in the Transistor Amp is there to give the amplifier power gain and to reduce the effect of the noise from the 125-Ω output resistor, plus the base spreading resistance, on the noise factor. If we remove it, its noise figure increases from 5 dB to about 13 dB. According to the sensitivity in cell J40, the cascade noise figure should therefore increase by [0.095 (8 dB) =] 0.76 dB. If we make the change in module noise figure in the spreadsheet, we actually see an increase of 1.74 dB, the inaccuracy being due to the large size of the change, as can be seen in Fig. 3.21.

Removing that transformer would have an even more important effect on gain, decreasing it by almost 12 dB. Based on sensitivity, this would increase the cascade noise figure by [−0.666 (−11.8 dB) =] 7.87 dB. Again, if we make the change we see a larger increase, 10.3 dB.

The total cascade noise figure increase, due to both effects, would be 10.5 dB, which is less than the sum of the two effects, again a result of the relatively large change. If we decrease the module gain only 1 dB or increase its noise figure only 1 dB, we obtain cascade noise figure increases of 0.685 and 0.103 dB

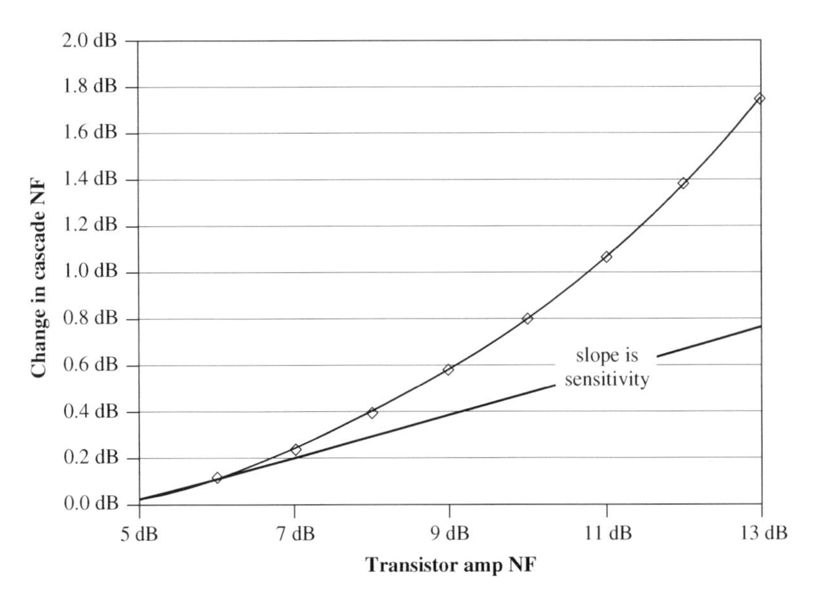

Fig. 3.21 Change in cascade noise figure with change in Transistor Amp noise figure.

respectfully. If we make both changes, we get a resulting change in cascade noise figure of 0.773, within 2% of the sum of the individual changes. This shows the importance of small changes for accuracy. In spite of the inaccuracy for large changes, however, the sensitivities do point out the relative importance of this module and the order of the changes to be expected.

3.13 GAIN CONTROLS

3.13.1 Automatic Gain Control

Example 3.9 Gain Determines Input Traditional automatic gain control (AGC) incorporates an adjustment of gain to bring the signal level at the cascade output to a desired level. Figure 3.22 is a modification of Fig. 3.3 in which only mean parameters have been retained. A target output level has been added at cell B31. Cell B32 shows the input signal level for which that target output level will be attained. A box has been drawn about cell B10 to indicate that it is the cell where gain is changed to attain the target level. Of course, the input level in cell B32 will respond to changes in any of the chain parameters that affect gain. One can vary the module gain in cell B10 and record the corresponding input level although, in practice, it is the input level that causes a change in module gain. This represents a control loop of at least type 1, since there is no error in the output level, relative to the target, regardless of the input level. The input level is easily computed from the cumulative gain and the target level. A type 0 loop would have some error, which would change proportionally to the input.

	A	B	C	D
2	T = 290 K assumed		SWR	
3		Gain	at out	$\lvert a_{RT}\rvert$
4	Module 1	12.0 dB	1.5	
5	Cable 1	–1.5 dB	1.5	0.0283
6	Module 2	10.0 dB	2	
7	Attenuator	–8.0 dB	1.5	0.0106
8	Module 3	7.0 dB	2.8	
9	Cable 2	–0.8 dB	3.2	0.2064
10	Gain Control	30.0 dB		
11			DERIVED	
12		Mean Gain		
13	Module 1	12.00 dB		
14	Cable 1	–1.50 dB		
15	Module 2	10.00 dB		
16	Attenuator	–8.00 dB		
17	Module 3	7.00 dB		
18	Cable 2	–0.61 dB		
19	Gain Control	30.00 dB		
20			CUMULATIVE	
21		Mean Gain		
22	at output of			
23	Module 1	12.00 dB		
24	Cable 1	10.50 dB		
25	Module 2	20.50 dB		
26	Attenuator	12.50 dB		
27	Module 3	19.50 dB		
28	Cable 2	18.89 dB		
29	Gain Control	48.89 dB		
30				
31	Target out:	–50 dBm		
32	Input Level:	–98.89 dBm		

Fig. 3.22 AGC with input level indicator.

Example 3.10 Input Determines Gain The spreadsheet in Fig. 3.23 provides similar information but is a better model of the cascade. It is designed so the gain (cell B11) of the Gain Control module changes in response to the input level given in cell B34. The required gain is the difference between the target output level and the input level. The gain that is required in the Gain Control (cell B35) is the difference between this required gain and the cumulative gain for all the preceding modules. The gain of the Gain Control (cell B11) is set equal to that value unless it is out of the range given by cells C11 and D11. (Module gains do have limits.) If it is out of range, the Gain Control gain goes to the nearest limit.

	A	B	C	D
2	T = 290 K assumed		SWR	
3		Gain	at out	$\lvert a_{RT}\rvert$
4	Module 1	12.0 dB	1.5	
5	Cable 1	−1.5 dB	1.5	0.0283
6	Module 2	10.0 dB	2	
7	Attenuator	−8.0 dB	1.5	0.0106
8	Module 3	7.0 dB	2.8	
9	Cable 2	−0.8 dB	3.2	0.2064
10			min	max
11	Gain Control	21.1 dB	10 dB	50 dB
12				
13	DERIVED			
14		Gain		
15	Module 1	12.00 dB		
16	Cable 1	−1.50 dB		
17	Module 2	10.00 dB		
18	Attenuator	−8.00 dB		
19	Module 3	7.00 dB		
20	Cable 2	−0.61 dB		
21	Gain Control	30.00 dB		
22	CUMULATIVE			
23		Gain		
24	at output of			
25	Module 1	12.00 dB		
26	Cable 1	10.50 dB		
27	Module 2	20.50 dB		
28	Attenuator	12.50 dB		
29	Module 3	19.50 dB		
30	Cable 2	18.89 dB		
31	Gain Control	48.89 dB		
32				
33	Target out:	−50 dBm		
34	Input Level:	−90 dBm		
35	Required Gain Control:	21.1 dB		

Fig. 3.23 AGC with specified input level.

3.13.2 Level Control

Figure 3.24 shows another type of gain control, one we might call Level Control. Its object is to keep the output noise level fixed. This might be used in conjunction with a circuit that is set to detect signals that surpass the received noise level by a given amount. In the system, the output noise power is somehow measured in a manner to exclude signal power. The measured value is compared to the desired level, and the gain of the Gain Control is adjusted to minimize the difference. This could be done either manually or automatically.

	A	B	C	D	E
2	T = 290 K assumed		SWR		
3		Gain	at out	$\lvert a_{RT} \rvert$	NF
4	Module 1	12.0 dB	1.5		2.0 dB
5	Cable 1	–1.5 dB	1.5	0.0283	
6	Module 2	10.0 dB	2		4.0 dB
7	Attenuator	–8.0 dB	1.5	0.0106	
8	Module 3	7.0 dB	2.8		3.0 dB
9	Cable 2	–0.8 dB	3.2	0.2064	
10			min G	max G	
11	Gain Control	39.35 dB	10 dB	50 dB	5.0 dB
12					
13	DERIVED				
14		Gain			NF
15	Module 1	12.00 dB			2.00 dB
16	Cable 1	–1.50 dB			1.54 dB
17	Module 2	10.00 dB			4.00 dB
18	Attenuator	–8.00 dB			8.06 dB
19	Module 3	7.00 dB			3.00 dB
20	Cable 2	–0.61 dB			0.93 dB
21	Gain Control	39.35 dB			5.00 dB
22	CUMULATIVE				
23		Gain			NF
24	at output of				
25	Module 1	12.00 dB			2.00 dB
26	Cable 1	10.50 dB			2.07 dB
27	Module 2	20.50 dB			2.42 dB
28	Attenuator	12.50 dB			2.54 dB
29	Module 3	19.50 dB			2.67 dB
30	Cable 2	18.89 dB			2.68 dB
31	Gain Control	58.25 dB			2.74 dB
32					
33					
34	Bandwidth:	2 MHz			
35	Noise Into Gain Control:	–89.35 dBm			
36	Target out:	–50 dBm			
37	Required in Gain Control:	39.35 dB			
38	Set Gain Control:	0.00 dB	(0 dB for Automatic)		

Fig. 3.24 Level control.

Example 3.11 Open-Loop Control In the spreadsheet (Fig 3.24) thermal noise in the specified bandwidth is computed and multiplied by the noise factor and the cumulative gain to the input of the last module. This total is subtracted from the target noise output to give the required gain in the last module, the Gain Control. The Gain Control is given that gain if it is within the allowed limits (cells C11 and D11) and if cell B38 contains zero. If cell B38 does not contain zero, the Gain Control gain is set to the value in cell B38. This allows the gain to be either specified or automatically controlled. The value of zero was chosen to set automatic level control because it is well out of the range of gains that would be specified.

Example 3.12 Closed-Loop Control There is sometimes another reason to provide the ability to set the gain manually. If the noise factor of the last module should vary with its gain (this could be incorporated in the formula for cell E11, for example) or if the Gain Control module should not be the last module in the cascade, the control process would become iterative because the noise figure could change with gain. The spreadsheet will execute a settable number of iterations, but it might be necessary to set some reasonable value of gain initially to permit the final value to be achieved. An example of such a spreadsheet is shown in Fig. 3.25 where the computed output noise level is partially determined by the variable that is being adjusted, the gain of the Gain Control module.

These same processes can easily be implemented for multiple conditions (e.g., maximum NF and minimum gain) on the same spreadsheet.

Advantages of building in the automatic gain adjustment include being more easily able to see the overall effect of a change in a module parameter, for example, the change in cascade noise figure that occurs when the gain of some module changes, or to see if the Gain Control module goes out of its allowed range as a result of some parameter change. (A conditional warning to that effect has been incorporated in cells C37 and D37 in Fig. 3.24.)

3.14 SUMMARY

- Noise factor f is the noise at the output of a module or cascade relative to what would be there if only the amplified theoretical noise of the source, at a temperature of 290 K, were present.
- In this book, noise figure F is f expressed in dB.
- For a cascade, $(f - 1)$ is the sum of noise contributions from the cascade's elements, each represented by $(f - 1)$ for the element divided by the preceding gain.
- Source impedance can influence module noise factor. Theoretically, f for a module is measured with the same driving impedance that the module sees in the cascade.
- Commonly, f is measured with standard interface impedance.
- This commonly measured f is appropriate for use in our "standard cascade" model where unilateral modules are interconnected by cables of standard impedances.

	A	B	C	D	E
2	T = 290 K assumed		SWR		
3		Gain	at out	$\lvert a_{RT} \rvert$	NF
4	Module 1	12.0 dB	1.5		2.0 dB
5	Cable 1	−1.5 dB	1.5	0.02832	
6	Module 2	10.0 dB	2		4.0 dB
7	Attenuator	−18.0 dB	1.5	0.00106	
8			min G	max G	
9			5 dB	35 dB	
10	Gain Control	25.5 dB	2.8		3.0 dB
11	Cable 2	−0.8 dB	3.2	0.20638	
12	Module 4	29.0 dB			10.0 dB
13					
14	DERIVED				
15		Gain			NF
16	Module 1	12.00 dB			2.00 dB
17	Cable 1	−1.50 dB			1.54 dB
18	Module 2	10.00 dB			4.00 dB
19	Attenuator	−18.00 dB			18.01 dB
20	Gain Control	25.51 dB			3.00 dB
21	Cable 2	−0.61 dB			0.93 dB
22	Module 4	29.00 dB			10.00 dB
23	CUMULATIVE				
24		Gain			NF
25	at output of				
26	Module 1	12.00 dB			2.00 dB
27	Cable 1	10.50 dB			2.07 dB
28	Module 2	20.50 dB			2.42 dB
29	Attenuator	2.50 dB			3.62 dB
30	Gain Control	28.01 dB			4.56 dB
31	Cable 2	27.40 dB			4.56 dB
32	Module 4	56.40 dB			4.59 dB
33					
34	Bandwidth:	2 MHz			
35	Noise Out:	−50 dBm			
36	Target out:	−50 dBm			
37	Required Gain Control:	25.5 dB			
38	Gain Error:	0.0E+00	<-enter cmd+= to iterate		
39	Set Gain Control:	99.00 dB	(99 dB for Automatic)		

Fig. 3.25 Level control with iteration.

- The noise factor for an attenuator at T_0 equals its attenuation ($f = 1/g$).
- Interconnect cables have effective noise factors that depend on the output SWR of the driving module.
- The variance of cascade noise figure, due to variations in individual module noise figures, can be conveniently computed by extending the cascade one element at a time.
- The effective noise factor of a mixer must account for the addition of image noise.
- Noise factor for a cascade of voltage amplifiers can be given in terms that are more convenient when power gain is difficult to use.
- Noise factor can be obtained for cascades consisting of sections of standard cascades, simple cascades, and voltage-amplifier cascades.
- Sensitivities are useful in analysis of the effects of module gains and noise figures on cascade noise figure.
- The signal level corresponding to a module gain can be indicated for automatic gain control (AGC), or the gain can be set in response to a given input level.
- Level control, to standardize the output noise level, can be incorporated in the spreadsheet.

ENDNOTES

[1] $4.0038\ldots \times 10^{-21}$ or $-173.9753\ldots$ dBm/Hz.

[2] Simple bandpass filters are reflective at out-of-band frequencies. A diplexer consists of two parallel filters whose interaction has been taken into account. We would pass the signal through one of these filters and terminate the input to the other so the mixer would see a proper impedance match over both the desired and image bands. Ideally, a triplexer, which provides three parallel filters, could provide proper termination in the RF passband and at frequencies above and below the passband. Similar considerations may be even more important at the other mixer ports (see Section 7.2.2).

CHAPTER 4

NONLINEARITY IN THE SIGNAL PATH

In this chapter we consider how to represent nonlinearities in modules and cascades. Nonlinearities produce additional signals that are often objectionable in RF systems. Some of these can be removed by filtering. Some cannot. To effectively design RF systems, we must be able to predict at what frequencies these spurious signals will occur and their expected magnitudes.

4.1 REPRESENTING NONLINEAR RESPONSES

Figure 4.1 shows a typical curve of output voltage plotted against input voltage. Ideally the curve would be a straight line extending indefinitely, but, practically, it will have some curvature and eventually saturate. However, we can usually represent a curve such as this by a Taylor series,

$$v_{\text{out}} = a_0 + a_1 v_{\text{in}} + a_2 v_{\text{in}}^2 + a_3 v_{\text{in}}^3 + a_4 v_{\text{in}}^4 + a_5 v_{\text{in}}^5 + \cdots, \tag{4.1}$$

where v_{in} is the change in input voltage from the operating point (the point about which the series is written) and the a_i are real. (Phase shift and frequency sensitivity can be accounted for in functions preceding or following the nonlinearity.)

The first term is a bias term and not of interest here. The second term is the desired linear term, a_1 being the linear voltage gain a. The other terms represent the curvature of the gain curve, and they create undesired components at other frequencies. If only one signal is present, the undesired components will be harmonics of the fundamental, but, if there are more signals in v_{in}, signals will be produced with frequencies that are mathematical combinations of the frequencies of the input signals (e.g., three times the frequency of one signal less

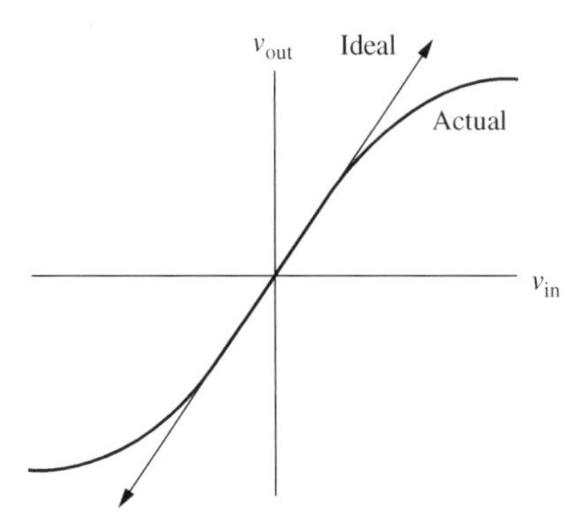

Fig. 4.1 Voltage transfer curve with straight-line approximation.

the frequency of the second), called intermodulation products or intermods. It is instructive to study the results when there are two input signals (although we will eventually consider large numbers of signals). This shows the formation of harmonics as well as intermods.

We write the input signal as

$$v_{\text{in}} = A\cos\varphi_a(t) + B\cos\varphi_b(t), \tag{4.2}$$

where

$$\varphi_a(t) = \omega_a t + \theta_a \tag{4.3}$$

and

$$\varphi_b(t) = \omega_b t + \theta_b. \tag{4.4}$$

At times we will drop the explicitly shown time dependence, writing φ for $\varphi(t)$.

4.2 SECOND-ORDER TERMS

The even-order term of primary interest will be the second-order term, that is, the one arising from v_{in}^2. Combining the third term on the right of Eq. (4.1) with Eq. (4.2), we obtain

$$v_2 = a_2[A\cos\varphi_a(t) + B\cos\varphi_b(t)]^2 \tag{4.5}$$

$$= a_2[A^2\cos^2\varphi_a(t) + 2AB\cos\varphi_a(t)\cos\varphi_b(t) + B^2\cos^2\varphi_b(t)] \tag{4.6}$$

$$= a_2\left\{\frac{A^2}{2}[1+\cos 2\varphi_a(t)] + AB\left\{\begin{array}{l}\cos[\varphi_a(t)-\varphi_b(t)]\\ \quad +\cos[\varphi_a(t)+\varphi_b(t)]\end{array}\right\}\right.$$

$$\left. + \frac{B^2}{2}[1+\cos 2\varphi_b(t)]\right\} \tag{4.7}$$

$$= a_2 \left(\begin{array}{l} \dfrac{A^2 + B^2}{2} + \dfrac{A^2}{2} \cos 2\varphi_a(t) + \dfrac{B^2}{2} \cos 2\varphi_b(t) \\ + AB\{\cos[\varphi_a(t) - \varphi_b(t)] + \cos[\varphi_a(t) + \varphi_b(t)]\} \end{array} \right), \tag{4.8}$$

where trigonometric identities have been employed to obtain Eq. (4.7) from (4.6).

The first term in Eq. (4.8) is a direct current (DC) term, essentially detection. The second and third terms are second harmonics of the two signals, shown at d and f in Fig. 4.2, where the fundamentals are at a and b. The last terms are the difference frequency term, at c in Fig. 4.2, and the sum frequency term, which has frequency between those of the two harmonics, at e. These last two terms are intermods. When the amplitudes of the input signals are equal, they are 6 dB greater than the harmonics, as can be seen from Eq. (4.8) and is suggested in Fig. 4.2.

4.2.1 Intercept Points

We can plot (Fig. 4.3) the powers of these undesired signals on the same plot with the power in the desired output fundamental, all against the power of each input signal. At low levels, all of these curves are straight lines (we will discuss the curvature at high levels presently). Since the second-order products increase twice as fast as the desired fundamental, the straight lines cross. At the crossing point, either for the intermod or the harmonic, the fundamental and the second-order product have equal output powers. Since the slopes of the straight lines are known, these crossing points, called intercept points (IPs), define the second-order products at low levels. They are called by terms such as the "second-order intermod output intercept point," for the power out at the intersection of the intermod and fundamental curves, and represented by shortcuts such as OIP2_{IM}. For the input power where the harmonic curve crosses the fundamental, this would be IIP2_H. Since an IP lies on the linear response curve, an OIP is higher than the corresponding IIP by the linear gain. Typically, the larger of the input or output intercept points is specified; so amplifiers use OIPs and mixers use

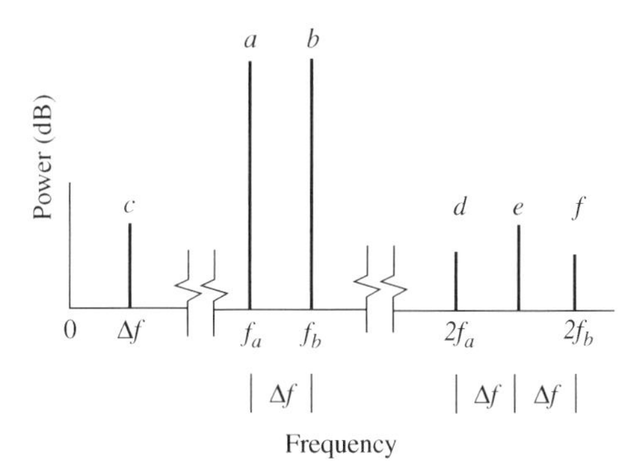

Fig. 4.2 Spectrum of second-order products from two equal-amplitude signals.

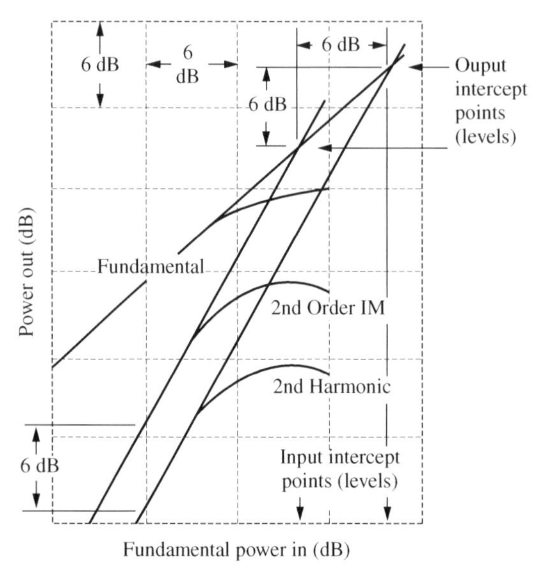

Fig. 4.3 Output powers of fundamentals and second-order products, two equal-power inputs.

IIPs (which makes sense from a marketing viewpoint, large numbers being more desirable). Some may even add the power of the two fundamentals, increasing the value of the IP by 3 dB — we will not!

Refer to Fig. 4.3. If an output is x dB below the second-harmonic IP, its second harmonic will be $2x$ dB below that IP. Similarly, if two equal-amplitude signals are x dB below the second-order IM intercept point, their IMs will be $2x$ dB below that IP. This implies that the difference (separation) between two equal-amplitude fundamentals and their harmonics or IMs is the same as the difference between those fundamentals and the corresponding IP. In other words, the signal level is midway between the IP level and the corresponding harmonic or IM level.

Example 4.1 Second Harmonic See Fig. 4.4. The output second-harmonic IP (OIP2_H) is at 17 dBm and the output signal power is −8 dBm, *25 dB* below the intercept point. Therefore, the second harmonic is another *25 dB* down, at −33 dBm, (2 × 25 *dB* =) 50 dB below the intercept point. We also know, from the *25-dB* difference between the IIP2_H and the input signal power, that the harmonic is *25 dB* below the signal at the output.

If the amplitude of only one input signal changes, we see from Eq. (4.8) that the harmonic of the changing signal will change by twice as many dB as does the input, but the other harmonic will be unaffected. The intermods' amplitudes change by the sum of the changes in the two input signals; so, if only one

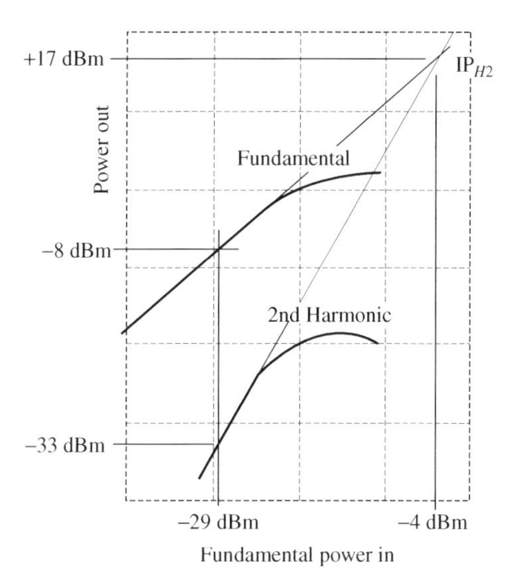

Fig. 4.4 Example 4.1.

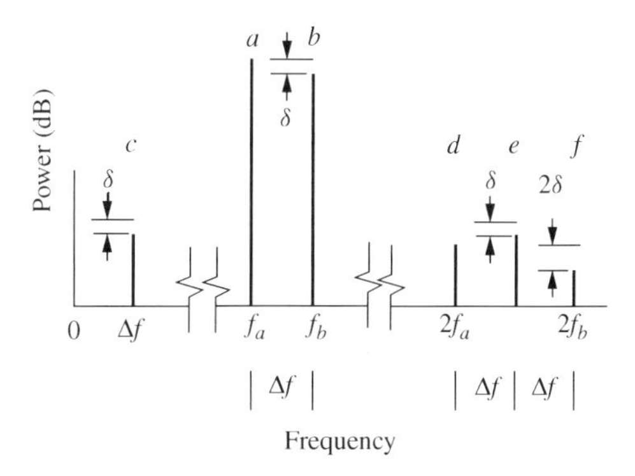

Fig. 4.5 Spectrum of second-order products from two unequal signals.

fundamental changes, the IMs will change by the same amount. This is illustrated in Fig. 4.5.

4.2.2 Mathematical Representations

Now we will express mathematically what we have just described, based on Eq. (4.8). The second-harmonic output power $p_{\text{out},H2}$ is related to the fundamental output power $p_{\text{out},F}$ and the second-order harmonic output intercept point OIP2_H by

$$p_{\text{out},H2} = \frac{p_{\text{out},F}^2}{p_{\text{OIP2},H}}. \tag{4.9}$$

Our symbol $p_{\text{OIP2},H}$ is perhaps more proper than simply OIP2_H, although the latter is generally considered a power also, even though it is termed a "point."

The output power $p_{\text{out,IM2}}$ in either second-order IM is related to the powers in the two fundamental outputs, $p_{\text{out},F1}$ and $p_{\text{out},F2}$, and to the second-order IM intercept point OIP2_{IM}, by

$$p_{\text{out,IM2}} = \frac{p_{\text{out},F1}\, p_{\text{out},F2}}{p_{\text{OIP2,IM}}}. \tag{4.10}$$

Here, $F1$ represents the fundamental a or b in Figs. 4.2 and 4.5 and $F2$ is the other fundamental.

The ratio between the second harmonic of the fundamental F and the fundamental at the output can be obtained by dividing Eq. (4.9) by $p_{\text{out},F}$:

$$\frac{p_{\text{out},H2}}{p_{\text{out},F}} = \frac{p_{\text{out},F}}{p_{\text{OIP2},H}}. \tag{4.11}$$

The ratio of either second-order IM to the output power in fundamental number 1 is similarly obtained by dividing Eq. (4.10) by the power in that fundamental:

$$\frac{p_{\text{out,IM2}}}{p_{\text{out},F1}} = \frac{p_{\text{out},F2}}{p_{\text{OIP2,IM}}}. \tag{4.12}$$

All these expressions can be related to equivalent input parameters by dividing the variables by the gain at the fundamental, for example:

$$\frac{p_{\text{in,IM2}}}{p_{\text{in},F1}} = \frac{p_{\text{in},F2}}{p_{\text{IIP2,IM}}}. \tag{4.13}$$

The variables here are: equivalent input power level for the second-order IM, $p_{\text{in,IM2}}$; power input at fundamentals 1 and 2, $p_{\text{in},F1}$ and $p_{\text{in},F2}$, respectively; and input intercept point for second-order IMs, $p_{\text{IIP2,IM}}$. The equivalent input power for a harmonic or IM is the input power that would have generated that signal had it been linearly amplified from the input rather than being created within the module.

These expressions can all be written in dB also. For example, Eq. (4.9) becomes

$$P_{\text{out},H2} = 2P_{\text{out},F} - P_{\text{OIP2},H}. \tag{4.14}$$

See Appendix H for a compilation of these various forms.

From Eqs. (4.1), (4.2), and (4.8), and the definition of intercept point, it is apparent that the output amplitude at IP2_{IM} satisfies

$$A_{\text{OPI2,IM}} = |a_1| A_{\text{IIP2,IM}} = |a_2| A^2_{\text{IIP2,IM}}, \tag{4.15}$$

so

$$A_{\text{IIP2,IM}} = |a_1/a_2|, \tag{4.16}$$

implying a power, dissipated in the resistance R across which the voltage appears, of

$$p_{\text{IIP2,IM}} = \frac{1}{2R}\left(\frac{a_1}{a_2}\right)^2. \tag{4.17}$$

Note from Eq. (4.1) that a_1 is unitless and a_2 has inverse-voltage units, giving Eq. (4.17) power units.

4.2.3 Other Even-Order Terms

The fifth term on the right side of Eq. (4.1) contains v_{in}^4 and will look like Eq. (4.8) squared, except for the coefficient a_4 instead of a_2. This will produce additional harmonic and intermodulation terms. Since the slope of the output power versus fundamental power for these, and higher-order, terms will be steeper than for the second-order terms, they will become negligible (compared to the second-order terms) at sufficiently low signal levels. Their influence on the second-order terms is of interest, however, because they, and other even-order terms, account for the curvature in the second-order curves of Fig. 4.3 at high levels. Note that the DC term in Eq. (4.8), when multiplied by another copy of Eq. (4.8), as occurs when the fourth-order term is formed, produces terms with frequencies identical to those of the second-order term. The same thing happens when higher even-order terms are expanded. Thus, the second harmonic and second-order IMs are proportional to $c_2a_2C_2 + c_4a_4C_4 + c_6a_6C_6 + \cdots$, where a_i is from Eq. (4.1), c_i is another constant, and C_i is some product A^jB^{i-j}, ranging from A^i to B^i, depending on the particular IM or harmonic. When $A = B$, for example, $C_i = A^jA^{i-j} = A^i$. Only the first (lowest order) term is significant at low levels of A and B, leading to the straight-line characteristic at low levels in Fig. 4.3, but the other terms become significant at high levels. The combination of all these terms at high levels produces the flattening of the curve there. It is conceivable that a higher slope could occur at high levels, corresponding to these higher powers in C_i, but we must remember that the values of the set of coefficients in Eq. (4.1) are an effect of the true curve, not its cause.

4.3 THIRD-ORDER TERMS

The third-order term in Eq. (4.1) is $a_3v_{\text{in}}^3$. It can be obtained by multiplying Eq. (4.8) by Eq. (4.2), excepting that a_2 is replaced by a_3. The result is

$$a_3v_{\text{in}}^3 = a_3\left\{\begin{array}{l}\dfrac{A^2+B^2}{2} + \dfrac{A^2}{2}\cos 2\varphi_a + \dfrac{B^2}{2}\cos 2\varphi_b \\ + AB[\cos(\varphi_a - \varphi_b) + \cos(\varphi_a + \varphi_b)]\end{array}\right\}(A\cos\varphi_a + B\cos\varphi_b) \tag{4.18}$$

$$
= a_3 \left\{ \begin{array}{l} \dfrac{A^2 + B^2}{2}(A \cos \varphi_a + B \cos \varphi_b) \\ + \dfrac{A^2}{4}[A \cos \varphi_a + A \cos 3\varphi_a + B \cos(2\varphi_a - \varphi_b) + B \cos(2\varphi_a + \varphi_b)] \\ + \dfrac{B^2}{4}[A \cos(\varphi_a - 2\varphi_b) + A \cos(\varphi_a + 2\varphi_b) + B \cos \varphi_b + B \cos 3\varphi_b] \\ + \dfrac{AB}{2}\left[\begin{array}{l} A \cos(2\varphi_a - \varphi_b) + 2A \cos \varphi_b + A \cos(2\varphi_a + \varphi_b) \\ + B \cos(\varphi_a - 2\varphi_b) + 2B \cos \varphi_a + B \cos(\varphi_a + 2\varphi_b) \end{array} \right] \end{array} \right\} \tag{4.19}
$$

$$
= \frac{a_3}{4} \left\{ \begin{array}{l} (3A^3 + 6AB^2) \cos \varphi_a + (3B^3 + 6A^2 B) \cos \varphi_b \\ + 3[A^2 B \cos(2\varphi_a - \varphi_b) + AB^2 \cos(\varphi_a - 2\varphi_b)] \\ + 3[A^2 B \cos(2\varphi_a + \varphi_b) + AB^2 \cos(\varphi_a + 2\varphi_b)] \\ + A^3 \cos 3\varphi_a + B^3 \cos 3\varphi_b \end{array} \right\}, \tag{4.20}
$$

where $\varphi_a \equiv \varphi_a(t)$ and $\varphi_b \equiv \varphi_b(t)$.

The first line in Eq. (4.20) contains signals at the fundamental frequencies, but their amplitudes are nonlinear functions of the input amplitudes (when A and B are equal, for example, they are proportional to the cube of the input amplitudes). They will contribute to the nonlinear shape of the fundamental gain at high levels. The second and third lines contain the IM terms and the last line has the third harmonics. At low levels, IMs and harmonics that contain n times a frequency have amplitudes that are proportional to the nth power of the corresponding fundamental amplitudes. At high levels, other terms with powers of $n + 2i$, where i is an integer, become appreciable and produce curvature in the (dB) response plots.

Figure 4.6 shows the third-order frequency spectrum with two inputs of the same level, and Fig. 4.7 shows changes to that spectrum when the amplitude of only one of the signals changes. Third-order IMs that are close to the desired signals (containing terms with frequency differences) are particularly troublesome because of the difficulty in filtering them.

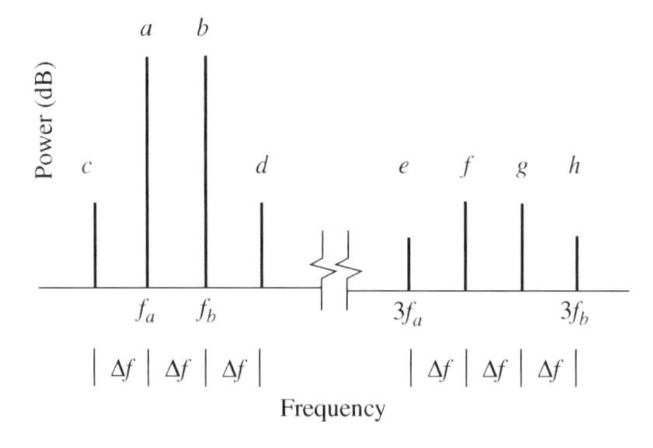

Fig. 4.6 Spectrum of third-order products from two equal-power inputs.

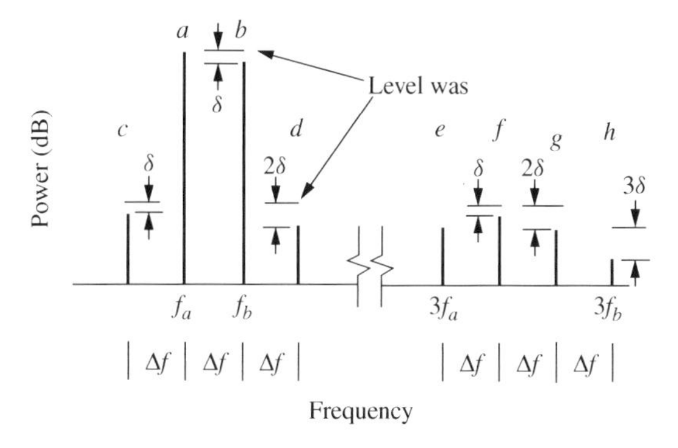

Fig. 4.7 Spectrum of third-order products from two unequal signals.

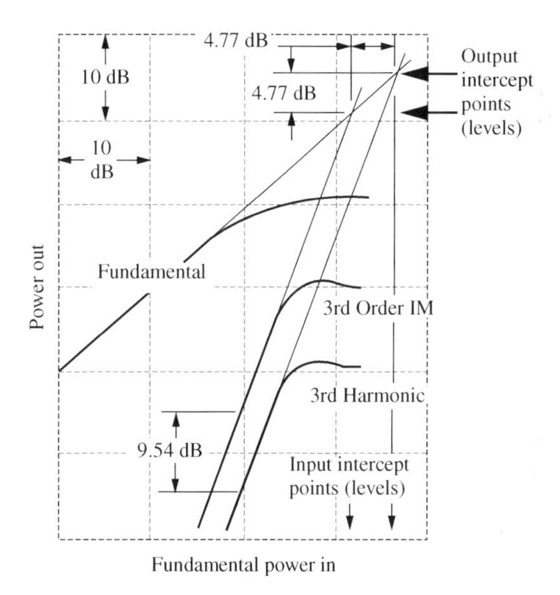

Fig. 4.8 Output powers of fundamentals and third-order products.

4.3.1 Intercept Points

Figure 4.8 shows the straight-line changes of fundamental, harmonic, and intermod powers with input power at low levels and their extensions to third-order IPs. This is similar to what is shown in Fig. 4.3 for second-order products, but the slopes for the third-order products are steeper since they represent cubic nonlinearities rather than squares. IMs and harmonics change 3 dB for each dB change in the inputs and fundamental outputs. Their ratios to the desired fundamentals change 2 dB per dB of changes in the latter. The same variations in the manner of specifying IP2s that were discussed in Section 4.2.1 apply here for IP3s.

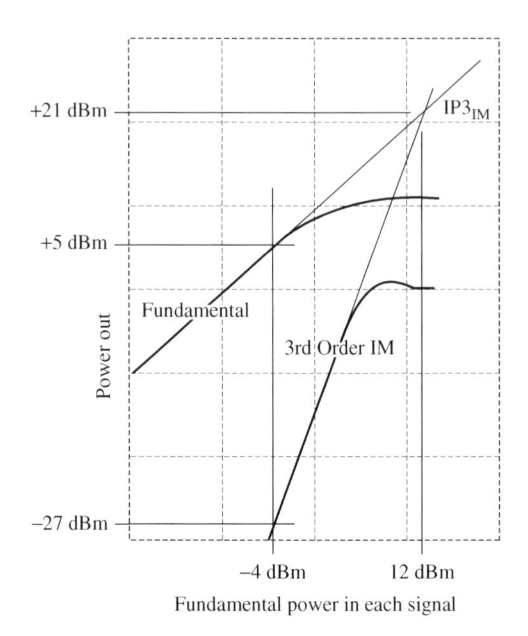

Fig. 4.9 Example 4.2.

Example 4.2 Third-Order IM See Fig. 4.9. The output third-order-IM IP (OIP3_{IM}) is at 21 dBm, and the output power for both signals is +5 dBm, *16 dB* below the intercept point. Therefore, the third-order IMs are twice *16 dB* below the signal, at -27 dBm, which is also thrice *16 dB* below the IP. We can also ascertain this 32-dB difference between fundamental and IM from the 16-dB difference between input signal and IIP3_{IM}.

4.3.2 Mathematical Representations

The following can be discerned by examination of Eq. (4.20). The third-harmonic output power $p_{\text{out},H3}$ is related to the fundamental output power $p_{\text{out},F}$, and the third-order harmonic output intercept point OIP3_H by

$$p_{\text{out},H3} = \frac{p^3_{\text{out},F}}{p^2_{\text{OIP3},H}}. \tag{4.21}$$

The IM output power $p_{\text{out,IM3}}$ at a frequency $\pm 2f_1 \pm f_2$ is related to the powers in the two fundamental outputs, $p_{\text{out},F1}$ at f_1 and $p_{\text{out},F2}$ at f_2, and to the third-order IM intercept point OIP3_{IM}, by

$$p_{\text{out,IM3}}(\pm 2f_1 \pm f_2) = \frac{p^2_{\text{out},F1}\, p_{\text{out},F2}}{p^2_{\text{OIP3,IM}}}. \tag{4.22}$$

Note that $p_{\text{out},F1}$ is the power of the fundamental whose frequency is doubled in the formula for the IM's output frequency. Which of the two frequencies is f_1 will be different for different IMs. In the region near the fundamental frequencies (which, for this discussion, we assume to be relatively close), each IM uses the frequency of its nearest fundamental twice in Eq. (4.20) so $p_{\text{out},F1}$ is the power in the nearest fundamental. In the region near the third harmonics, $p_{\text{out},F1}$ is the power of the fundamental that produces the nearest harmonic. For example, in Fig. 4.7, the power in a is squared in obtaining the power in c and in f, whereas the power in b is used only once.

The ratio of the third harmonic to the desired fundamental output can be obtained by dividing Eq. (4.21) by $P_{\text{out},F}$:

$$\frac{p_{\text{out},H3}}{p_{\text{out},F}} = \frac{p^2_{\text{out},F}}{p^2_{\text{OIP3},H}}. \tag{4.23}$$

The ratio of an IM3 to its nearest fundamental, or to the fundamental that produces the nearest third harmonic, is, from Eq. (4.22),

$$\frac{p_{\text{out,IM3}}(\pm 2f_1 \pm f_2)}{p_{\text{out},F1}} = \frac{p_{\text{out},F1}\, p_{\text{out},F2}}{p^2_{\text{OIP3,IM}}}. \tag{4.24}$$

These expressions can be written in terms of input quantities or in dB in the same manner as demonstrated in Section 4.2. See Appendix H for a compilation of these forms.

From Eqs. (4.1), (4.2), and (4.20), and the definition of IP, it is apparent that the output amplitude at IP3_{IM} satisfies

$$A_{\text{OIP3,IM}} = |a_1| A_{\text{IIP3,IM}} = \left(\tfrac{3}{4}\right) |a_3| A^3_{\text{IIP3,IM}}, \tag{4.25}$$

so

$$A^2_{\text{IIP3,IM}} = \left(\tfrac{4}{3}\right) |a_1/a_3|, \tag{4.26}$$

implying a power, dissipated in the resistance R across which the voltage appears, of

$$p_{\text{IIP3,IM}} = \frac{2}{3R}\left|\frac{a_1}{a_3}\right|. \tag{4.27}$$

Note from Eq. (4.1) that a_1 is unitless and a_3 has units of inverse-voltage squared, giving Eq. (4.27) power units.

4.3.3 Other Odd-Order Terms

The last term shown if Eq. (4.1) can be obtained, except for a change from $a_2 a_3$ to a_5, by multiplying Eq. (4.20) by Eq. (4.8). Again, as with even orders, the DC term in Eq. (4.8) causes all of the frequencies in the third-order term to reappear in the fifth-order term and, as in that case, the third-order magnitudes take on the form $d_3 a_3 C_3 + d_5 a_5 C_5 + d_7 a_7 C_7 + \cdots$, where the first term dominates at low levels but the others explain the curvature at high levels.

In a similar manner, since Eq. (4.20) contains signals at the fundamental frequencies, the fundamental contained within this and additional odd-order terms explain the curvature of the fundamental response at high levels. The curvature of the gain curve for $\cos\varphi_a(t)$ is due to A^n terms originating in nth-order nonlinearities. However, the multiplier for $\cos\varphi_a(t)$ also contains powers of B. Thus, when a strong signal is present at $\varphi_b(t)$, the gain for $\cos\varphi_a(t)$ can be seriously affected by the amplitude of the signal at $\varphi_b(t)$. This can cause *desensitization*, a reduction of the signal strength, and thus the sensitivity, for one signal in the presence of another, strong, signal (sometimes called a *blocker*) (Domino et al., 2001). It can also cause *cross modulation* because, when the strong signal is amplitude modulated, the gain for the weaker signal will change as the amplitude of the strong signal changes, causing amplitude modulation (AM) to be transferred from the strong to the weak signal (Rohde and Bucher, 1988, pp. 72–75).

Appendix P contains additional mathematical development that can be applied to higher order IMs, as well as those discussed in this chapter.

4.4 FREQUENCY DEPENDENCE AND RELATIONSHIP BETWEEN PRODUCTS

Frequency dependence can cause the observed powers to differ from the powers that enter or leave the nonlinearity. IMs that are expected to have identical amplitudes may have different levels as a result. This can be accommodated in our model by preceding and/or following the nonlinearity by filters. (In the case of a feedback amplifier, for example, a following low-pass filter might account for gain rolloff after the nonlinearity while a preceding high-pass filter might account for a compensating drive increase with frequency. The effects of the two filters would ideally cause the gain to be frequency independent, but the IMs would still increase with frequency due to the increasing signal seen by the nonlinearity.)

To the degree that the frequency response after the nonlinearity is flat, IMs may be predictable from harmonics. We can see from Fig. 4.3 that the second-order IMs are 6 dB higher than the harmonics, as are the corresponding IP2s. From Fig. 4.8 we see that the third-order IMs are 9.54 dB greater than the third harmonics and that the IP3_H exceeds the IP3_{IM} by 4.77 dB.

In addition, we may be able to relate the -1-dB compression level to the IP3. We can see, from Eq. (4.20), that, when only one signal is present, the amplitude of the fundamental term, $(\frac{3}{4})a_3 A^3$, is thrice the amplitude of the third harmonic. It is also equal to the amplitudes of the IMs that occur when a second signal of equal amplitude (i.e., $A = B$) is added. The fundamental from the first-order product [Eq. (4.1)] is $a_1 A$. If we assume that the signs of these two fundamental terms oppose and that there are no other significant terms at the fundamental, 1-dB compression occurs when the third-order fundamental term reduces the sum with the first-order term by 1 dB:

$$\frac{|a_1|A_{-1\text{ dB}} - \frac{3}{4}|a_3|A^3_{-1\text{ dB}}}{|a_1|A_{-1\text{ dB}}} = 1 - \frac{3}{4}\left|\frac{a_3}{a_1}\right| A^2_{-1\text{ dB}} = 10^{-1/20}, \tag{4.28}$$

$$\frac{3}{4}\left|\frac{a_3}{a_1}\right| A^2_{-1\text{ dB}} = 0.10875. \tag{4.29}$$

Substituting for a_3/a_1 from Eq. (4.26), we obtain

$$\frac{A^2_{-1\text{ dB}}}{A^2_{\text{IIP3,IM}}} = \frac{p_{\text{in},-1\text{ dB}}}{p_{\text{IIP3,IM}}} = 0.10875 \Rightarrow -9.64\text{ dB}. \tag{4.30}$$

Since there is a 1-dB gain reduction at the 1-dB compression level, the *output* power at this level is

$$P_{\text{out},-1\text{ dB}} = (P_{\text{IIP3,IM}} + G - 1\text{ dB}) - 9.64\text{ dB} = P_{\text{OIP3,IM}} - 10.64\text{ dB}. \tag{4.31}$$

If we measure an IM that is near the signal (in frequency) to obtain OIP3_{IM}, the frequency response could easily be the same for the signal and the IM, removing some risk from our flat-response assumption. However, we cannot know that higher-odd-order terms will not make significant contributions to the compression.[1] While we can reduce signal levels to eliminate higher order terms in relating harmonics to IMs, a low signal level does not apply at the compression point. If we are going to measure the harmonics or IMs at the high level required for compression, we might as well measure the compression level directly. The development above does provide some theoretical basis for the 1-dB output compression point being 10.6 dB below the OIP3_{IM}, but that depends on the third-order product being the only significant contributor to compression. We can find many amplifiers that are within a few dB of demonstrating this relationship, but we can also find some that deviate from it significantly. Perhaps this relationship is best used as an estimate in early design based on the hope that we will be able to find an amplifier for which it does hold if we need to.

4.5 NONLINEAR PRODUCTS IN THE CASCADES

Throughout our development of intercept points for cascades, we will assume that gain is the same for all of the signals of interest, intermods and fundamentals. We do this to simplify the expressions. We could develop a general expression with different gains for each signal but, if the gains differ significantly, it may be as well to rely on our ability to compute each intermod in each module, applying the appropriate gains to each signal both before and after the point of generation and appropriately adding the resulting intermods at the output or, equivalently, at the input. We can still develop an overall intercept point based on the signal strengths and the basic equations such as Eq. (4.10). For the case where the frequency response is sufficiently flat, however, and to aid our general understanding, it will be useful to have a simple expression relating the cascade intercept points to the module intercept points.

We begin by determining the intercept point for two modules in cascade before attacking the more general case.

4.5.1 Two-Module Cascade

Multiplying the third-order intermod power at the output of the first module, as given by Eq. (4.22), by the gain g_2 of a second module, we obtain its power at the output of the second module:

$$g_2 p_{\text{out,IM3,1}}(\pm 2f_1 \pm f_2) = g_2 \frac{p^2_{\text{out},F1,1} p_{\text{out},F2,1}}{p^2_{\text{OIP3,IM,1}}}. \tag{4.32}$$

The last subscripts refer to the number of the module where the intermod is generated or where the gain occurs. To this will be added the intermod generated in the output module (number 2). If the phase relationship between the two intermods is random, the mean power (over all possible phases) will be the sum of the individual powers:

$$p_{\text{out,IM3,casc}}(\pm 2f_1 \pm f_2) = g_2 \frac{p^2_{\text{out},F1,1} p_{\text{out},F2,1}}{p^2_{\text{OIP3,IM,1}}} + \frac{p^2_{\text{out},F1,2} p_{\text{out},F2,2}}{p^2_{\text{OIP3,IM,2}}} \tag{4.33}$$

$$= \frac{(g_2 p_{\text{out},F1,1})^2 (g_2 p_{\text{out},F2,1})}{(g_2 p_{\text{OIP3,IM,1}})^2} + \frac{p^2_{\text{out},F1,2} p_{\text{out},F2,2}}{p^2_{\text{OIP3,IM,2}}} \tag{4.34}$$

$$= p^2_{\text{out},F1,\text{casc}} p_{\text{out},F2,\text{casc}} \left[\frac{1}{(g_2 p_{\text{OIP3,IM,1}})^2} + \frac{1}{p^2_{\text{OIP3,IM,2}}} \right]. \tag{4.35}$$

Here we have recognized that power at the output of module 2 equals power at the output of the cascade, as does the output power of module 1 multiplied by the g_2. From this we see that we can write an expression for the cascade in the form of Eq. (4.22) if

$$\frac{1}{p^2_{\text{OIP3,casc}}} = \frac{1}{(g_2 p_{\text{OIP3,1}})^2} + \frac{1}{p^2_{\text{OIP3,2}}}, \tag{4.36}$$

which, therefore, defines the relationship between the cascade and module third-order intercept points for the case where phases add randomly. In the worst case, where the intermods add in phase, we must add voltages rather than powers, so we would change Eqs. (4.33)–(4.36) to read

$$p^{1/2}_{\text{out,IM3,casc}}(\pm 2f_1 \pm f_2) = g_2^{1/2} \frac{p_{\text{out},F1,1} p^{1/2}_{\text{out},F2,1}}{p_{\text{OIP3,IM,1}}} + \frac{p_{\text{out},F1,2} p^{1/2}_{\text{out},F2,2}}{p_{\text{OIP3,IM,2}}} \tag{4.37}$$

$$= \frac{(g_2 p_{\text{out},F1,1})(g_2 p_{\text{out},F2,1})^{1/2}}{g_2 p_{\text{OIP3,IM},1}} + \frac{p_{\text{out},F1,2}\, p_{\text{out},F2,2}^{1/2}}{p_{\text{OIP3,IM},2}} \tag{4.38}$$

$$= p_{\text{out},F1,\text{casc}}\, p_{\text{out},F2,\text{casc}}^{1/2} \left(\frac{1}{g_2 p_{\text{OIP3,IM},1}} + \frac{1}{p_{\text{OIP3,IM},2}} \right). \tag{4.39}$$

Squaring this expression and comparing to Eq. (4.22), we find that

$$\frac{1}{p_{\text{OIP3,casc}}} = \frac{1}{g_2 p_{\text{OIP3},1}} + \frac{1}{p_{\text{OIP3},2}}. \tag{4.40}$$

The IM subscripts have been left off Eqs. (4.36) and (4.40) because they apply both to IMs and harmonics, which we can easily see by repeating the developments for harmonics, referring to Eq. (4.21) rather than to Eq. (4.22).

If we follow a similar process for second-order intermods, referring to Eq. (4.10) rather than (4.22) or Eq. (4.9) rather than to Eq. (4.21), we obtain

$$\frac{1}{p_{\text{OIP2,casc}}} = \frac{1}{g_2 p_{\text{OIP2},1}} + \frac{1}{p_{\text{OIP2},2}} \tag{4.41}$$

for random phases and

$$\frac{1}{p_{\text{OIP2,casc}}^{1/2}} = \frac{1}{(g_2 p_{\text{OIP2},1})^{1/2}} + \frac{1}{p_{\text{OIP2},2}^{1/2}} \tag{4.42}$$

for coherent addition of the intermods.

Similar equations can be written for cascade input IPs by referring all of the module IPs to the cascade input. For example, we can multiply Eq. (4.42) by the square root of the cascade gain $g_1 g_2$ to give

$$\frac{1}{p_{\text{IIP2,casc}}^{1/2}} = \frac{1}{(p_{\text{OIP2},1}/g_1)^{1/2}} + \frac{1}{[p_{\text{OIP2},2}/(g_1 g_2)]^{1/2}} \tag{4.43}$$

$$= \frac{1}{p_{\text{IIP2},1}^{1/2}} + \frac{1}{(p_{\text{IIP2},2}/g_1)^{1/2}}. \tag{4.44}$$

4.5.2 General Cascade

The ratio of two output powers (or voltages), such as in Eqs. (4.11) or (4.24), or of two input powers, as in Eq. (4.13), does not change as we move through the stages in a cascade. That is, if the ratio exists at some point in the cascade, it will be the same when both quantities are amplified by subsequent stages or when the signals are referred to the input, assuming the same frequency response for the two powers forming the ratio. For example, if we divide both sides of

Eq. (4.32) by $g_2 p_{\text{out},F1,1}$, we obtain

$$\frac{g_2 p_{\text{out,IM3},1}(\pm 2f_1 \pm f_2)}{g_2 p_{\text{out},F1,1}} = \frac{p_{\text{out},F1,1}\, p_{\text{out},F2,1}}{p_{\text{OIP3},1}^2}, \tag{4.45}$$

which is independent of g_2. Thus, the output (or the input by reference) will contain many intermodulation products at the same frequency, each produced in a different stage of the cascade, and each with the same power, relative to the signal, that it had when it was created. Here we will find a single equivalent IP, to represent the entire cascade, that produces the same result. The development will be similar to the previous section but more general.

The general form of the expressions that we have obtained for nth-order intermods or harmonics at a module output can be written

$$\frac{p_{\text{out},n}}{p_{\text{out},F1}} = \frac{k_{\text{out}}}{p_{\text{OIP}n}^{n-1}}, \tag{4.46}$$

where

$$k_{\text{out}} = \prod_{m=1}^{n-1} p_{\text{out},Fi} \tag{4.47}$$

and $i = 1$ or $i = 2$ (e.g., $p_{\text{out},F1} p_{\text{out},F2}$ or $p_{\text{out},F1}^3$), is some product of output powers that depends on the particular IM or harmonic being considered. Here n indicates the order of the nonlinearity. We have dropped the designator H or IM; the development will apply to either. Compare to Eqs. (4.11), (4.12), (4.23), and (4.24). For the whole cascade this would be written

$$\frac{p_{\text{out},n,\text{cas}}}{p_{\text{out},F1,\text{cas}}} = \frac{k_{\text{out,cas}}}{p_{\text{OIP}n,\text{cas}}^{n-1}}. \tag{4.48}$$

4.5.3 IMs Adding Coherently

In the worst case, all q of the products generated in the q stages of a cascade are in phase; so their voltage amplitudes add at the output, as do the q ratios of the product to one of the fundamental voltages:

$$\left(\frac{v_{\text{out},n}}{v_{\text{out},F1}}\right)_{\text{cas}} = \sum_{i=1}^{q} \left(\frac{v_{\text{out},n}}{v_{\text{out},F1}}\right)_i. \tag{4.49}$$

Adding those ratios is the same as adding the intermod voltages because the fundamental is the same for each of the addends at the output or when referred to any given point in the cascade.

We can write Eq. (4.46) for the cascade as

$$\left(\frac{k_{\text{out}}}{p_{\text{OIP}n}^{n-1}}\right)_{\text{cas}} = \left(\frac{p_{\text{out},n}}{p_{\text{out},F1}}\right)_{\text{cas}} = \left(\frac{v_{\text{out},n}}{v_{\text{out},F1}}\right)_{\text{cas}}^2 \tag{4.50}$$

and combine it with Eq. (4.49), after taking the square root, to give

$$\left(\frac{k_{\text{out}}}{p_{\text{OIP}n}^{n-1}}\right)_{\text{cas}}^{1/2} = \sum_{i=1}^{q}\left(\frac{v_{\text{out},n}}{v_{\text{out},F1}}\right)_i. \tag{4.51}$$

Now using Eq. (4.46) for the modules, this becomes

$$\left(\frac{k_{\text{out}}}{p_{\text{OIP}n}^{n-1}}\right)_{\text{cas}}^{1/2} = \sum_{i=1}^{q}\left(\frac{k_{\text{out}}}{p_{\text{OIP}n}^{n-1}}\right)_i^{1/2}. \tag{4.52}$$

Dividing both sides by $k_{\text{out,cas}}^{1/2}$ we obtain

$$\frac{1}{p_{\text{OIP}n,\text{cas}}^{(n-1)/2}} = \frac{1}{k_{\text{out,cas}}^{1/2}} \sum_{i=1}^{q} \frac{k_{\text{out},i}^{1/2}}{p_{\text{OIP}n,i}^{(n-1)/2}}. \tag{4.53}$$

Since k_{out} is the product of $(n-1)$ output power levels [see Eq. (4.47)], the ratio of $k_{\text{out,cas}}$ to $k_{\text{out},i}$ is the product of $n-1$ power gains from the output of module i to the cascade output. Using that equivalence, we have

$$\frac{1}{p_{\text{OIP}n,\text{cas}}^{(n-1)/2}} = \sum_{i=1}^{q} \frac{1}{(g_{i+1,q}\, p_{\text{OIP}n,i})^{(n-1)/2}}, \tag{4.54}$$

where

$$g_{k,q} \triangleq \prod_{j=k}^{q} g_j \tag{4.55}$$

is the gain for modules k through q and, thus, the gain from the input to module k to the cascade output. Thus $g_{i+1,q}\, p_{\text{OIP}n,i}$ just represents the intercept point of module i amplified or referred to the output, where it is combined with the other amplified module intercept points.

We can change this relationship between OIPs to one between IIPs by dividing the denominator by $g_{1,q}^{(n-1)/2}$ to obtain

$$\frac{1}{p_{\text{IIP}n,\text{cas}}^{(n-1)/2}} = \sum_{i=1}^{q}\left(\frac{g_{1,i}}{p_{\text{OIP}n,i}}\right)^{(n-1)/2} = \sum_{i=1}^{q}\left(\frac{g_{1,i-1}}{p_{\text{IIP}n,i}}\right)^{(n-1)/2} \tag{4.56}$$

The IM3s that are close to the desired signals tend to receive the same phase shift as the fundamental because their frequencies are close. Consider this scenario, which is tabulated in Table 4.1. Suppose that, at some point, two fundamental signals have phases θ_1 and θ_2 and a third-order product is created with time-varying phase of $2\varphi_1(t) - \varphi_2(t)$, implying a frequency of $2\omega_1 - \omega_2$ (close to the signal at ω_1) plus a phase $2\theta_1 - \theta_2$. In traveling to another module, these signals

TABLE 4.1 Phases of Close (in frequency) Signals and IMs Formed at Two Different Locations

Frequency of signal	f_1	f_2	$f_{\mathrm{IM3}} = 2f_1 - f_2$
Phase at module 1	θ_1	θ_2	$2\theta_1 - \theta_2$
Phase at module 2	$\theta_1 + \Delta\theta$	$\theta_2 + \Delta\theta$	$2\theta_1 - \theta_2 + \Delta\theta$
Phase of new IM3 at module 2			$2(\theta_1 + \Delta\theta) - \theta_2 - \Delta\theta =$ $2\theta_1 - \theta_2 + \Delta\theta$

pick up a phase shift of $\Delta\theta$ so the fundamentals have phase shift $\theta_1 + \Delta\theta$ and $\theta_2 + \Delta\theta$ and the third-order IM has phase of $2\theta_1 - \theta_2 + \Delta\theta$. Then an IM is created at the same frequency in the second module. If the process is similar to what produced the first IM, the new IM will have phase $2(\theta_1 + \Delta\theta) - (\theta_2 + \Delta\theta) = 2\theta_1 - \theta_2 + \Delta\theta$. But this is the same phase that the first IM has on arrival in the second module; so the IMs created in two different locations add in phase.

Therefore, it would not be surprising to find that IM3s close to the signals add in phase and Eq. (4.53), while being worst case, may also be close to typical for third-order IMs near the signals (Maas, 1995). For third-order IMs, Eq. (4.54) becomes

$$\frac{1}{p_{\mathrm{OIP3,cas}}} = \sum_{i=1}^{q} \frac{1}{g_{i+1,q}\, p_{\mathrm{OIP3},i}}; \tag{4.57}$$

so the cascade IP3 is the reciprocal of the sum of the reciprocals of individual IP3s. They add much as do parallel resistances.

We could also write [Eq. (4.56)] as

$$\frac{1}{p_{\mathrm{IIP3,cas}}} = \sum_{i=1}^{q} \frac{g_{1,i-1}}{p_{\mathrm{IIP3},i}}. \tag{4.58}$$

4.5.4 IMs Adding Randomly

If the IMs have random phase, we expect the *powers* to add. Equation (4.53) still gives the worst case, but we expect the results to be closer to a value given by adding powers rather than voltages. Proceeding as before, but using powers, Eq. (4.50) leads to

$$\left(\frac{k_{\mathrm{out}}}{p_{\mathrm{OIP}n}^{n-1}}\right)_{\mathrm{cas}} = \sum_{i=1}^{q} \left(\frac{p_{\mathrm{out},n}}{p_{\mathrm{out},Fj}}\right)_i = \left(\frac{p_{\mathrm{out},n}}{p_{\mathrm{out},Fj}}\right)_{\mathrm{cas}} \tag{4.59}$$

$$= \sum_{i=1}^{q} \left(\frac{k_{\mathrm{out}}}{p_{\mathrm{OIP}n}^{n-1}}\right)_i, \tag{4.60}$$

which can be written

$$\frac{1}{p_{\mathrm{OIP}n,\mathrm{cas}}^{n-1}} = \sum_{i=1}^{q} \frac{1}{(g_{i+1,q}\, p_{\mathrm{OIP}n,i})^{n-1}}. \tag{4.61}$$

This differs from Eq. (4.54) for coherent addition in that the powers, $n-1$, were there divided by 2.

We could also write the relationship in terms of input intercept points:

$$\frac{1}{p_{\mathrm{IIP}n,\mathrm{cas}}^{n-1}} = \sum_{i=1}^{q} \left(\frac{g_{1,i-1}}{p_{\mathrm{IIP}n,i}} \right)^{n-1}. \tag{4.62}$$

These equations are more appropriate for products that are not close to the desired signal, such as the third-order products near the third harmonics and all of the second-order products. For these, the phase shifts produced by time delays in traveling will differ because, for a given delay, the phase shifts are proportional to frequency. In addition, multiplications that are not of the form $m\theta_1 - (m-1)\theta_2$, as they are for the close-in IM3s, will not produce the same phase shift, relative to the fundamental, when they occur at two different locations.

For $n = 2$, Eq. (4.61) gives

$$\frac{1}{p_{\mathrm{OIP2,cas}}} = \sum_{i=1}^{q} \frac{1}{g_{i+1,q}\, p_{\mathrm{OIP2},i}}, \tag{4.63}$$

which has the same form as Eq. (4.57). Therefore, this form may be appropriate for both second-order and close-to-signal third-order products.

4.5.5 IMs That Do Not Add

Third-order products, such as c and d in Fig. 4.6, follow the signal through frequency translations since they are always separated from the signals by a fixed offset. Thus, IM3s of this type, generated before a frequency translation (heterodyning in a mixer), add to the same type generated after a translation. The pretranslation set is translated to the frequencies at which the posttranslation IMs occur.

Other products shown in Fig. 4.6 and the second-order products in Fig. 4.2 do not possess this property. Their separations from the signal depend on the signal frequency. Therefore, pretranslation IMs will occur at different frequencies after the translation than will the IMs created there.

The IM3s (c and d in Fig. 4.6) that follow the signals (a and b in Fig. 4.6) are most important because of their ability to sneak through filters along with signals and because they reinforce across translations. The next most important IMs are usually second-order since they are closer to the signals than the other third-order IMs. They are most important in video bands, that is, bands that include zero frequency or have a very large ratio of upper- to lower-edge frequencies. Such bands can contain fundamentals (a and b in Figs. 4.2 and 4.6) and their harmonics and signals near the harmonics (d–f in Fig. 4.2 and e–h in Fig. 4.6) as well as difference-frequency signals (c in Fig. 4.2).

Filtering may determine whether certain IMs are generated throughout the system or only in certain parts. Two signals that are very closely spaced and that are well within all filter bands will generate IM3s that add everywhere. As the separation between the signals increases, one or both of the signals or one or both of the IMs may be reduced by filtering. Then the amplitudes of the IMs in the cascade become a function of the separation between the signals (Synder, 1978).

4.5.6 Effect of Mismatch on IPs

In Chapter 3, we showed how the noise factor, as usually measured, was appropriate for use in our standard cascade. We used an isolated noise source in that development, but we also considered, in Section 3.8, how the driving impedance could affect the value of the noise source and how to use such information when available. There are similar considerations for the use of IPs in cascade.

As long as the IPs, as measured with standard impedances, do not change, they are appropriate for use in a standard cascade. The gain that references them to the cascade input is appropriate for relating the output measured during test (v_{ojT} in Fig. 2.5) to the system input. However, the accuracy of the cascade analysis suffers from any dependence of IP on mismatch at the module output. If the reflection from the module output $S_{22(k-)}$ were linear, the relative IMs transmitted through the interconnect would be the same as in v_{ojT}, but, in many cases, this would not be true. Since reflected signals can cause an effective change in the load of the output amplifier, reflections can significantly influence the IP. For example, higher impedance would cause a larger voltage swing and thus produce distortion at a lower module input level.

Commonly, the only information available on module IPs would have been obtained into standard impedance, and these would be used as best estimates of performance in the actual cascade. However, for greater accuracy, we could measure the IPs with the load that the module sees in the cascade or an equivalent load (Fig. 4.10). Since we are measuring the forward wave in the cable, the gain from the forward wave at the input of the module under test to the forward wave at the cable output can also be determined, avoiding the usual use of a range of effective cable gains.

In general, however, we are assuming that the load impedance is unknown, except for its standing-wave ratio (SWR). We would therefore vary the phase of the mismatch in Fig. 4.10, one having the specified SWR (perhaps by using a line stretcher in the interconnect), and measure the IPs as a function of phase

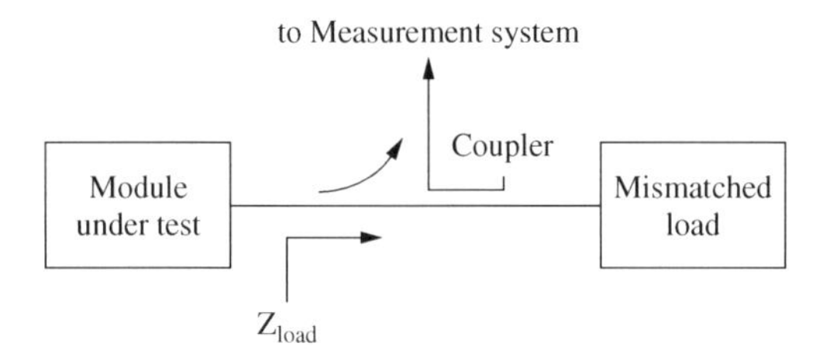

Fig. 4.10 Measuring IPs with mismatches.

at fixed SWR. This would be referred to the module input or output to give the range of IPs to be expected with the specified load SWR. We should correlate the measured IP with the observed effective cable gain so we do not use an inappropriate combination in our cascade calculations (e.g., low IP at low gain when the two do not occur simultaneously).

4.6 EXAMPLES: SPREADSHEETS FOR IMs IN A CASCADE

Example 4.3 Computing IMs of a Cascade Figure 4.11 is a spreadsheet for a standard cascade with second- and third-order OIPs added in cells F4–G10. The cumulative IIP2s and IIP3s for the cascade are given in cells F24–K30 for the cascade ending with each module or interconnect. They could be obtained from Eq. (4.63) multiplied by the cascade gain to give input IPs. However, as was true for noise figures, we will find it more convenient to use formulas for two elements, the first of which consists of a cascade composed of all previous modules. Therefore, we use Eqs. (4.40) and (4.41), multiplying them by the gain $g_1 g_2$ for the cascade being considered. Since both equations have the same form, we can just write

$$\frac{1}{p_{\mathrm{OIP,casc}}/g_1 g_2} = \frac{1}{p_{\mathrm{OIP},1}/g_1} + \frac{1}{p_{\mathrm{OIP},2}/g_1 g_2}, \tag{4.64}$$

$$= \frac{1}{p_{\mathrm{IIP,casc}}}, \tag{4.65}$$

where the module or interconnect under consideration has index 2 and the preceding cascade has index 1.

Note how severely the input IPs can be affected by gain variations (compare cells F30–H30 and also I30–K30).

The products being computed in this spreadsheet are IMs, rather than harmonics, according to the label in cells F2–G2, although only the OIP values would be changed for harmonics. We just need to label the spreadsheet to show what we are computing.

A display of sensitivities of the cascade IPs to module IPs and gains can be valuable. These would be obtained, as they were for gains (Section 2.5) and noise figures (Section 3.11), by listing changes in the overall IPs obtained by experimentally changing component parameters on the spreadsheet and normalizing to a 1-dB component parameter change.

Example 4.4 Frequency Conversion and IMs That Do Not Add The problem, discussed in Section 4.5.5, of IP2s that do not add, is illustrated in Fig. 4.12. Here a mixer changes the signal frequency in the midst of the cascade and thus changes the frequencies of the second-order products. Line 32 has been added and labeled "mixer from in." It refers to the second-order IMs at the sum or difference of two input frequencies. These are translated by the mixer in the same manner

	A	B	C	D	E	F	G	H	I	J	K
2		Gain	Gain	SWR		IMs					
3		nom	+/−	at out	$\lvert a_{RT} \rvert$	OIP3	OIP2				
4	Module 1	12.0 dB	1.0 dB	1.5		0.0 dBm	10.0 dBm				
5	Cable 1	−1.5 dB		1.5	0.0283						
6	Module 2	8.0 dB	2.0 dB	2		10.0 dBm	23.0 dBm				
7	Cable 2	−1.0 dB		2	0.0883						
8	Module 3	2.0 dB	2.0 dB	2.8		10.0 dBm	22.0 dBm				
9	Cable 3	−0.8 dB		3.2	0.2064						
10	Module 4	15.0 dB	2.0 dB			24.0 dBm	35.0 dBm				
11	DERIVED										
12		Gain									
13		mean	max	min	±						
14	Module 1	12.00 dB	13.00 dB	11.0 dB	1.00 dB						
15	Cable 1	−1.50 dB	−1.25 dB	−1.74 dB	0.25 dB						
16	Module 2	8.00 dB	10.00 dB	6.00 dB	2.00 dB						
17	Cable 2	−0.97 dB	−0.20 dB	−1.73 dB	0.77 dB						
18	Module 3	2.00 dB	4.00 dB	0.00 dB	2.00 dB						
19	Cable 3	−0.61 dB	1.21 dB	−2.43 dB	1.82 dB						
20	Module 4	15.00 dB	17.00 dB	13.00 dB	2.00 dB						
21	CUMULATIVE										
22		Gain				IIP3 with			IIP2 with		
23	at output of	mean	max	min	±	mean gain	max gain	min gain	mean gain	max gain	min gain
24	Module 1	12.00 dB	13.00 dB	11.00 dB	1.00 dB	−12.00 dB	−13.00 dB	−11.00 dB	−2.00 dB	−3.00 dB	−1.00 dB
25	Cable 1	10.50 dB	11.75 dB	9.26 dB	1.25 dB	−12.00 dB	−13.00 dB	−11.00 dB	−2.00 dB	−3.00 dB	−1.00 dB
26	Module 2	18.50 dB	21.75 dB	15.26 dB	3.25 dB	−13.60 dB	−15.43 dB	−12.03 dB	−2.88 dB	−4.39 dB	−1.54 dB
27	Cable 2	17.54 dB	21.55 dB	13.52 dB	4.01 dB	−13.60 dB	−15.43 dB	−12.03 dB	−2.88 dB	−4.39 dB	−1.54 dB
28	Module 3	19.54 dB	25.55 dB	13.52 dB	6.01 dB	−15.04 dB	−18.50 dB	−12.60 dB	−3.99 dB	−7.00 dB	−1.96 dB
29	Cable 3	18.93 dB	26.76 dB	11.09 dB	7.83 dB	−15.04 dB	−18.50 dB	−12.60 dB	−3.99 dB	−7.00 dB	−1.96 dB
30	Module 4	33.93 dB	43.76 dB	24.09 dB	9.83 dB	−16.21 dB	−22.19 dB	−12.84 dB	−5.17 dB	−10.98 dB	−2.18 dB

Fig. 4.11 Spreadsheet giving IPs for standard cascade.

	A	B	C	D	E	F	G	H
2		Gain	Gain	SWR		IMs		
3		nom	+/–	at out	$\lvert a_{RT} \rvert$	OIP3	OIP2a	OIP2b
4	Module 1	12.0 dB	1.0 dB	1.5		9.0 dBm	19.0 dBm	
5	Cable 1	-1.5 dB		1.5	0.0283			
6	Module 2	15.0 dB	1.5 dB	2		27.0 dBm	40.0 dBm	
7	Cable 2	-1.0 dB		2	0.0883			
8	Module 3 (mixer)	-9.0 dB	1.5 dB	2.8		19.0 dBm	52.0 dBm	57.0 dBm
9	Cable 3 (diplexer)	-7.0 dB		1.5	0.0189	24.0 dBm		60.0 dBm
10	Module 4	15.0 dB	1.0 dB	2		26.0 dBm		37.0 dBm
11	Cable 4	-0.8 dB		1.7	0.0719			
12	Module 5	6.0 dB	1.5 dB	3		30.0 dBm		44.0 dBm
13	DERIVED							
14		Gain						
15		mean	max	min	±			
16	Module 1	12.00 dB	13.00 dB	11.00 dB	1.00 dB			
17	Cable 1	-1.50 dB	-1.25 dB	-1.74 dB	0.25 dB			
18	Module 2	15.00 dB	16.50 dB	13.50 dB	1.50 dB			
19	Cable 2	-0.97 dB	-0.20 dB	-1.73 dB	0.77 dB			
20	Module 3 (mixer)	-9.00 dB	-7.50 dB	-10.50 dB	1.50 dB			
21	Cable 3 (diplexer)	-7.00 dB	-6.83 dB	-7.16 dB	0.16 dB			
22	Module 4	15.00 dB	16.00 dB	14.00 dB	1.00 dB			
23	Cable 4	-0.78 dB	-0.15 dB	-1.40 dB	0.63 dB			
24	Module 5	6.00 dB	7.50 dB	4.50 dB	1.50 dB			
25	CUMULATIVE							
26		Gain				IIP3 with	IIP2 with	IIP2 with
27	at output of	mean	max	min	±	mean gain	mean gain	max gain
28	Module 1	12.00 dB	13.00 dB	11.00 dB	1.00 dB	-3.00 dBm	7.00 dBm	6.00 dBm
29	Cable 1	10.50 dB	11.75 dB	9.26 dB	1.25 dB	-3.00 dBm	7.00 dBm	6.00 dBm
30	Module 2	25.50 dB	28.25 dB	22.76 dB	2.75 dB	-4.32 dBm	6.29 dBm	4.98 dBm
31	Cable 2	24.54 dB	28.05 dB	21.02 dB	3.51 dB	-4.32 dBm	6.29 dBm	4.98 dBm
32	mixer from in	15.54 dB	20.55 dB	10.52 dB			6.28 dBm	4.97 dBm
33	mixer at out	15.54 dB	20.55 dB	10.52 dB	5.01 dB	-4.99 dBm	41.46 dBm	36.45 dBm
34	Cable 3 (diplexer)	8.54 dB	13.72 dB	3.36 dB	5.18 dB	-5.03 dBm	41.46 dBm	36.45 dBm
35	Module 4	23.54 dB	29.72 dB	17.36 dB	6.18 dB	-5.74 dBm	13.45 dBm	7.28 dBm
36	Cable 4	22.76 dB	29.57 dB	15.96 dB	6.80 dB	-5.74 dBm	13.45 dBm	7.28 dBm
37	Module 5	28.76 dB	37.07 dB	20.46 dB	8.30 dB	-6.53 dBm	11.25 dBm	4.09 dBm

Fig. 4.12 Spreadsheet, with frequency conversion, giving IP2s.

as are the signals. The next line is labeled "mixer at out" and refers to the IM at the sum or difference of two output signal frequencies. Lines 33–37 contain IIP2 data for the part of the cascade after the mixer. Line 33 begins anew with the OIP2 of the mixer output, not combining the previous IIP2s since they are at different frequencies. Thus cells G32–H32 contain IIPs for one set of IMs and cells G37–H37 contain IIPs for another set at different frequencies.

We could use multiple columns if we were interested in more than one IM that did not add through the frequency translation and that were characterized by different sets of IPs. Note that cells G8 and H8 contain different OIP2 values for the mixer. There are separate OIP2s for the input frequencies and for the output frequencies. The value in G8 is related to the 1×2(LO $\times$ 2RF) mixer spurs, involving the second-order products of the input signal (RF). The value in H8 is related to the 2×2 products (2LO $\times$ 2RF), for example, the second harmonic of the desired 1×1 mixer output (IF). Another of the 2×2s appears at the mixer output at the frequency difference between the two signals, but the product that was created at the same frequency at the input has there been translated to a different frequency.

"Cable 3" is actually a diplexer (or, perhaps, a triplexer), a filter for removing undesired products while presenting a proper interface impedance both within and without the passband. Treating it as a cable assumes perfect match, at its

	A	B	C	D	E	F	G	H
2		Gain	Gain	SWR		IMs		
3		nom	+/−	at out	$\|a_{RT}\|$	OIP3	OIP2a	OIP2b
4	Module 1	12.0 dB	1.0 dB	1.5		9 dBm	19 dBm	
5	Cable 1	−1.5 dB		1.5	0.0283			
6	Module 2	15.0 dB	1.5 dB	2		27 dBm	40 dBm	
7	Cable 2	−1.0 dB		2	0.0883			
8	Module 3 (mixer)	−9.0 dB	1.5 dB	2.8		19 dBm	52 dBm	57 dBm
9	Cable 3 (diplexer)	−7.0 dB		1.5	0.0189	24 dBm		60 dBm
10	Module 4	15.0 dB	1.0 dB	2		26 dBm		37 dBm
11	Cable 4	−0.8 dB		1.7	0.0719			
12	Module 5	6.0 dB	1.5 dB	3		30 dBm		44 dBm

	A	B	C	D	E	F
13	DERIVED					
14		Gain				
15		mean	max	min	±	
16	Module 1	12.00 dB	13.00 dB	11.00 dB	1.00 dB	
17	Cable 1	−1.50 dB	−1.25 dB	−1.74 dB	0.25 dB	
18	Module 2	15.00 dB	16.50 dB	13.50 dB	1.50 dB	
19	Cable 2	−0.97 dB	−0.20 dB	−1.73 dB	0.77 dB	
20	Module 3 (mixer)	−9.00 dB	−7.50 dB	−10.50 dB	1.50 dB	
21	Cable 3 (diplexer)	−7.00 dB	−6.83 dB	−7.16 dB	0.16 dB	
22	Module 4	15.00 dB	16.00 dB	14.00 dB	1.00 dB	
23	Cable 4	−0.78 dB	−0.15 dB	−1.40 dB	0.63 dB	
24	Module 5	6.00 dB	7.50 dB	4.50 dB	1.50 dB	
25	CUMULATIVE					
26		Gain	IIP3		IIP2 with	
27	at output of	mean	coherent	non-coherent	coherent	non-coherent
28	Module 1	12.00 dB	−3.00 dBm	−3.00 dBm	7.00 dBm	7.00 dBm
29	Cable 1	10.50 dB	−3.00 dBm	−3.00 dBm	7.00 dBm	7.00 dBm
30	Module 2	25.50 dB	−4.32 dBm	−3.26 dBm	3.94 dBm	6.29 dBm
31	Cable 2	24.54 dB	−4.32 dBm	−3.26 dBm	3.94 dBm	6.29 dBm
32	mixer from in	15.54 dB			3.74 dBm	6.28 dBm
33	mixer at out	15.54 dB	−4.99 dBm	−3.35 dBm	41.46 dBm	41.46 dBm
34	Cable 3 (diplexer)	8.54 dB	−5.03 dBm	−3.35 dBm	39.08 dBm	41.46 dBm
35	Module 4	23.54 dB	−5.74 dBm	−3.50 dBm	13.02 dBm	13.45 dBm
36	Cable 4	22.76 dB	−5.74 dBm	−3.50 dBm	13.02 dBm	13.45 dBm
37	Module 5	28.76 dB	−6.53 dBm	−3.73 dBm	8.04 dBm	11.25 dBm

Fig. 4.13 Spreadsheet giving IPs with both coherent and noncoherent addition.

terminals, to the standard impedance. An alternative, which would allow the inclusion of SWRs for the diplexer, would be to treat it as a unilateral module with cables on either side. That would be an approximation also, depending on its attenuation and the matches at the cable ends to give effective unilaterality.

Example 4.5 Coherent and Noncoherent Addition Figure 4.13 shows both coherent and noncoherent addition of IMs, according to Eq. (4.56) and (4.62), respectively (modified for only two levels, as before). Only mean gains are used.

4.7 ANOMALOUS IMs

Occasionally, we may find a module with IMs at the frequencies expected for IM3s but that vary with amplitude like IM2s (i.e., as in Fig. 4.3). Such an anomaly is illustrated in Fig. 4.14. This can occur when the transfer curve has hysteresis, due to the presence of magnetic circuits. For example, power amplifiers often use ferrite cores in making baluns, transformers, and combiners, and filters may use such transformers in matching. The theory we have used to this point was based on the Taylor series of Eq. (4.1), representing a curve such as is shown in Fig. 4.1, but this does not describe a transfer function containing hysteresis, in which the curve differs for increasing and decreasing input values and changes with the magnitude of the signals. Two such curves are shown in Fig. 4.15. The magnetic flux density B, which is ultimately proportional to a voltage within the

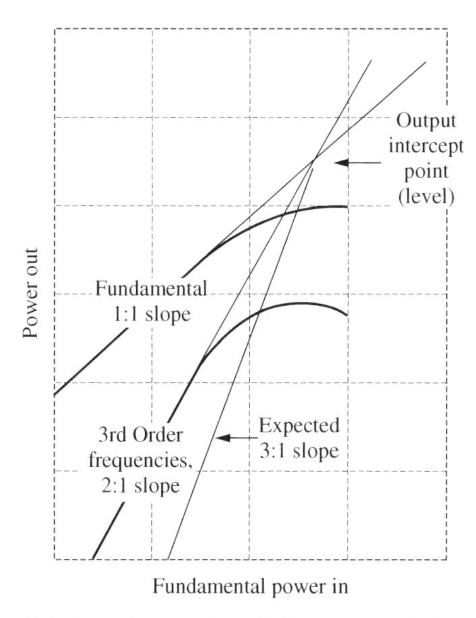

Fig. 4.14 Anomalous IMs have frequencies of third-order IMs but amplitude dependence of second-order IMs.

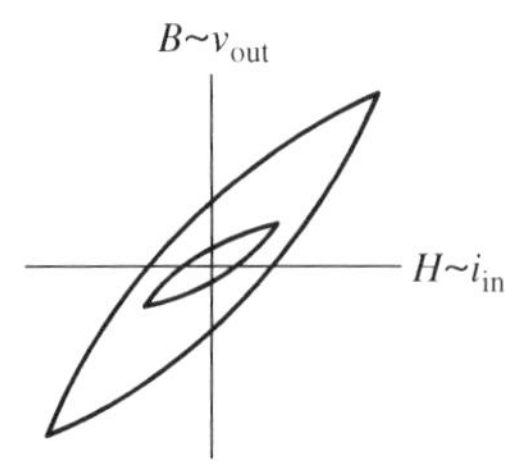

Fig. 4.15 Hysteresis.

device, and which at least influences the module output voltage, is here plotted against the magnetic field intensity, which is proportional to the input current and, thus, to the input voltage. The change in shape between the two curves shown, one of which represents larger peak values than the other, is apparent.

Two 1935 studies (Kalb and Bennett, 1935; Latimer, 1935–36) obtained the response for such a circuit, assuming the hysteresis curves consisted of back-to-back parabolas (Snelling, 1988). The term representing the hysteresis produces a third harmonic (in the case of one signal) and is proportional to the *square* of the peak value of H. For two equal-amplitude driving signals, the ratios of output IMs and harmonics to the fundamentals have been shown to be proportional to the input voltage, in the fashion of a second-order IM, not to its square, as with third orders.

Both the curvature at high levels and the possibility of anomalies suggest the measurement of multiple points, preferably in the region of expected input powers, to confidently establish the intercept points. Perhaps of more concern is the possibility of a mix of both normal and anomalous third-order IMs, as illustrated in Fig. 4.16, since there is a greater possibility of the anomaly being undetected when a few points on a third-order slope are measured at higher levels.

4.8 MEASURING IMs

Figure 4.17 shows the setup for measuring harmonic IP (Barkley, 2001). Ideally, with the switches up, we measure the power of the signal from the module under test and the relative level of the harmonic using a calibrated spectrum analyzer. Then, as we change the power from the generator, we plot these, as in Fig. 4.3 or 4.8, and determine the power where the two extended lines cross. If we know we are in the desired straight-line regions, measurements will only be necessary at one power level. However, generally we should verify that the straight line continues through the region representing the powers that are of interest in the system.

A potential problem arises from the possible generation of the same harmonics we are trying to measure in some part of the measurement system. Even high-quality signal generators have surprisingly high harmonic levels. If the signal generator harmonic level, measured without the module in place, is significant,

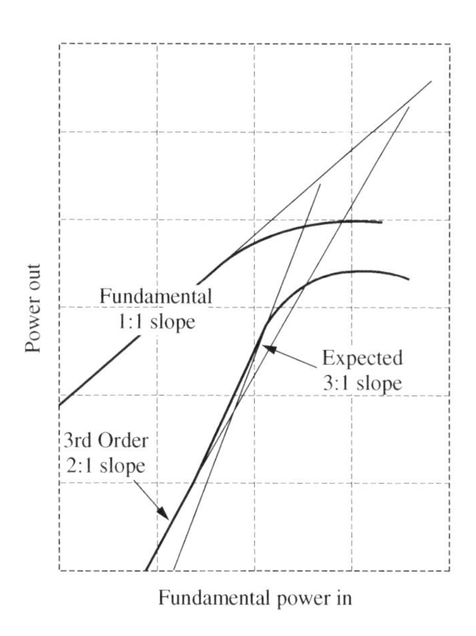

Fig. 4.16 Normal and anomalous third-orders are both present.

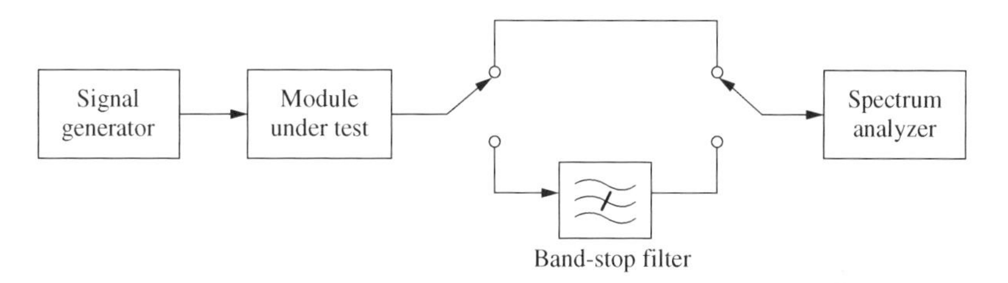

Fig. 4.17 Harmonic test.

it will have to be filtered before driving the module under test. The spectrum analyzer is also suspect as a source of harmonics. If a change in its input attenuator does not change both the fundamental and the harmonic by the change in attenuation, some of the harmonic power is being generated in the analyzer. In that case, a filter must be inserted before the analyzer to attenuate the fundamental so it will not generate harmonics in the analyzer. Of course, filter losses must be accounted for in our calculations.

We assume that the switch is either free from significant harmonics and IMs or, more likely, that it actually represents reconnection of cables, there not being an actual switch. Figure 4.18 shows the setup for measuring IP_{IM}. The additions are a second generator and a power combiner to add the two signals. In addition to the potential problems encountered in the harmonic test, we now must be concerned about the generation of IMs in these two added components. A signal leaking through the power combiner from one generator to the other can generate

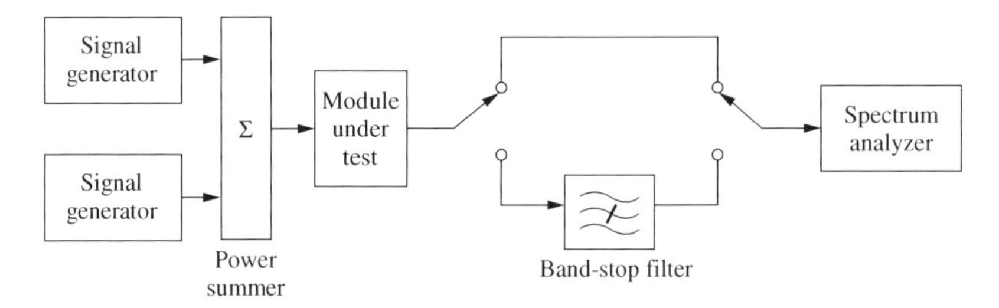

Fig. 4.18 IM test.

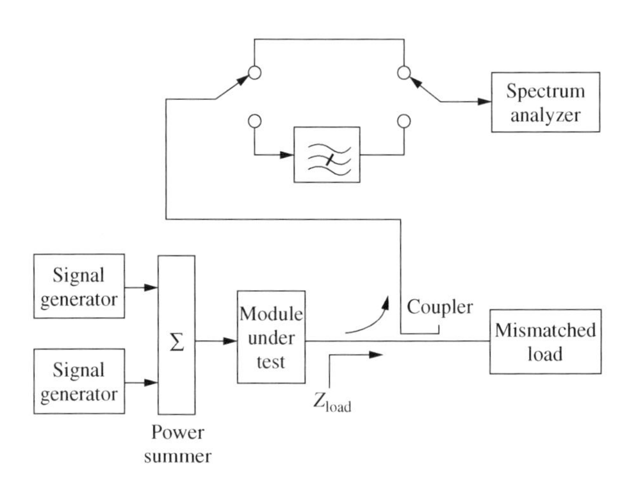

Fig. 4.19 IM test with mismatches.

IMs in the generator. Some generators are rated for IM generation under these conditions. It may be necessary to add isolators or attenuators at the generator outputs. In addition, the power combiner is a possible source for IMs.

Testing the input signal to the module for IMs may be difficult; another amplifier may be needed to get the significant IM level high enough to observe without the module under test. If filtering is required to eliminate the fundamentals before the analyzer, it must now filter out two signals while passing the IMs.

Figure 4.19 is a combination of Figs. 4.18 and 4.10, which shows the testing of IMs with a mismatched load more completely than does Fig. 4.10. The equivalent diagram for testing harmonics would be a similar combination of Figs. 4.17 and 4.10.

In Section 4.4 we discussed the relationship between IMs and harmonics that would enable us to predict the former from the latter under certain restrictions. This has the advantage of only requiring one signal source. One restriction was that the same gain should apply at the harmonics and the IMs. Another concerned small contribution from higher-order products so the harmonics and IMs are of the lowest order (second or third), as our relationships assume. If we measure harmonic power versus input power over a range, we can ensure that the curve is following the theoretical relationship for the lowest order IM and use data in that region. Modulated signals have also been used to measure distortion (Heutmaker et al., 1997).

While we have emphasized IMs in modules, we should be aware that apparently innocuous components, such as cable assemblies (Deats and Hartman, 1997), can produce IMs that are significant at high power levels.

4.9 COMPRESSION IN THE CASCADE

The output power level that is 1 dB below the level expected using the small-signal gain is a measure of gain compression at high signal levels. It is called the 1-dB compression level, $P_{\text{out},-1\text{ dB}}$, and is illustrated in Fig. 4.20. To predict the 1-dB compression level for a cascade, we would have to multiply (add gains in dB of) all of the transfer curves in the compression region. We could then find the 1-dB compression level from the composite, as in Fig. 4.20. Because of the complexity of this process, we consider two approximate processes in the following example.

Example 4.6 Refer to Fig. 4.21. In cells I25–K31, we show the equivalent cascade input compression point for each module, obtained by dividing the module 1-dB compression power by the preceding power gain and adding 1 dB (to account for the 1-dB gain compression in the module). These numbers are not accurate unless all of the preceding modules are in their linear region, but they allow us to compare the potential effects of all of the compressions of the various modules. They tell us the effect of each module if it alone were in compression.

We show the lowest of these values in cells I33–K33. The first module is dominant except at max gain, where the last module becomes so.

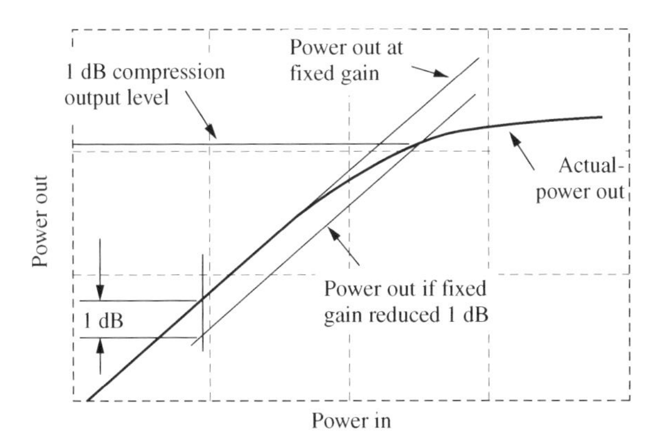

Fig. 4.20 1-dB compression.

	A	B	C	D	E	F	G	H	I	J	K
2		Gain	Gain	SWR			1-dB compression				
3		nom	+/–	at out	$\|a_R\|_T$	OIP3	output				
4	Module 1	12.0 dB	1.0 dB	1.5		0.0 dBm	10.0 dBm				
5	Cable 1	–1.5 dB		1.5	0.0283						
6	Module 2	8.0 dB	2.0 dB	2		10.0 dBm	23.0 dBm				
7	Cable 2	–1.0 dB		2	0.0883						
8	Module 3	2.0 dB	2.0 dB	2.8		10.0 dBm	22.0 dBm				
9	Cable 3	–0.8 dB		3.2	0.2064						
10	Module 4	15.0 dB	2.0 dB			24.0 dBm	35.0 dBm				
11						DERIVED					
12			Gain								
13		mean	max	min	±						
14	Module 1	12.00 dB	13.00 dB	11.00 dB	1.00 dB						
15	Cable 1	–1.50 dB	–1.25 dB	–1.74 dB	0.25 dB						
16	Module 2	8.00 dB	10.00 dB	6.00 dB	2.00 dB						
17	Cable 2	–0.97 dB	–0.20 dB	–1.73 dB	0.77 dB						
18	Module 3	2.00 dB	4.00 dB	0.00 dB	2.00 dB						
19	Cable 3	–0.61 dB	1.21 dB	–2.43 dB	1.82 dB						
20	Module 4	15.00 dB	17.00 dB	13.00 dB	2.00 dB		Highlight compressions within			5 dB	of minimum
21						CUMULATIVE					
22									Input at 1-dB compression in module		
23			Gain				IIP3 with		if no compression in previous stages		
24	at output of	mean	max	min	±	mean gain	max gain	min gain	mean gain	max gain	min gain
25	Module 1	12.00 dB	13.00 dB	11.00 dB	1.00 dB	–12.00 dBm	–13.00 dBm	–11.00 dBm	**–1.00 dBm**	–2.00 dBm	**0.00 dBm**
26	Cable 1	10.50 dB	11.75 dB	9.26 dB	1.25 dB	–12.00 dBm	–13.00 dBm	–11.00 dBm			
27	Module 2	18.50 dB	21.75 dB	15.26 dB	3.25 dB	–13.60 dBm	–15.43 dBm	–12.03 dBm	5.50 dBm	2.25 dBm	8.74 dBm
28	Cable 2	17.54 dB	21.55 dB	13.52 dB	4.01 dB	–13.60 dBm	–15.43 dBm	–12.03 dBm			
29	Module 3	19.54 dB	25.55 dB	13.52 dB	6.01 dB	–15.04 dBm	–18.50 dBm	–12.60 dBm	**3.46 dBm**	–2.55 dBm	9.48 dBm
30	Cable 3	18.93 dB	26.76 dB	11.09 dB	7.83 dB	–15.04 dBm	–18.50 dBm	–12.60 dBm			
31	Module 4	33.93 dB	43.76 dB	24.09 dB	9.83 dB	–16.21 dBm	–22.19 dBm	–12.84 dBm	**2.07 dBm**	**–7.76 dBm**	11.91 dBm
32		1-dB compression if		**11 dB**	higher than IP3 at output:					MINIMUM	
33			Input Power at 1-dB Compression:			–4.2 dBm	–10.2 dBm	–0.8 dBm	–1.0 dBm	–7.8 dBm	0.0 dBm
34			Output Power at 1-dB Compression:			28.7 dBm	32.6 dBm	22.3 dBm	31.9 dBm	35.0 dBm	23.1 dBm

Fig. 4.21 Spreadsheet for 1-dB compression.

To aid our analysis, we have set a level, in cell J20, that indicates a range of equivalent input compression levels to be highlighted. In this example we have chosen 5 dB so any level within 5 dB of the minimum (row 33) will be displayed as bold. For mean gain (cells I25–I31), this results in three highlighted values that are within 5 dB of the minimum (cells I33–K33). With min or max gain, no other modules are this close to the minimum, showing that the modules are better matched for compression effects at mean gain.

We might see a relatively fixed relationship between IP3 and $P_{\text{out},-1\text{ dB}}$ levels for the various modules that make up the cascade (see Section 4.4). This is probably due to the dependence of both parameters on the same distortion. (For example, we might expect a very overdriven amplifier to produce something like a square wave. It has many odd harmonics, whose amplitudes are proportional to the amplitude of the square wave and thus, to the limiting level.) We have set an approximate amount by which $P_{\text{out},-1\text{ dB}}$ exceeds OIP3_{IM} in cell D32, and this is added to the cascade IIP3s (cells F31–H31) in cells F33–H33 to obtain estimates of the equivalent input level at 1-dB compression.

Line 34 shows equivalent output 1-dB compression powers. They are obtained by adding cascade gain (cells B31–D31) less 1 dB to each input level in line 33.

4.10 OTHER NONIDEAL EFFECTS

Amplitude modulation to phase modulation (AM-to-PM) conversion (Toolin, 2000; Laico, 1956) can be significant in some cascades, and frequency modulation to amplitude modulation (FM-to-AM) conversion can be produced whenever the passband is not flat. When the frequencies of two signals become close, the variation in supply current at their difference frequency may become too low for effective bypassing by a particular bias network, causing supply voltage variation and AM at the difference frequency. Time-varying heat dissipation due to such beats or due to other transient effects can cause parameter variations (Yang et al., 2000).

4.11 SUMMARY

- Second-order harmonics and IMs increase in power 2 dB for each dB increase in the fundamental power. Therefore, their ratio to the fundamental increases 1 dB per dB of increase in the fundamental.
- Third-order harmonics and IMs increase in power 3 dB for each dB increase in the fundamental power. Therefore, their ratio to the fundamental increases 2 dB per dB of increase in the fundamental.
- Second- and third-order harmonics and IMs can be represented by intercept points.
- Some third-order IMs remain close in frequency to the fundamentals that caused them. These are particularly important because they often cannot be filtered out.
- Third-order IMs that are close to the signal in frequency tend to add coherently. The reciprocal of the cascade p_{IP3} equals the sum of the reciprocals of the individual module p_{IP3}s, all referenced to the cascade input, assuming coherent addition.
- Second-order IMs tend to add randomly. The reciprocal of the cascade p_{IP2} is the sum of the reciprocals of the individual module p_{IP2}s, all referenced to the cascade input, assuming random phases.
- Frequency conversions change the frequency offset of the second-order IMs but close in third-order IMs generated before the conversion continue to add to those generated after the conversion.
- Anomalous IMs, with third-order frequencies and second-order amplitude responses, can be caused by the hysteresis in ferrites.
- In measuring IPs, care must be taken to avoid harmonics and IMs generated by the measurement equipment.
- Computation of 1-dB compression for a cascade is awkward; so we look at the affect of individual modules on the cascade, highlighting the significant contributors, or relate the 1-dB compression level to the IP3.

ENDNOTE

[1]Winder (1993) has shown that, if the fifth harmonic is at least 41 dB weaker than the third, the effect of fifth-order products on the third-order relationships will be small, less than 25% in voltage.

CHAPTER 5

NOISE AND NONLINEARITY

Here we consider the interaction between the noise and the nonlinearities that were discussed in the previous chapters. First, we will consider the intermodulation products that are produced by noise or by noiselike signals. Then we will consider how the noise figure and nonlinearities combine to establish a dynamic range. We will conclude with a discussion of spreadsheets that combine these effects and of certain enhancements that can be introduced into them.

5.1 INTERMODULATION OF NOISE

Noise is also affected by nonlinearities and dense signals [e.g., frequency division multiplex (FDM)] are sometimes characterized as noise for analysis. We can extend the work we have done for two signals to many signals with a specified amplitude distribution, writing them as a summation of individual signals and formulating the results of multiplying these summations in the process of obtaining v_{in}^2 or v_{in}^3 in Eq. (4.1). Then we must determine how the resulting power distribution is described, taking into account coherence or noncoherence of the various components. We then allow the number of signals to approach infinity as each represents the power in a bandwidth that approaches differential width. This can be an arduous process. We will attempt to gain an understanding of this process by doing it here for a simple case using a simplified process, leaving greater detail and rigor to works on communication theory (Rice 1944, 1945; Blachman 1966; Davenport and Root 1958; Schwartz et al., 1966).[1] Nevertheless, we will look at several practical applications of the results we obtain.

5.1.1 Preview

For a preview of the kind of results to be expected, refer to Fig. 5.1, which shows results from the second-order v_{in}^2 term. The input power spectrum is shown in dashed lines. [In this section we use two-sided power spectral density (PSD) where the power is evenly divided between positive and negative frequencies.] At Fig 5.1*a*, no signal is present. The second-order term produces three triangular distributions of noise power, corresponding to the IMs, *c* and *e* in Fig. 4.2. As the number of signals increases, the number of IMs grows faster than the number of harmonics, and the IMs come to dominate. While the sum-frequency response may appear to be equal to the difference-frequency response in Fig. 4.2 but not in Fig. 5.1, the total bandwidth, considering both positive and negative frequencies,

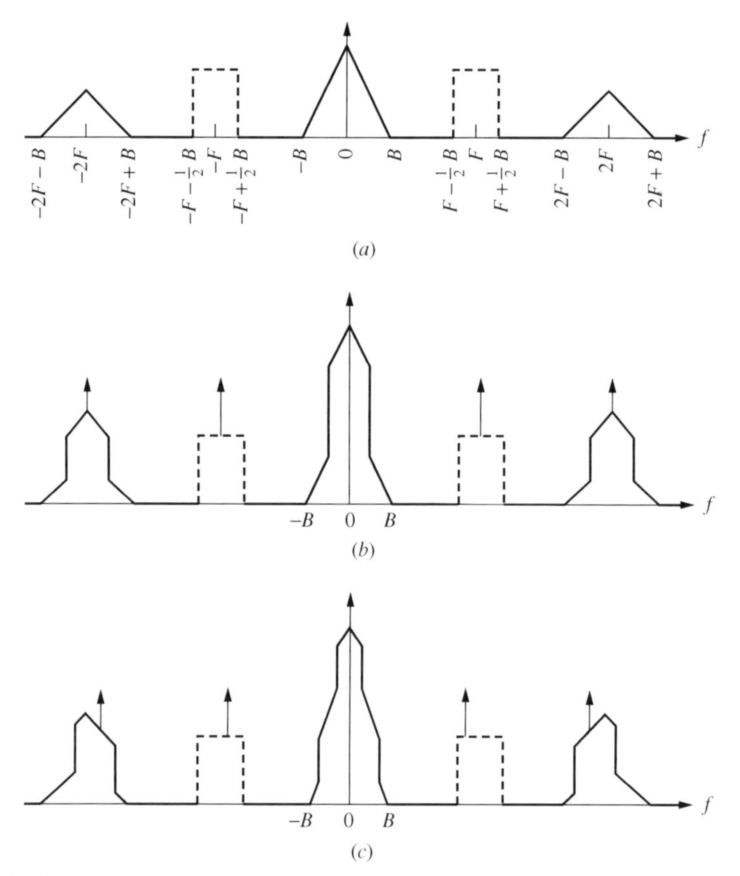

Fig. 5.1 Output power spectral density (shown solid) of a square-law device whose input is Gaussian noise with a rectangular power spectral density (shown dashed) plus a sinusoid. The powers of the sinusoids are represented by arrows. At (*a*) the sinusoidal input has zero amplitude. At (*b*) it is centered in the noise spectrum. At (*c*) it is eccentric. (From Blachman, 1966, p. 91; used with permission.)

is twice as great for the sum frequencies as for the difference frequencies in Fig. 5.1. The powers in the two regions are equal, but it is more spread out at the sum frequencies. Note that, in Fig. 4.2, if a and b represented the extremes of a band of many signals, the difference frequencies would extend from zero to the frequency of c, whereas the sum frequencies would extend over a band twice that wide, from d to f.

When a signal is added (in Fig. 5.1b), the shape of the power density changes due to mixing between the discrete signal and the noise. The display at Fig. 5.1c shows what happens when that signal moves from the center of the band.

We will develop the details for a case where no signal is present, similar to Fig. 5.1a but will also include third-order products.

5.1.2 Flat Bandpass Noise

Figure 5.2 shows white noise with power p over a bandwidth of B centered at f_c. The two-sided PSD is, therefore, $S_0/2 = p/(2B)$; S_0 is the one-sided PSD, applicable when negative frequencies are not used, and $S_0/2$ is the two-sided PSD, where half of the power is at positive frequencies and half at negative frequencies. The linear term will have the same PSD multiplied by a_1^2 [Eq. (4.1)], since the voltage is multiplied by a_1.

5.1.3 Second-Order Products

Second-order terms arise from multiplication of the input by itself. Initially, assume that the input consists of a large number of cosines, closely and evenly spaced in frequency, with voltage amplitudes corresponding to the power density represented by Fig. 5.2. Since multiplication of time waveforms implies convolution of their transforms, the Fourier transform of the second-order term arises from the convolution of the finely spaced impulses, representing cosines, and having power spectral density shown in Fig. 5.2 (Bracewell, 1965, pp. 79–80, 24–40). The convolution is obtained by integrating the product of the transform with another image of the transform, flipped about the zero frequency axis (which makes no difference for transforms of cosines, since they are even functions of frequency) and shifted by f. This result gives the transform of the product at frequency f. It is equivalent to what would be obtained by writing a sum of

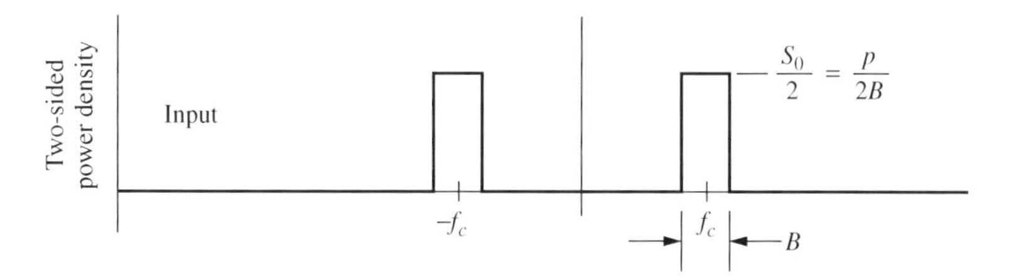

Fig. 5.2 Input noise spectrum.

infinitesimally spaced cosines [e.g., $v_i = A_i \cos 2\pi(f_{\text{start}} + i\delta)$] and adding the products of the various components that produce signals at a frequency f (e.g., $v_i v_{i+k}$ if $f = k\delta$). This process produces a triangle in the frequency domain with maximum amplitude at zero frequency, where the two factors are aligned.

5.1.3.1 *DC Term* The mean-square voltage in a differential bandwidth at any frequency x within the rectangles of Fig. 5.2 is

$$\tilde{e}_x^2 dx = \frac{pR}{2B} dx, \tag{5.1}$$

where R is the resistance in which the power is dissipated and across which the voltage appears. Here x *is* used for frequency to differentiate it from the independent variable f of the convolution, which represents the offset of the two multiplied spectrums. The transform of the second-order term in Eq. (4.1) is obtained by the convolution of two copies of this spectrum multiplied by a_2. Its value, for $f = 0$, is

$$v(0) = a_2 \left[\int_{-f_c-B/2}^{-f_c+B/2} \tilde{e}_x^2 \, dx + \int_{f_c-B/2}^{f_c+B/2} \tilde{e}_x^2 \, dx \right] = a_2 \tilde{e}_x^2 2B = a_2 pR. \tag{5.2}$$

This is a zero-frequency, or DC, term and the corresponding power is

$$p_{20} = \frac{v^2(0)}{R} = a_2^2 p^2 R = \left(\frac{a_2}{a_1} \right)^2 (a_1^2 p) pR. \tag{5.3}$$

Writing $a_1^2 p$ as the output power p_1 and substituting from Eq. (4.17), we obtain

$$p_{20} = \frac{p}{2 p_{\text{IIP2,IM}}} p_1 = \frac{p_1^2}{2 p_{\text{OIP2,IM}}}, \tag{5.4}$$

where we multiplied numerator and denominator by $|a_1|^2$ to get the last term. This power is represented by an impulse at zero frequency of value (area) p_{20}. It corresponds to the DC terms in Eq. (4.8). If the originally assumed cosines now are allowed to become noiselike, having random phases, this term is not changed because, since $f = 0$, each member of the summation is being multiplied by an identical member. The fundamental and DC outputs are shown in Fig. 5.3.

5.1.3.2 *Density* Except for this case with exactly zero-frequency offset between the factors, the convolution represents a summation of voltages, taken from various parts of the original distribution, whose frequencies differ by f. If we now no longer assume a series of cosines but, rather, a series of sinusoids with random phases (noise), integration will be the summation of noncoherent sinusoids, sinusoids whose phases are randomly related, except where some special consideration shows them to be coherent.

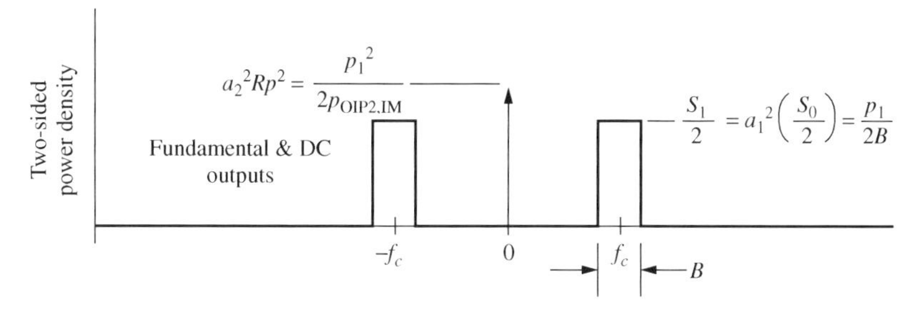

Fig. 5.3 Fundamental and second-order DC outputs.

Looking at the sum a little closer, we find that it is made up of coherent pairs since, for each product $e_i e_j$ that occurs in multiplying one set of voltages by the other, there is a second identical product $e_j e_i$. One pair arises from e_i in the first distribution and e_j in the second, and the other arises from e_j in the first distribution and e_i in the second.[2] The voltages add for the two members of each coherent pair whereas the powers add when the various pairs, which are not coherent with each other (i.e., their phase relationships are random), are added. Therefore, the result has twice the power density that it would if everything were noncoherent. [Adding the two members of a pair gives a power four times the individual powers, rather than two times, as noncoherent addition would give. Thus the summation of voltage that occurred for coherent signals (cosines) implies twice the summation of powers.] Therefore, we integrate the product of input PSDs to get a resulting PSD and double the result in recognition of the coherent pairs:

$$\frac{S_2'(f)}{2} = 2\frac{S_0(f)}{2} \bigstar \frac{S_0(f)}{2}, \tag{5.5}$$

where correlation is indicated by the pentagram and the prime differentiates this function from the same function after it has been multiplied by appropriate constants. This correlation is the same as convolution, since $S(f) = S(-f)$.

As f approaches zero, the correlation becomes equal to the integral of a constant $[S_0(0)/2]^2$ over a width of $2B$, so Eq. (5.5) gives

$$\frac{S_2'(0)}{2} = 4B\left(\frac{S_0}{2}\right)^2 \tag{5.6}$$

at the center.[3] As f increases, the two rectangle pairs shift relative to each other, decreasing the region over which a nonzero product exists, and leading to a function that decreases linearly to zero value at $f = B$. (As f approaches B, the number of components e_j whose frequencies differ by f approaches zero.) This is the middle triangle shown in Fig. 5.4*a*.

As f approaches $2f_c$, one of the rectangles begins to overlap the other, producing the additional result shown near $\pm 2f_c$. The amplitude of this triangle is

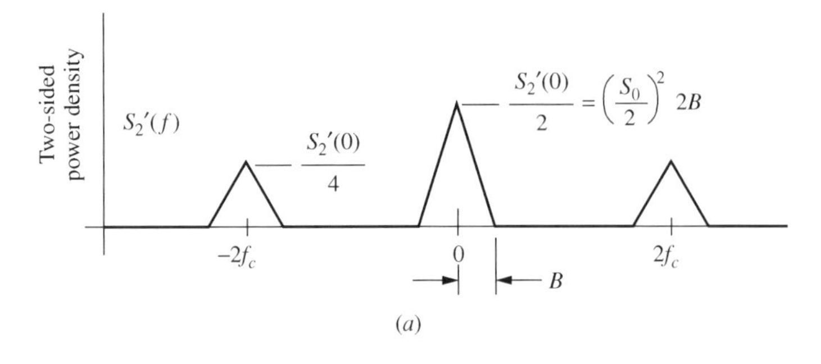

(*a*)

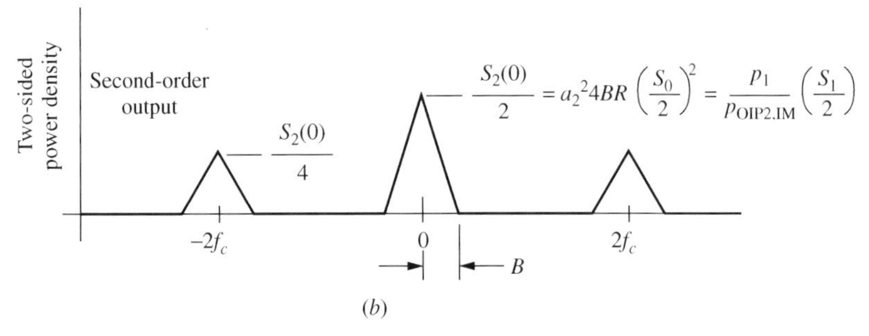

(*b*)

Fig. 5.4 Second-order noise products.

only half that of the middle triangle because only one set of products is involved, whereas the products of both rectangles were summed for the middle triangle.

Multiplying Eq. (5.6) by $a_2^2 R$, we obtain the PSD for the second-order term in Eq. (4.1),[4]

$$\frac{S_2(0)}{2} = a_2^2 R \frac{S_2'(0)}{2} = a_2^2 4BR\left(\frac{S_0}{2}\right)^2. \tag{5.7}$$

We can replace the input power density multiplied by bandwidth by the equivalent power p,

$$\frac{S_2(0)}{2} = \left(\frac{a_2}{a_1}\right)^2 2R\left(\frac{S_0}{2}\right) a_1^2 p. \tag{5.8}$$

Using Eq. (4.17) and writing power gain times input power as output power, this is

$$\frac{S_2(0)}{2} = \frac{p_1}{p_{\text{IIP2,IM}}}\left(\frac{S_0}{2}\right) = \frac{p_1}{p_{\text{OIP2,IM}}}\left(\frac{S_1}{2}\right), \tag{5.9}$$

which is shown in Fig. 5.4*b*.

5.1.3.3 Effect of a Signal with the Noise The DC power represented by the impulse in Fig. 5.3 is proportional to the total power squared; so it increases when a signal is present. Since this term is proportional to p^2, it is proportional to

$$(p_{\text{signal}} + p_{\text{noise}})^2 = p_{\text{signal}}^2 + 2p_{\text{signal}}p_{\text{noise}} + p_{\text{noise}}^2, \tag{5.10}$$

rather than to just the sum of the two squared powers. The middle term is a signal–cross–noise ($s \times n$) term, and it increases the response to the signal. In addition, there are $s \times n$ noise density terms that increase the noise in the presence of the signal, as seen in Fig. 5.1. The ratio of the $s \times n$ noise power to the $n \times n$ power in the middle of that figure (a region that would be low passed for detection) is twice the input S/N:

$$\left(\frac{p_{s\times n}}{p_{n\times n}}\right)_{\text{low-frequency}} = 2\left(\frac{S}{N}\right)_{\text{in}} \tag{5.11}$$

(Davenport and Root, 1958, pp. 257–263). Thus, if a signal pulse appears whose power equals the noise power, the mean ("DC") value of the detected output will increase to four times what it was before the pulse, according to Eq. (5.10), while the noise ($p_{s\times n} + p_{n\times n}$) will also increase, to three times its former value, according to Eq. (5.11). If a detection threshold has been set to minimize false detections due to the noise existing in the absence of the signal, the first effect helps to bring the output over that threshold when a signal arrives. However, the second effect introduces an uncertainty as to whether the pulse will exceed the threshold, since it increases the noise during the pulse. It may cause a pulse that would otherwise be too small to break threshold to do so, or it might cause a larger pulse to be below threshold.

5.1.3.4 Crystal Video Receiver with Preamplification A crystal video receiver consists of an RF filter, a detector, and a video filter. In Fig. 5.5, these are preceded by an amplifier (Klipper, 1965). The RF filter determines the range of frequencies admitted, and the video filter is made wide enough to pass a detected pulse with required fidelity. (We show a *bandpass* video filter so the DC component due to noise alone will be rejected.) Thus an RF band of width B_r can be observed without tuning. The noise from the square-law detector [i.e., using the second-order term in Eq. (4.1)] has a distribution shaped like Fig. 5.4 with peak value given by Eq. (5.7) with B designated as B_r. Here $S_0 = N_0 f_{\text{pre}} g_{\text{pre}}$, where N_0 is the one-sided input thermal noise power density and f_{pre} and g_{pre} are the cascade noise factor and gain of the preamplifier and filters that precede the detector.

Therefore, the peak PSD is

$$\frac{S_2(0)}{2} = a_2^2 R \left(\frac{N_0}{2}\right)^2 4B_r f_{\text{pre}}^2 g_{\text{pre}}^2. \tag{5.12}$$

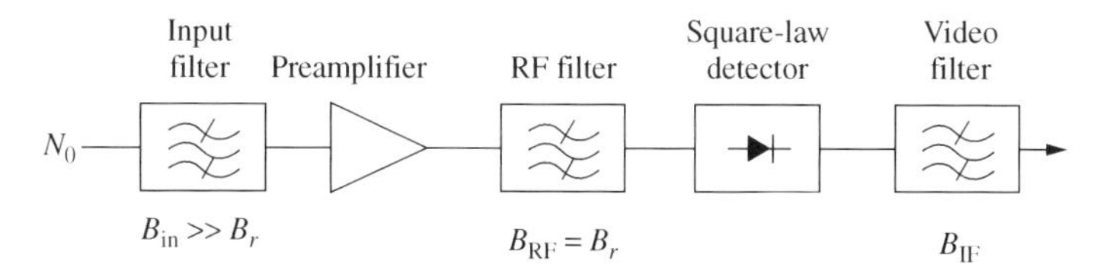

Fig. 5.5 Crystal video receiver.

The video band extends from zero (approximately) to B_v, and the average height of the triangle in that range is (see Fig. 5.4*b*)

$$\frac{S_{2,\text{avg}}}{2} = \frac{S_2(0)}{2}\left(1 - \frac{B_v}{2B_r}\right), \tag{5.13}$$

leading to a noise power of

$$p_n = \int_0^{B_v} S_2(f)\,df = S_{2,\text{avg}} B_v = S_2(0)\left(1 - \frac{B_v}{2B_r}\right) B_v \tag{5.14}$$

$$= a_2^2 R N_0^2 f_{\text{pre}}^2 g_{\text{pre}}^2 2\left(B_r B_v - \frac{B_v^2}{2}\right). \tag{5.15}$$

Note that, while narrowing B_v reduces the noise, the noise power still depends on the RF bandwidth B_r. This unusual dependence of noise power on the bandwidths will be seen in expressions for noise in this type of receiver when there is no signal. The addition of a signal creates additional power proportional to the signal voltage and terms resulting from multiplication of the signal by the noise (Klipper, 1965).

5.1.4 Third-Order Products

5.1.4.1 Density Spectrum If we convolve the rectangular voltage spectrum of the input (corresponding to Fig. 5.2) with that for the triangular second-order response (corresponding to Fig. 5.4*a*), we obtain the voltage spectrum for e^3, shaped as shown in Fig. 5.6. This time, since we are multiplying three copies of the set of cosines, we find that the result at a given frequency consists of noncoherent groups of six coherent pairs.[5] Therefore, the transform of the cube of the PSD is

$$\frac{S_3'(f)}{2} = 6\frac{S(f)}{2} \bigstar \frac{S(f)}{2} \bigstar \frac{S(f)}{2}. \tag{5.16}$$

When the rectangular input spectrum (Fig. 5.2) is shifted by $\pm f_c$, the rectangle multiplies the center of the second-order PSD plus the half-size triangle at $2f_c$

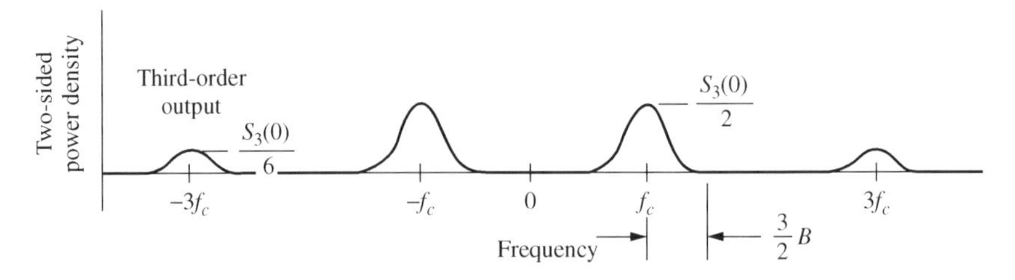

Fig. 5.6 Third-order noise products.

(Fig. 5.4*b*). The peak of the third-order response is

$$\frac{S_3(0)}{2} = 6a_3^2 \left(\frac{S_0}{2}\right)^3 2BR^2 \left(\frac{3}{4}\right) B \left(\frac{3}{2}\right). \tag{5.17}$$

The factor 6 comes from the number of voltages in a coherent group. Coefficient a_3 comes from Eq. (4.1). The next term and the factor $2B$ is from the product of the $S_0/2$ and $S_2'(0)/2$, the density of the input (Fig. 5.2) and the density at the center to the second-order triangle (Fig. 5.4*a*). The term R^2 results from the conversion of S_0 to a mean-squared voltage (generating R^3) and reconversion of the *product* to a power density (generating $1/R$), much as occurred in Eq. (5.12). The factor $\frac{3}{4}$ represents the ratio of average-to-peak value for the triangle in the region $\pm B/2$ so that multiplication by $\frac{3}{4}B$ amounts to integration over that bandwidth. The factor $\frac{3}{2}$ adds the product of the smaller triangle and rectangle to that of the larger triangle and rectangle. This can be simplified to

$$\frac{S_3(0)}{2} = a_3^2 R^2 \frac{27}{2} \left(\frac{S_0}{2}\right)^3 B^2 = \frac{27}{8} \left(\frac{a_3}{a_1}\right)^2 R^2 p^2 \left(\frac{S_1}{2}\right) \tag{5.18}$$

$$= \frac{27}{8} \left(\frac{2}{3Rp_{\mathrm{IIP3,IM}}}\right)^2 R^2 p^2 \left(\frac{S_1}{2}\right) = \frac{3}{2} \left(\frac{S_1}{2}\right) \left(\frac{p_1}{p_{\mathrm{OIP3,IM}}}\right)^2. \tag{5.19}$$

The shape of this curve can be determined without great difficulty by analysis of the correlation process.

5.1.4.2 Third-Order Terms at Input Frequencies Since there are terms in Eq. (4.20) at the frequency of the input, we might expect to see them also when working with densities. Appendix T shows that there is an additional output PSD at the input frequencies of

$$\varepsilon = 4 \left(\frac{S_1}{2}\right) \left[\mathrm{sign}\left(\frac{a_3}{a_1}\right) \left(\frac{p}{p_{\mathrm{IIP3,IM}}}\right) + \left(\frac{p}{p_{\mathrm{IIP3,IM}}}\right)^2 \right]. \tag{5.20}$$

This modifies $S_1/2$ by a small amount as long as $p \ll p_{\mathrm{IIP3,IM}}$. It is included in Fig. 5.7, which shows a composite of all the spectrum components that we have discussed.

5.1.4.3 NPR Measurement Noise power ratio (NPR) is a parameter used to determine whether a system is sufficiently noise free and distortion free to handle frequency-division-multiplex (FDM) traffic. The test is performed by creating a rectangular noise spectrum that emulates the FDM channels and removing a narrow slot, representing one channel, by filtering (Fong et al., 1986). Third-order nonlinearities will fill in the slot (Fig. 5.8). The depth of the slot after the spectrum has passed through the system is a measure of the amount of the noise that can be expected in a channel due, for one thing, to power in adjacent

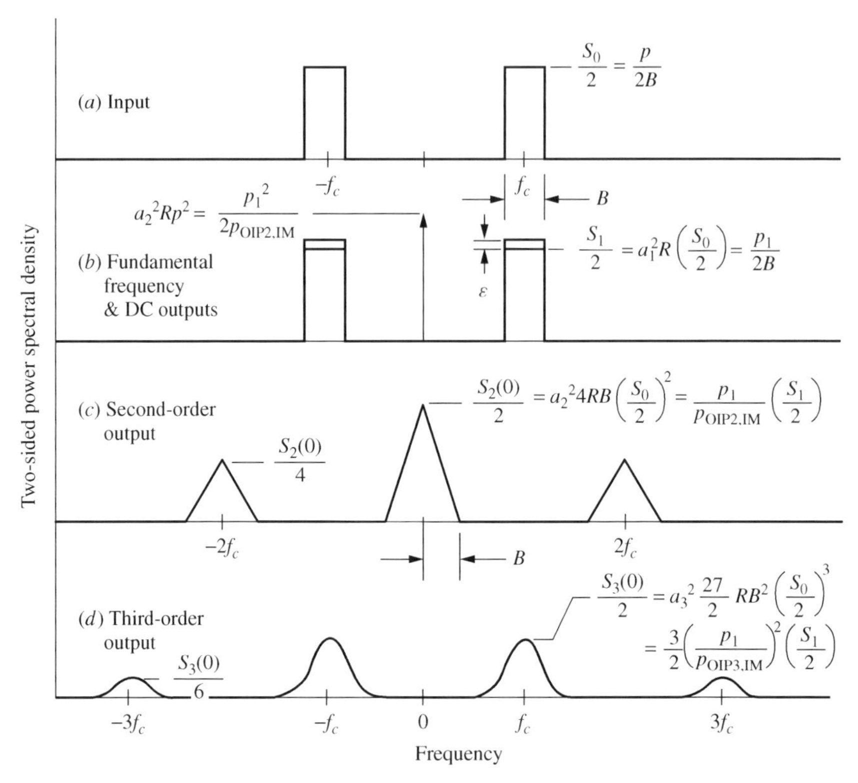

Fig. 5.7 Second- and third-order noise outputs. The impulse shown at (*b*) is a second-order product and ε is a third-order product that is coherent with first-order response; ε can be negative.

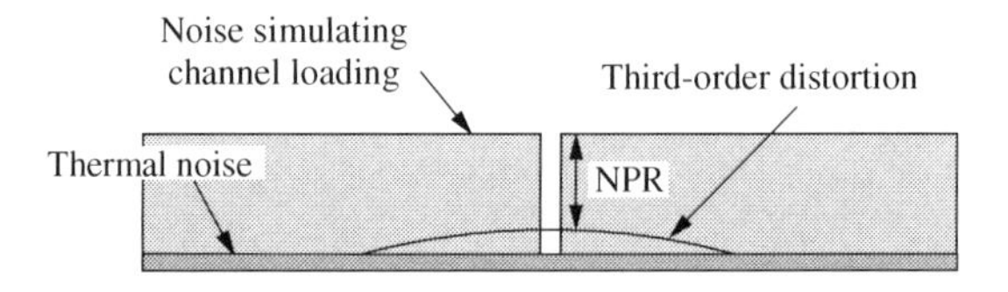

Fig. 5.8 NPR noise loading and distortion.

channels. A slot that is narrow compared to the noise band will have little effect on the third-order products produced, in which case Eq. (5.19) will apply at midband, enabling us to compute the NPR there due to third-order products.

Example 5.1 NPR An FDM system has $OIP3_{IM} = 29$ dBm. What total signal power at the output will permit 50 dB NPR for any channel due to IMs? Since the maximum third-order product is in the center of the input band (Fig. 5.7), the required output power is the level that will cause that density to be 50 dB lower than the first-order output density. Using Eq. (5.19) (assuming for now, that we can ignore ε), we have

$$\frac{S_3(0)}{2} = 10^{-50/10}\left(\frac{S_1}{2}\right) = \frac{3}{2}\left(\frac{S_1}{2}\right)\left(\frac{p_1}{10^{29/10}\text{mW}}\right)^2, \tag{5.21}$$

$$10^{-5} = \frac{3}{2}\left(\frac{p_1}{10^{2.9}\text{mW}}\right)^2, \tag{5.22}$$

$$p_1 = \sqrt{\frac{2}{3}}10^{-2.5}10^{2.9}\text{mW} = 2.05\ \text{mW}. \tag{5.23}$$

We will now check the assumption that the modification of the signal strength by ε is negligible. From Eq. (5.20),

$$\varepsilon = 4\left(\frac{S_1}{2}\right)\left[\pm\left(\frac{2.05\ \text{mW}}{10^{2.9}\text{mW}}\right) + \left(\frac{2.05\ \text{mW}}{10^{2.9}\text{mW}}\right)^2\right], \tag{5.24}$$

$$|\varepsilon| \le 4\left(\frac{S_1}{2}\right)[2.6\times 10^{-3} + 6.7\times 10^{-6}] = 0.010\left(\frac{S_1}{2}\right). \tag{5.25}$$

Thus the signal PSD is changed by 1%, modifying the NPR by only 0.04 dB.

5.2 COMPOSITE DISTORTION

Cable television (CATV) systems are sensitive to a type of interference consisting of spurs produced by the influence of nonlinearities on the many visual (picture) carriers (Thomas, 1995). Due to the presence of many evenly spaced channels in these systems, interference can be produced in a given channel by multiple spurious signals, all appearing at the same frequency and caused by various combinations of carriers. This interference is called composite. The two types of primary concern are composite second-order (CSO) distortion, caused by second-order nonlinearities, and composite triple beat (CTB) distortion, caused by third-order nonlinearities. In the HRC CATV system, carriers occur at multiples of 6 MHz, beginning at 54 MHz, while, in the IRC system, they are offset from these 6-MHz multiples, being higher by 1.25 MHz. The most common, or Standard, system is similar to the IRC system except that carriers at 73.25, 79.25, and 85.25 MHz are replaced by carriers at 77.25 and 83.25 MHz. Most of the channels in the Standard system are thus the same as for the IRC system, and we will ignore the deviations from that scheme for simplicity. In the HRC system all of the in-band interferers fall on carrier frequencies. The situation is more complicated for the other, offset, systems. Second-order products of offset (by 1.25 MHz) carriers will occur at sum frequencies, making them higher by 1.25 MHz than the nearest channel frequency:

$$(6n + 1.25) + (6m + 1.25) = (6q + 1.25) + 1.25 \tag{5.26}$$

or at difference frequencies, making them 1.25 MHz low:

$$(6n + 1.25) - (6m + 1.25) = (6q + 1.25) - 1.25. \tag{5.27}$$

Third-order products of offset carriers will be at carrier frequencies,

$$(6m + 1.25) + (6n + 1.25) - (6p + 1.25) = (6q + 1.25), \tag{5.28}$$

or offset by 2.5 MHz,

$$(6m + 1.25) + (6n + 1.25) + (6p + 1.25) = (6q + 1.25) + 2.5\text{ MHz}, \tag{5.29}$$

$$(6m + 1.25) - (6n + 1.25) - (6p + 1.25) = (6q + 1.25) - 2.5\text{ MHz}. \tag{5.30}$$

While the interferers are very close to each other in frequency, their relative phases wander over time so the average sum of spurious powers is measured. The RF bandwidth is usually 30 kHz so only the responses at one offset are summed. Our development for intermodulation of noise spectrums in the previous section began by considering a large number of evenly spaced discrete signals whose spacing was then allowed to shrink to zero. Here we are faced with a large number of evenly spaced signals whose spacing does not shrink to zero, but we may be able to approximate them as a continuous spectrum and use the previous development to determine the resulting spurious spectrum, given the IP2 and IP3. Practically, there are many things that will limit the accuracy of this approach. The amplifiers may operate at total powers that are higher than the power where the intercept points accurately predict IM levels. Output powers are generally not flat (which interferes with the application of our particular development, which assumed flat spectrums) and IPs are often frequency sensitive. Nevertheless, even a limited ability to relate CSO and CTB distortion to IPs can be of value. Figure 5.9 is the same as Fig. 5.7 but redrawn for a 110-channel IRC (or Standard, approximately) CATV system. Each 6-MHz frequency segment represents the power in one carrier centered in that segment (thus the edges extend 3 MHz beyond the end carriers). One thing we note is that the parts of the spectrum that are at negative frequencies now produce IMs with positive frequencies, and visa versa. Note the apparent similarity between the third-order output at positive frequencies in Fig. 5.9 and the calculated density of CTBs in Fig. 5.10.

5.2.1 Second-Order IMs (CSO)

Note, in Fig. 5.9*c*, that the maximum value of $S_2(0)/2$ is almost equal to the value at the first system carrier frequency, 55.25 MHz. It is only 1/11 of the way from the peak of the 666-MHz-wide sloped region and less than 0.5 dB from the peak. Therefore, we will take the peak to be the worst case for CSO. While the larger central response contains difference frequencies, the smaller (half height) responses contain sum frequencies, and thus the actual discrete frequencies are at different offsets. Even if they did add, the maximum would not be

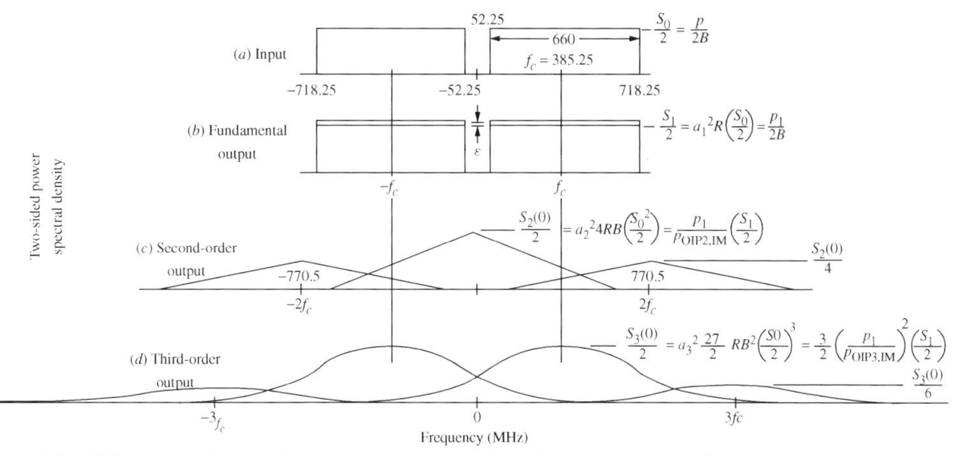

Fig. 5.9 Second- and third-order power density for 110-channel CATV video carriers. Carriers are spaced at 6 MHz so each has been approximated as spread over ±3 MHz.

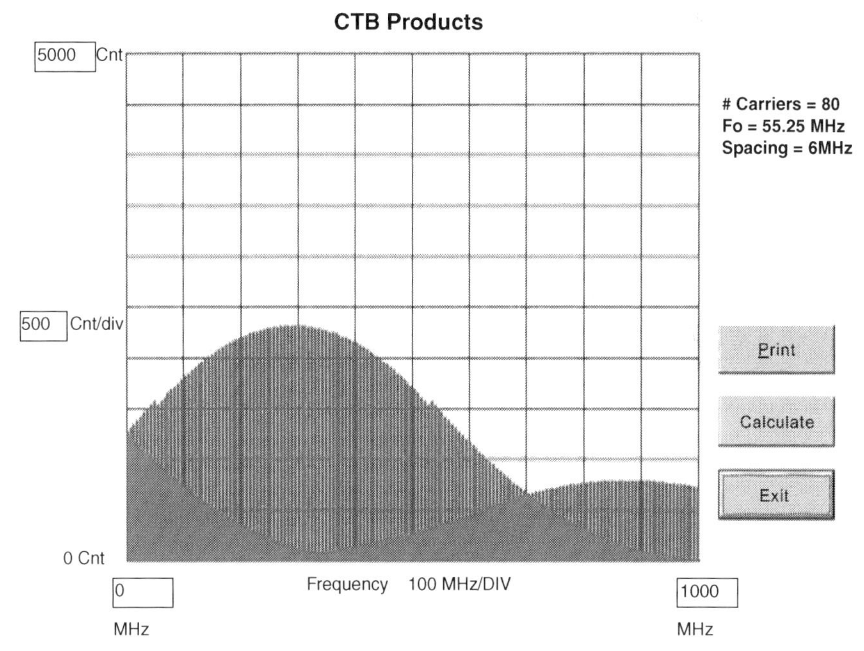

Fig. 5.10 Number of CTB products versus frequency for an 80-carrier IRC CATV system. (From Cain, 1999; used with permission.)

changed because the smaller responses go to zero where the larger one peaks. The maximum magnitude of the second-order density relative to the fundamental is

$$\frac{S_2(0)}{S_1} = \frac{p_1}{p_{OIP2,IM}}. \tag{5.31}$$

Since we are representing both the CSO distortion and the carrier by densities integrated over 6 MHz, we can multiply both numerator and denominator by 6 MHz to obtain the equivalent composite distortion and carrier, respectively. Therefore, this ratio is also the maximum CSO to carrier ratio:

$$\mathrm{CSO}_{\mathrm{relative}} < \frac{p_1}{p_{\mathrm{OIP2,IM}}}. \tag{5.32}$$

5.2.2 Third-Order IMs (CTB)

Similarly, the main third-order responses will not occur at the same frequencies as do the spill-over from negative frequencies [the positive and negative frequencies for a given carrier are separated by $2(6n + 1.25)$ MHz $= (6q + 1.25)$ MHz $+$ 1.25 MHz] or as the spectrum at three times the frequency, but these would not contribute significantly at the peak anyway. By a procedure similar to what we used for CSO,

$$\mathrm{CTB}_{\mathrm{relative}} \le \frac{S_3(0)}{S_1} = \frac{3}{2}\left(\frac{p_1}{p_{\mathrm{OIP3,IM}}}\right)^2. \tag{5.33}$$

5.2.3 CSO and CTB Example

Example 5.2 Let us see how well this theory agrees with the typical values for a CATV amplifier, one whose data sheet provides all of the values needed for computation, the RF Micro-Devices (2001) model RF2317. It is tested with 110 carriers, each at an input voltage of +10 dBmV in a 75-Ω system. The nominal gain is 15 dB so the output power is -23.8 dBm per signal:

$$15\ \mathrm{dB} + 10\ \mathrm{dB} + 10\ \mathrm{dBW}\log\left[\frac{(10^{-3}\ \mathrm{V})^2/75\ \Omega}{1\ \mathrm{W}}\right] \tag{5.34}$$

$$= 25\ \mathrm{dB} - 78.75\ \mathrm{dBW} = -23.75\ \mathrm{dBm}. \tag{5.35}$$

Total output power for 110 carriers is

$$p_1 = -23.75\ \mathrm{dBm} + 10\ \mathrm{dB}\log(110) = -3.34\ \mathrm{dBm}. \tag{5.36}$$

The OIP2 is given at +63 dBm. Substituting these last two numbers into Eq. (5.32), we obtain

$$\mathrm{CSO}_{\mathrm{relative}} \le -66\ \mathrm{dBc}. \tag{5.37}$$

The highest CSO given on the data sheet is -63 dBc at 1.25 MHz below the lowest carrier. That location agrees with the theoretical maximum but the level is 3 dB higher.

Typical OIP3 is +40 dBm at 500 MHz and goes to +42 at 100 MHz and +38 at 900 MHz. Equation (5.33) at 40 dBm OIP3 and -3.3 dBm p_1 gives

$$\mathrm{CTB}_{\mathrm{relative}} \le 10\ \mathrm{dB}\log(1.5) - 2(3.3\ \mathrm{dBm} + 40\ \mathrm{dBm}) = -84.8\ \mathrm{dBc}. \tag{5.38}$$

The data sheet gives CTB as −85 dBc at 331.25, and 547.25 MHz and 1 dB lower at 55.25 MHz, which very closely matches our estimate. These agreements are probably closer than we should expect given the variations in parameters with power and frequency.[6]

5.3 DYNAMIC RANGE

Dynamic range is the range of signal power levels over which a system will operate properly. The lower limit is generally set by noise and the upper limit is set by some undesirable phenomenon.

5.3.1 Spurious-Free Dynamic Range

We can set a threshold or lower limit P_T at which signals can be detected without excessive interference by noise. This will form the lower limit of an acceptable range of signal powers. As the power of input signals, say a pair of them, increases, spurs will eventually be created. If the spur power rises above that of the noise in the processing, or analysis, bandwidth B_p, signals at P_T will begin to see interference at a level greater than what we have defined as acceptable. The bandwidth B_p is the noise bandwidth in which the signal is ultimately observed or processed so the level of interference depends on the noise power in that bandwidth. (Actually, when the spurs are just at the noise level the *total* interference will have been increased. We will still consider P_T the acceptable minimum signal level. Perhaps we will take into consideration the possibility of interference due to both noise and equal-power spurs when we choose P_T, or perhaps we will disregard the degradation from the spurs because they occur less often than the noise, which is continuous.) The input level P_M that produces spurs at levels equal to the noise power is the upper limit of the range of acceptable signal powers. The difference between the minimum level P_T and the maximum level P_M is called the spur-free dynamic range (SFDR). This is sometimes called the instantaneous SFDR (ISFDR) to differentiate it from a system in which variable attenuators permit reception of strong signals at one time and weak signals at another time. Usually the spurs considered are close-in third-order IMs, since it is difficult or impossible to eliminate them by filtering.

To relate the ISFDR to the IP3 and the third-order IM level (Tsui, 1985, pp. 28–31; Tsui, 1995, pp. 204–205), we write the relationship illustrated in Fig. 4.8, using Eq. (28) in Appendix H for two equal-power input signals (see Fig. 5.11), as

$$P_{\text{in,IM3}} = 3P_{\text{in},F} - 2P_{\text{IIP3,IM}} \tag{5.39}$$

and rearrange to obtain

$$3(P_{\text{in},F} - P_{\text{in,IM3}}) = 2(P_{\text{IIP3,IM}} - P_{\text{in,IM3}}) \tag{5.40}$$

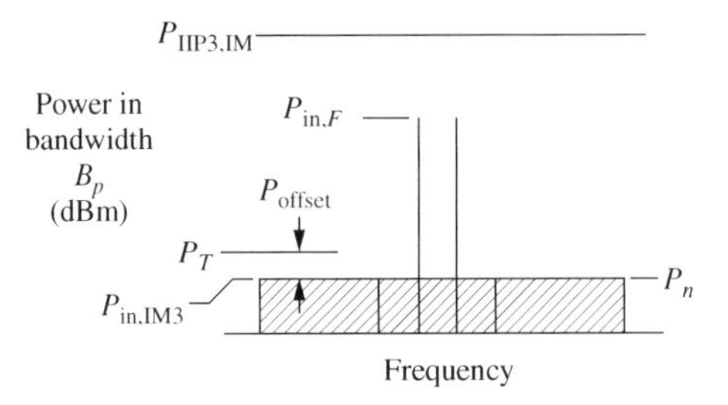

Fig. 5.11 SFDR.

or

$$(P_{\text{in},F} - P_{\text{in,IM3}}) = \tfrac{2}{3}(P_{\text{IIP3,IM}} - P_{\text{in,IM3}}). \tag{5.41}$$

This says that the separation between the signal and the IM3 spur is two thirds of the separation between the IP3 and that spur, as can be seen in Fig. 4.8.

Since the IM power, when the input level is P_M, is equal to the noise level, we have there

$$P_{\text{in,IM3}} = P_n, \tag{5.42}$$

and Eq. (5.41) becomes

$$(P_M - P_n) = \tfrac{2}{3}(P_{\text{IIP3,IM}} - P_n). \tag{5.43}$$

The ISFDR is equal to the difference between P_M and P_n, as given by Eq. (5.43), reduced by the amount P_{offset} by which P_T exceeds P_n:

$$\text{ISFDR} = \tfrac{2}{3}(P_{\text{IIP3,IM}} - P_n) - P_{\text{offset}}, \tag{5.44}$$

where P_n is given by

$$P_n = 10 \text{ dB} \log_{10}(kTB_p) + F \tag{5.45}$$

$$= 10 \text{ dB} \log_{10}(B_p/\text{Hz}) + F - 174 \text{ dBm}. \tag{5.46}$$

It is not unusual to set $P_{\text{offset}} = 0$ in order to obtain a measure that is independent of the particular processing on which P_{offset} depends.

Note how heavily ISFDR depends on B_p [Eqs. (5.44) and (5.46)]. The same cascade can have vastly different ISFDRs for different processing bandwidths, a parameter that may not be inherent in the cascade.

Example 5.3 ISFDR The third-order input intercept point IIP3 is -3 dBm and the noise figure is 8 dB. Find the ISFDR for a 40-MHz processing bandwidth. Find it for a 4-kHz processing bandwidth. Use $P_{\text{offset}} = 0$.

From Eq. (5.46), the noise level in 40 MHz is

$$P_n = 76 \text{ dB} + 8 \text{ dB} - 174 \text{ dBm} = -90 \text{ dBm}.$$

Using this in Eq. (5.44), we obtain

$$\text{ISFDR}|_{40 \text{ MHz}} = \tfrac{2}{3}(-3 \text{ dBm} + 90 \text{ dBm}) = 58 \text{ dB}.$$

For a 4-kHz bandwidth, we obtain $P_n = -130$ dBm and, as a result,

$$\text{ISFDR}|_{4 \text{ kHz}} = \tfrac{2}{3}(-3 \text{ dBm} + 130 \text{ dBm}) = 84.7 \text{ dB}.$$

For the wider bandwidth, the maximum signal is -32 dBm, 58 dB above the noise level of -90 dBm. With the narrower bandwidth, the signal is only -45.3 dBm, but this is 84.7 dB above the noise level of -130 dBm. Thus the maximum signal is 13.3 dB weaker (one third of the change in noise levels) when the dynamic range is 26.7 dB higher (two thirds of the change in noise levels). When the noise goes down, the maximum signal goes down also, but by a lesser amount, giving a larger separation between maximum signal and noise.

5.3.2 Other Range Limitations

The compression level (Section 4.9) can limit dynamic range, even for single signals. The resulting instantaneous dynamic range is the difference between the 1-dB compression level and the threshold P_T. If the IP3 is on the order of 10 dB higher than the compression level (Section 4.4), the ISFDR due to third-order spurs will be more limiting for ranges greater than about 20 dB. Nevertheless, in some applications single signals may be sufficiently more important or likely than multiple signals to make the limitation due to compression significant.

Dynamic range can also be limited by various spurs that are created in mixers (Chapter 7). These must be controlled through careful design of the frequency conversion, for which dynamic range is an important design parameter.

5.4 OPTIMIZING CASCADES

5.4.1 Combining Parameters on One Spreadsheet

We have seen how gain, noise factor, and intercept points can be included in spreadsheets. We will often include all of these on a single spreadsheet as we develop a design, enabling us to see, and to optimize, the trade-off between system intercept point and noise figure as we modify the distribution of gain. We will include them all here, first for an ideal standard cascade consisting of unilateral modules interconnected by cables that are well matched to the same standard impedance for which the modules are designed.

	A	B	C	D	E	F	G	H	I	J	K	L
2		Gain	Gain	SWR		IMs		specified NF		Temperature		
3		nom	+/–	at out	|a RT |	OIP3	mean	max	min	290 K		
4	Module 1	12.0 dB	1.0 dB	1.5		0.0 dBm	2.3 dB	2.8 dB	2.0 dB			
5	Cable 1	–1.5 dB		1.5	0.0283							
6	Module 2	8.0 dB	2.0 dB	2		10.0 dBm	3.0 dB	3.5 dB	2.5 dB			
7	Cable 2	–1.0 dB		2	0.0883							
8	Module 3	2.0 dB	2.0 dB	2.8		10.0 dBm	8.0 dB	9.2 dB	6.8 dB			
9	Cable 3	–0.8 dB		3.2	0.2064							
10	Module 4	15.0 dB	2.0 dB			24.0 dBm	5.0 dB	5.3 dB	4.7 dB			
11						DERIVED						
12			Gain				Noise Figure					
13						mean gain	min gain	max gain				
14		mean	min	max	±	mean NF	max NF	min NF				
15	Module 1	12.00 dB	11.00 dB	13.00 dB	1.00 dB	2.30 dB	2.80 dB	2.00 dB				
16	Cable 1	–1.50 dB	–1.74 dB	–1.25 dB	0.25 dB	1.54 dB	1.54 dB	1.54 dB				
17	Module 2	8.00 dB	6.00 dB	10.00 dB	2.00 dB	3.00 dB	3.50 dB	2.50 dB				
18	Cable 2	–0.97 dB	–1.73 dB	–0.20 dB	0.77 dB	1.08 dB	1.08 dB	1.08 dB				
19	Module 3	2.00 dB	0.00 dB	4.00 dB	2.00 dB	8.00 dB	9.20 dB	6.80 dB				
20	Cable 3	–0.61 dB	–2.43 dB	1.21 dB	1.82 dB	0.93 dB	0.93 dB	0.93 dB		processing bandwidth:		1.0E+5 Hz
21	Module 4	15.00 dB	13.00 dB	17.00 dB	2.00 dB	5.00 dB	5.30 dB	4.70 dB		threshold offset:		6.00 dB
22						CUMULATIVE						ISFDR
23			Gain				Noise Figure			IIP3 with		mean gain
24	at output of	mean	min	max	±		conditions as above		mean gain	min gain	max gain	mean NF
25	Module 1	12.00 dB	11.00 dB	13.00 dB	1.00 dB	2.30 dB	2.80 dB	2.00 dB	–12.00 dB	–11.00 dB	–13.00 dB	67.13 dB
26	Cable 1	10.50 dB	9.26 dB	11.75 dB	1.25 dB	2.37 dB	2.88 dB	2.06 dB	–12.00 dB	–11.00 dB	–13.00 dB	67.09 dB
27	Module 2	18.50 dB	15.26 dB	21.75 dB	3.25 dB	2.59 dB	3.19 dB	2.20 dB	–13.60 dB	–12.03 dB	–15.43 dB	65.87 dB
28	Cable 2	17.54 dB	13.52 dB	21.55 dB	4.01 dB	2.60 dB	3.21 dB	2.20 dB	–13.60 dB	–12.03 dB	–15.43 dB	65.87 dB
29	Module 3	19.54 dB	13.52 dB	25.55 dB	6.01 dB	2.81 dB	3.84 dB	2.27 dB	–15.04 dB	–12.60 dB	–18.50 dB	64.76 dB
30	Cable 3	18.93 dB	11.09 dB	26.76 dB	7.83 dB	2.82 dB	3.86 dB	2.27 dB	–15.04 dB	–12.60 dB	–18.50 dB	64.76 dB
31	Module 4	33.93 dB	24.09 dB	43.76 dB	9.83 dB	2.88 dB	4.18 dB	2.28 dB	–16.21 dB	–12.84 dB	–22.19 dB	63.94 dB

Fig. 5.12 Spreadsheet giving NF, IP3, and SFDR for standard cascade.

Example 5.4 Combined Parameters for a Standard Cascade Figure 5.12 shows such a spreadsheet in which cascade noise figures and third-order intercept points are obtained for several combinations of variations in the module parameters. The ISFDR is also given for mean gains and noise figures, based on Eqs. (5.44) and (5.46). Note that a combined spreadsheet is necessary for ISFDR since values are required for both noise figure and IP3.

Example 5.5 Combined Parameters for a Less Ideal Cascade In addition, we consider the less ideal circuit shown in Fig. 5.13 for which we make some approximations in order to fit the circuit to our standard cascade. The image filter, along with the cables on either end of it, is treated as a reflectionless interconnect. This is done because the filter cannot be realistically approximated as a unilateral module. The same kind of characterization is used for the diplexer. These approximations depend on well-matched components for accuracy. The mixer is characterized as a unilateral module. See Example 3.7.

The spreadsheet for this circuit is shown in Fig. 5.14. The effect of image noise has been included, but an image noise multiplier has been added to enable us to easily remove the image noise in order to observe its effect. Setting the multiplier (cell J5) to one includes the image noise in the cascade model while setting it to zero removes image noise. Cells F21–H21 contain the effective noise figure of the mixer according to Eq. (3.46). The term $f'_{B3}g'_{B3}$ is realized in cells I–K, 20 and 22. The process is the same as described in Example 3.5, but the fact that only two levels are involved makes that development overkill for this case. The noise figure for the two-element cascade between the filter and the mixer f'_{B3} can be represented by Eq. (3.14), where g_{pk} is just the gain of "amp 1," taken from cells B19–D19. This is then multiplied by g'_{B3}, which can be obtained either by summing the gains in rows 19 and 20 (columns B–D) or subtracting the cumulative gain at the filter output (cells B30–D30) from that at the mixer input (cells B32–D32).

The last line in Fig. 5.14 shows the change in the cascade parameters when the spreadsheet is simplified by removal of all reflections (SWR = 1 everywhere). We can see that the mismatches affect the extreme cascade parameters more than they affect mean values (see Section 2.3.2.1). This might lead us to expect that reflections at the filter or diplexer, which we have ignored, will have relatively little effect on the mean or typical performance. The effects of such missing

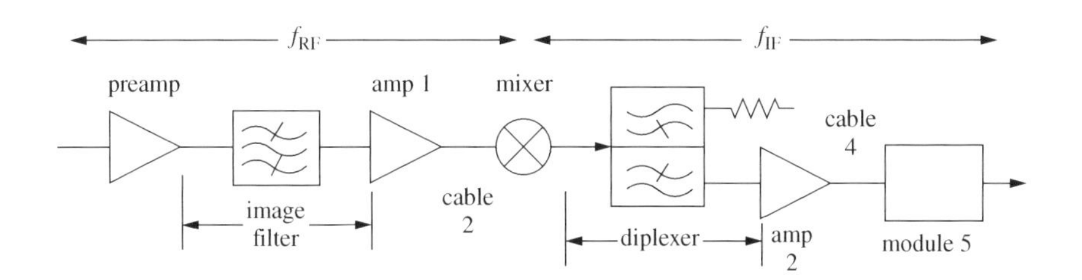

Fig. 5.13 Block diagram of cascade with frequency conversion.

	A	B	C	D	E	F	G	H	I	J	K
2		Gain	Gain	SWR		IMs	specified NF			Temperature	
3		nom	+/–	at out	|a RT |	OIP3	mean	max	min	290 K	
4	preamp (module 1)	12.0 dB	1.0 dB	1.5		0.0 dBm	2.3 dB	2.8 dB	2.0 dB	Image Noise Multiplier	
5	image filter (cable 1)	–1.5 dB		1.5	0.0283					1	
6	amp 1 (module 2)	8.0 dB	0.8 dB	2		10.0 dBm	3.0 dB	3.5 dB	2.5 dB		
7	cable 2	–1.0 dB		2	0.0883						
8	mixer (module 3)	–8.0 dB	1.2 dB	2.8		15.0 dBm	8.0 dB	9.2 dB	6.8 dB		
9	diplexer (cable 3)	–0.8 dB		3.2	0.2064						
10	amp 2 (module 4)	15.0 dB	1.0 dB	2.2		24.0 dBm	5.0 dB	5.3 dB	4.7 dB		
11	cable 4	–1.8 dB		2	0.0826						
12	module 5	6.0 dB	1.3 dB			30.0 dBm	5.0 dB	5.4 dB	4.6 dB		
13		DERIVED									
14		Gain				Noise Figure				Image Noise	
15						mean gain	min gain	max gain	mean gain	min gain	max gain
16		mean	min	max	±	mean NF	max NF	min NF	mean NF	max NF	min NF
17	preamp (module 1)	12.00 dB	11.00 dB	13.00 dB	1.00 dB	2.30 dB	2.80 dB	2.00 dB	Broadband Assumption: Parameters		
18	image filter (cable 1)	–1.50 dB	–1.74 dB	–1.25 dB	0.25 dB	1.54 dB	1.54 dB	1.54 dB	same at desired and image frequencies.		
19	amp 1 (module 2)	8.00 dB	7.20 dB	8.80 dB	0.80 dB	3.00 dB	3.50 dB	2.50 dB	Cumulative NF, amp 1 and cable 2		
20	cable 2	–0.97 dB	–1.73 dB	–0.20 dB	0.77 dB	1.08 dB	1.08 dB	1.08 dB	3.10 dB	3.60 dB	2.59 dB
21	mixer (module 3)	–8.00 dB	–9.20 dB	–6.80 dB	1.20 dB	11.94 dB	11.87 dB	12.29 dB	Plus gain, amp 1 and cable 2		
22	diplexer (cable 3)	–0.61 dB	–2.43 dB	1.21 dB	1.82 dB	0.93 dB	0.93 dB	0.93 dB	10.13 dB	9.07 dB	11.19 dB
23	amp 2 (module 4)	15.00 dB	14.00 dB	16.00 dB	1.00 dB	5.00 dB	5.30 dB	4.70 dB			
24	cable 4	–1.77 dB	–2.49 dB	–1.05 dB	0.72 dB	1.93 dB	1.93 dB	1.93 dB			
25	module 5	6.00 dB	4.70 dB	7.30 dB	1.30 dB	5.00 dB	5.40 dB	4.60 dB			
26		CUMULATIVE									
27		Gain				Noise Figure			IIP3 with		
28	at output of	mean	min	max	±	conditions as above			mean gain	min gain	max gain
29	preamp (module 1)	12.00 dB	11.00 dB	13.00 dB	1.00 dB	2.30 dB	2.80 dB	2.00 dB	–12.00 dB	–11.00 dB	–13.00 dB
30	image filter (cable 1)	10.50 dB	9.26 dB	11.75 dB	1.25 dB	2.37 dB	2.88 dB	2.06 dB	–12.00 dB	–11.00 dB	–13.00 dB
31	amp 1 (module 2)	18.50 dB	16.46 dB	20.55 dB	2.05 dB	2.59 dB	3.19 dB	2.20 dB	–13.60 dB	–12.31 dB	–14.96 dB
32	cable 2	17.54 dB	14.72 dB	20.35 dB	2.81 dB	2.60 dB	3.21 dB	2.20 dB	–13.60 dB	–12.31 dB	–14.96 dB
33	mixer (module 3)	9.54 dB	5.52 dB	13.55 dB	4.01 dB	3.17 dB	4.11 dB	2.57 dB	–13.66 dB	–12.34 dB	–15.05 dB
34	diplexer (cable 3)	8.93 dB	3.09 dB	14.76 dB	5.83 dB	3.23 dB	4.22 dB	2.60 dB	–13.66 dB	–12.34 dB	–15.05 dB
35	amp 2 (module 4)	23.93 dB	17.09 dB	30.76 dB	6.83 dB	3.76 dB	5.82 dB	2.75 dB	–13.84 dB	–12.39 dB	–15.65 dB
36	cable 4	22.16 dB	14.60 dB	29.71 dB	7.55 dB	3.77 dB	5.83 dB	2.75 dB	–13.84 dB	–12.39 dB	–15.65 dB
37	Cascade	28.16 dB	19.30 dB	37.01 dB	8.85 dB	3.79 dB	5.92 dB	2.76 dB	–13.95 dB	–12.41 dB	–16.21 dB
38	Δ with all SWRs = 1:	–0.26 dB	3.30 dB	–3.81 dB	–3.55 dB	0.02 dB	–0.82 dB	0.20 dB	0.02 dB	–0.13 dB	0.69 dB

Fig. 5.14 Spreadsheet with noise figure and IP3 for Fig. 5.13.

	A	B	C	D
1	SIMPLIFIED CASCADE SPREADSHEET			
2			Noise	IMs
3		Gain	Figure	OIP3
4	item 1	12.0 dB	2.3 dB	0.0 dBm
5	item 2	−1.5 dB	1.5 dB	
6	item 3	8.0 dB	3.0 dB	10.0 dBm
7	item 4	−1.0 dB	1.0 dB	
8	item 5	2.0 dB	8.0 dB	10.0 dBm
9	item 6	−0.8 dB	0.8 dB	
10	item 7	15.0 dB	5.0 dB	24.0 dBm
11		Cumulative Cascade		
12	at output of	Gain	NF	IIP3
13	item 1	12.00 dB	2.30 dB	−12.00 dB
14	item 2	10.50 dB	2.37 dB	−12.00 dB
15	item 3	18.50 dB	2.58 dB	−13.60 dB
16	item 4	17.50 dB	2.59 dB	−13.60 dB
17	item 5	19.50 dB	2.81 dB	−15.03 dB
18	item 6	18.70 dB	2.82 dB	−15.03 dB
19	item 7	33.70 dB	2.88 dB	−16.15 dB

Fig. 5.15 Simplified spreadsheet for cascade of Fig. 5.12.

reflections may be further countered by the fact that the SWRs that are included in the calculations are often specified maximums.

Example 5.6 Simplified Combined Spreadsheet Figure 5.15 is a very simple spreadsheet for the system analyzed in Fig. 5.12 in which all SWRs and variations are ignored. Compare the results in line 19 with the corresponding mean values on line 31 of Fig. 5.12. This spreadsheet is very easy to use and to expand [just insert any additional required lines for item parameters below line 10 and copy (present) line 19 below as many times as necessary]. Such a simple spreadsheet can be very useful for initial design calculations.

5.4.2 Optimization Example

Example 5.7 Figure 5.16 is the block diagram of a double-conversion receiver with the gain, noise figure, and $IIP3_{IM}$ [in (dBm)] plotted below. These cascade parameters were plotted from a simplified spreadsheet, such as that in Fig. 5.15, one that does not yet account for reflections. Gain is obtained as early in the cascade as possible so that the effect of subsequent noise figures will be minimized. The gain is limited, however, in order to preserve IIP3 by not driving the modules nearer to the output of the cascade too hard. Balancing noise figure and

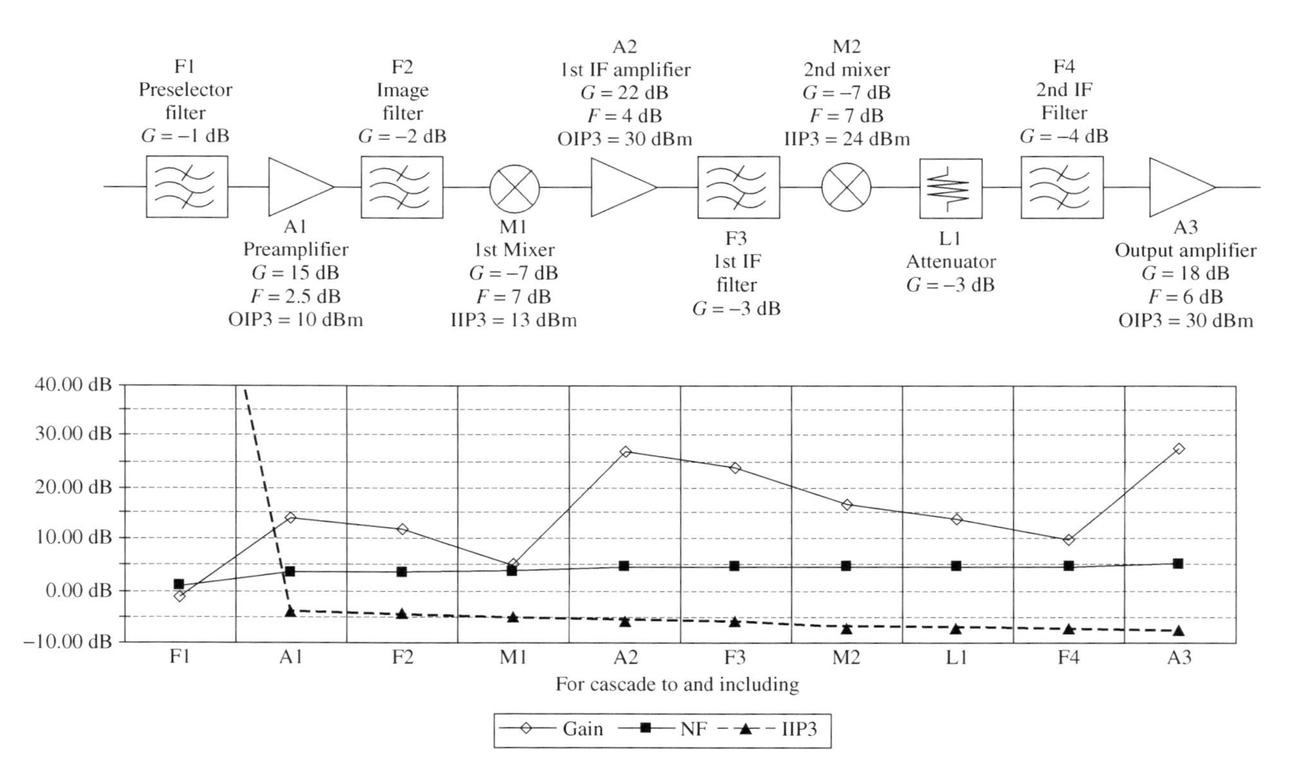

Fig. 5.16 Double conversion.

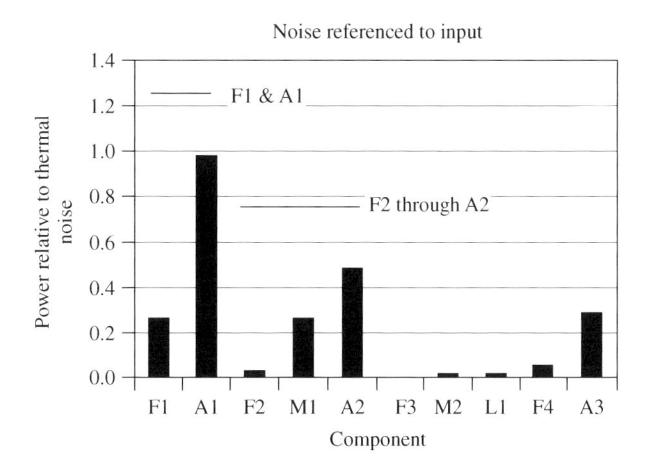

Fig. 5.17 Noise contributions of components.

IIP3 usually produces the seesawing gain that we see here as we move along the cascade. The resulting growth in noise figure and drop in IIP3 along the cascade can be seen in the figure.

Figure 5.17 shows the noise contribution, $f - 1$ divided by the preceding gain, of each module. Two horizontal lines show the contributions of the first two amplifiers combined with the directly preceding attenuations since the net effect is easily determined (Section 3.4). It is important to minimize losses before the preamplifier since they contribute directly to the cascade noise figure. Because of its gain, the preamplifier largely establishes the noise figure of the cascade, although, in order to keep the signal levels down, its gain is not so high that other components do not also make some contribution.

Figure 5.18 shows limitations due to component IIP3s, referenced to the cascade input. (Note that, whereas large values in Fig. 5.17 indicate significant contributions of cascade noise, in Fig. 5.18 small values indicate significant limitations on IIP3.) We see that the first amplifier also largely establishes the cascade IIP3. Higher power components may be used nearer to the output where the signal level has grown. For example, the second mixer M2 has a higher IIP3 than the first mixer M1. This can be accomplished by using a higher LO drive level in the second mixer. Notice that M2 still presents a greater limit to cascade IIP3 than does M1.

Maintaining a fairly constant gain tends to maximize the SFDR. If the three amplifiers were placed where A1 is in Fig. 5.16 (maintaining the same order as shown), noise figure would improve by about 1.7 dB but IIP3 would decrease by 39 dB, leading to a 20-dB degradation in SFDR (Table 5.1). If all three were placed at the output (again maintaining their order), IIP3 would improve 9 dB but noise figure would worsen by 24 dB, a devastating degradation for most receivers, and SFDR would be 10 dB worse than with the gain distributed as in Fig. 5.16.

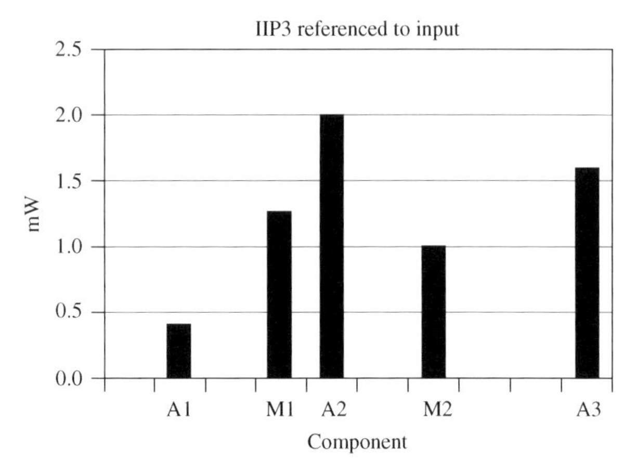

Fig. 5.18 IIP3 limitations of components.

TABLE 5.1 Effects of Redistributing Amplifiers

	NF (dB)	IIP3 (dBm)	ISFDR in 10 kHz (dB)
Distributed amplification	5.30	−7.4	80.90
All amps in front	3.62	−39.17	60.81
All amps in back	29.62	1.72	70.74

We can see from Figs. 5.17 and 5.18 that the first amplifier largely determines both the noise figure and the IIP3 and, therefore the dynamic range, for that configuration. The cascade SFDR is only 4.4 dB less than that of the first amplifier.

5.5 SPREADSHEET ENHANCEMENTS

There are many enhancements that can be usefully included, depending on the project. We have already seen how to include gain control. Here we list a few others, which may be added as the project develops and more data becomes available.

5.5.1 Lookup Tables

We may wish to represent the dependence of a module parameter on some other parameter, such as frequency or temperature or module gain. This other parameter can be entered manually or may be a module parameter. The dependent parameter can be taken from a table stored in some other part of the spreadsheet, perhaps on another page of a workbook, and its value can be interpolated from that table. Worksheet functions such as INDEX, MATCH, LOOKUP, VLOOKUP, HLOOKUP, and FORECAST can be useful in implementing these selections.

5.5.2 Using Controls

Buttons and other controls can be incorporated into a spreadsheet. We might use a button to sequence through various system configurations, displaying the identities of the configurations by using macros and lookup tables. Module or cable parameters can be keyed on the chosen configuration. We might use checkboxes for similar purposes or enter a number or a word in a cell as a control.

5.6 SUMMARY

- Noise also produces IM products. Although more difficult, methods used to determine IMs for discrete signals can also be applied with care to noise.
- Large numbers of discrete signals (e.g., FDM or CATV) can be approximated as noise.
- ISFDR is limited by spurs and noise. It depends on noise figure, intercept point (usually third-order), and processing bandwidth.
- Spreadsheets can incorporate harmonic and intercept point calculations along with gain and noise factors. These can be incorporated for various conditions and configurations and developed and refined as the project progresses.
- ISFDR can be included on a spreadsheet that incorporates noise figure and intercept point.
- Gain is needed at the front end of a cascade to reduce the contribution of subsequent components to the cascade noise figure.
- Excessive gain at the front end of a cascade reduces its input intercept points.
- Gain is usually kept fairly constant throughout the cascade to maximum ISFDR.

ENDNOTES

[1]The author is indebted to Dr. Nelson Blachman for private conversations and internal memos on this subject.
[2]Power is obtained from $e_i(x)e_j(f-x)^*$ but the spectrum is composed of odd imaginary terms and real even terms. The processes of conjugation and frequency negation effectively cancel each other for odd imaginary terms and have no effect for real even terms.
[3]This product is not valid at $f=0$; we have previously shown that coherence changes the results there. However, it is valid for any other value of f, no matter how small, and therefore $S_2(0)$ still represents the peak of the distribution.
[4]We multiply S_0 in Eq. (5.6) by R to convert from power to mean-square voltage, producing R^2. Then we multiply by a_2^2 to obtain the mean-square voltage from the second-order nonlinearity. Then we divide by R to obtain the corresponding power.
[5]For example, $(a+b+c+\ldots)^3 = (2ab+2bc+2ac+\ldots)(a+b+c+\ldots) = 2abc+2abc+2abc+\ldots$.
[6]Based on experiments, Germanov (1998) reduced estimates of multicarrier IMs by 3 dB below the levels that he had theoretically calculated from tests with two or three signals. He cited the lower voltage peaks, with a given total signal power, when there are many signals. In terms of Eq. (4.1), this may correspond to differing effects of higher order terms (which are responsible for

the curvature in the IM curves of Figs. 4.3 and 4.8) when the powers of individual signals decrease while the number of them increases. While it seems unlikely that a significant improvement in the linearity (in dB) of the relationship between IM and signal powers will occur as a result of simply decreasing the power per signal without decreasing the total power, neither is it apparent that the relationship is simply dependent on total RF power, independent of the number of signals over which it is spread. We would probably be most confident in the accuracy of predicted levels, based on IM level curves taken with two signals, when the *total* power of all signals does not exceed the total power for the two signals at the top of the linear range of those curves.

CHAPTER 6

ARCHITECTURES THAT IMPROVE LINEARITY

In this chapter we consider several architectures that can improve linearity by canceling IMs or harmonics that are produced in an amplifying component (Seidel et al., 1968). We begin with amplifier modules combined in parallel. We might note that this improves linearity inherently by reducing the power required from the individual combined modules. However, we will be concerned here with the cancellation of IMs that can occur, depending on the details of how the modules are combined.

Another way to improve linearity is the use of feedback, although its application is limited at higher frequencies due to potential instability associated with inherent delays. This problem is avoided in another method that we will consider, feedforward.

6.1 PARALLEL COMBINING

So far we have considered modules combined in cascade but modules are also combined in parallel. Amplifiers are often combined this way (Gonzalez, 1984, pp. 181–183) in order to obtain an RF power level that is beyond the capability of an individual amplifier. Some circuits that are used to combine and divide RF power[1] have unique properties that affect the performance parameters that we have studied. Internally these circuits often use transmission lines in interesting combinations to produce their unique properties (Sevick, 1987), but here we are concerned, not with these methods, but with the resulting external properties and their potential for linearity improvement. We assume that these properties are retained at all frequencies of interest although that may, at times, be problematical.

6.1.1 90° Hybrid

Figure 6.1 shows the ideal transfer characteristic of a 90° hybrid. There will be, in addition, a time delay that produces the same phase shift in each of the four paths without altering the ideality of the hybrid. The power of a signal entering one port is split into two equal parts, which appear at the two opposite ports, all ports being at the same impedance. Practically, there will be loss in the hybrids and undesired phase shifts, but we will study the ideal case to get an understanding of the general properties of circuits using 90° hybrids.

Simple 90° hybrids typically cover about an octave, but much wider bands are possible in designs that employ multiple sections.

6.1.1.1 Combining Amplifiers A typical use for the 90° hybrid is illustrated in Fig. 6.2, which shows a module that combines the power from two amplifiers. The upper amplifier receives the same signal as the lower one, but delayed 90°. When the signals are recombined, the output of the lower amplifier is delayed 90° in reaching the composite output so, if the amplifiers are identical, the two output signals combine in phase at the module output. Thus the powers of two amplifiers are added. This is a useful feature when one amplifier does not have sufficient power capacity. The scheme can be repeated for additional power increases.

The output termination receives the signal that passed through the lower amplifier plus the signal from the upper amplifier, which should be identical but shifted a total of 180°. Ideally these cancel, but the termination dissipates any power that results from differences in the two signals due to nonideal hybrids or mismatched amplifiers.

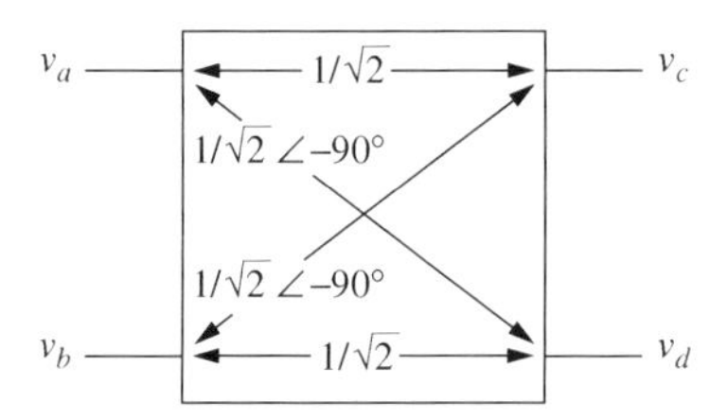

Fig. 6.1 90° hybrid.

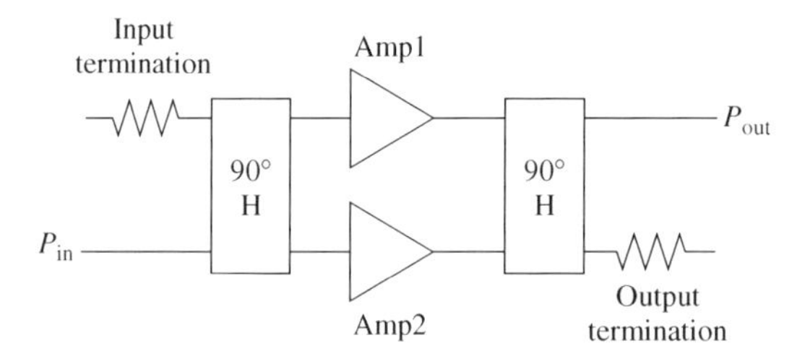

Fig. 6.2 Amplifiers combined using 90° hybrids.

There is some variation in power division across the hybrid's bandwidth. Thus, the 0° output may exceed the -90° output at some frequencies and conversely. Unfortunately, in Fig. 4.2, one signal path receives two 0° shifts from the hybrids, and the other receives two -90° shifts, tending to accentuate deviations from the ideal. If a sign reversal could be obtained in one of the amplifiers, the output port would be interchanged with the output termination port. If this could be done without degrading the match between the amplifiers, it would have the advantage of improving the match between the two signal paths because there would be one 0° and one -90° shift in each path.

6.1.1.2 Impedance Matching To the degree that the amplifiers are identical, the reflection coefficients at their inputs will be identical. Since the signal into the upper amplifier lags the lower one by 90°, its reflection will lag the lower reflection by 90° also. The upper reflection picks up another -90° going through the hybrid back to the module input, so, at the input, it is a total of 180° out of phase with the reflection from the lower amplifier. Thus, the reflections cancel at the module input. Tracing the phase of the reflection entering the input termination in the same way, we find that the two reflections are in phase there, so all of the reflected power is dissipated in the input termination. Thus, two poorly matched amplifiers can be combined to produce a well-matched amplifier module, if the individual amplifiers are identical.

The output port is well matched for the same reason. This is particularly important if Amp 1 and Amp 2 are not well matched to the standard impedance R_0. They may be just active devices with high output impedances. As long as the output impedances are identical, a signal sent into the output will end up in the output termination and not be reflected. Even if their impedances differ greatly from each other, if they are both much higher than R_0 they will produce nearly identical reflections that will cancel at the module output.

6.1.1.3 Intermods and Harmonics If second and third harmonics are generated in Amp 2, its output can be expressed as

$$v_{o2} = v_1 \cos\varphi(t) + v_2 \cos[2\varphi(t)] + v_3 \cos[3\varphi(t)]. \tag{6.1}$$

Similarly, the output from Amp 1 would be

$$v_{o1} = v_1 \cos[\varphi(t) - 90^\circ] + v_2 \cos\{2[\varphi(t) - 90^\circ]\} + v_3 \cos\{3[\varphi(t) - 90^\circ]\} \tag{6.2}$$

$$= v_1 \cos[\varphi(t) - 90^\circ] + v_2 \cos[2\varphi(t) - 180^\circ] + v_3 \cos[3\varphi(t) - 270^\circ]. \tag{6.3}$$

Output v_{o2} is delayed another 90°, producing

$$v_{o2d} = v_1 \cos[\varphi(t) - 90^\circ] + v_2 \cos[2\varphi(t) - 90^\circ] + v_3 \cos[3\varphi(t) - 90^\circ] \tag{6.4}$$

before adding to v_{o1} in the output. The sum voltage is

$$v_{oT} = (v_{o1} + v_{o2d})/\sqrt{2} \tag{6.5}$$

$$= \sqrt{2} v_1 \cos[\varphi(t) - 90^\circ] + v_2 \cos[2\varphi(t) - 135^\circ]. \tag{6.6}$$

The fundamentals have added, producing twice the power of a single fundamental. The second harmonic frequencies have added in quadrature, giving a 3-dB reduction relative to the fundamental. The third-harmonics have canceled. Ideally, this amplifier has no third harmonics. They are all sent to the output termination.

It is easy to show that second-order IMs act like second harmonics. When the fundamentals add, the IMs in v_{o1} contain -180°; when they subtract, they contain 0°. In either case they are in quadrature to the IMs in v_{o2d}, so the ratio of the second-order IMs to the fundamental is 3 dB lower in the output of the module than at the individual amplifiers.

Third-order intermods near the third harmonics (f and g in Fig. 4.6) result from the addition of frequencies and contain the same $3 \times 90^\circ$ that the third harmonics do. As a result they are canceled along with the harmonics. The more important third-order IMs (c and d in Fig. 4.6), those near the signals, however, act like the signals. Since their frequencies are the differences between one fundamental and the second harmonic of the other, their phases contain the same -90° that the fundamentals do, so these IMs from the two amplifiers add coherently.

6.1.1.4 Summary The 90° hybrids can be used to add the powers of two identical amplifiers. Ideally, the input and output ports of the composite amplifier will be reflectionless. The relative (to the signal) amplitudes of second-order harmonics and IMs will be reduced 3 dB (compared to their values in the individual amplifiers). Third harmonics and nearby third-order IMs will be eliminated while third-order IMs near the signals will not be reduced.

6.1.2 180° Hybrid

Figure 6.3 shows the ideal transfer characteristic of a 180° hybrid. Additional delay and loss will be present in practical hybrids, as noted for the 90° hybrid. The power of a signal entering one port is split in two equal parts, which appear at the two opposite ports, all ports being at the same impedance level. These devices are characteristically very broadband, sometimes covering two or three decades.

6.1.2.1 Combining Amplifiers The 180° hybrids can be used to combine identical amplifiers, as illustrated in Fig. 6.4. The input to the upper amplifier is delayed 180°, inverted, relative to the other. A similar operation at the output

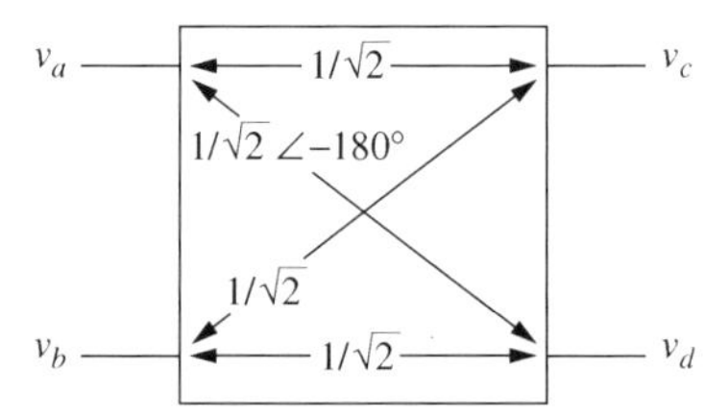

Fig. 6.3 180° hybrid.

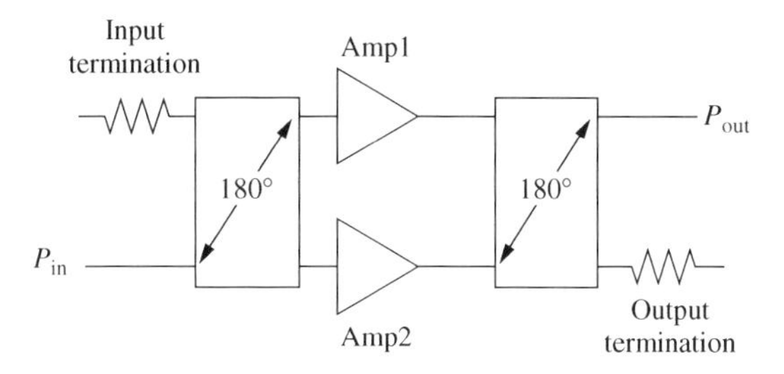

Fig. 6.4 Amplifiers combined using 180° hybrids.

recombines the signals in phase at the load, and any signal appearing in the output termination is due to imbalances.

6.1.2.2 Impedance Matching Reflections from the inputs or outputs of the individual amplifiers add at the module input or output, having made either a 0° or a 360° round trip, so there is no improvement in impedance matching.

6.1.2.3 Intermods and Harmonics If the output of Amp 2 is

$$v_{o2} = v_1 \cos\varphi(t) + v_2 \cos[2\varphi(t)] + v_3 \cos[3\varphi(t)], \tag{6.7}$$

the output from Amp 1 will be

$$\begin{aligned} v_{o1} &= v_1 \cos[\varphi(t) - 180^\circ] + v_2 \cos\{2[\varphi(t) - 180^\circ]\} \\ &\quad + v_3 \cos\{3[\varphi(t) - 180^\circ]\} \qquad (6.8) \\ &= v_1 \cos[\varphi(t) - 180^\circ] + v_2 \cos[2\varphi(t) - 360^\circ] \\ &\quad + v_3 \cos[3\varphi(t) - 540^\circ]. \qquad (6.9) \end{aligned}$$

Output v_{o2} is delayed another 180°, producing

$$v_{o2d} = v_1 \cos[\varphi(t) - 180^\circ] + v_2 \cos[2\varphi(t) - 180^\circ] + v_3 \cos[3\varphi(t) - 180^\circ], \tag{6.10}$$

before adding to v_{o1} in the output. The sum voltage is

$$\begin{aligned} v_{oT} &= (v_{o1} + v_{o2d})/\sqrt{2} \qquad (6.11) \\ &= \sqrt{2}\{v_1 \cos[\varphi(t) - 180^\circ] + v_3 \cos[3\varphi(t) - 180^\circ]\}. \qquad (6.12) \end{aligned}$$

The fundamentals have added, producing twice the power of each. The powers of odd-order harmonics likewise add at the output. Even-order harmonics cancel at the output, all their power going to the output termination.

IMs will have the same phase as harmonics of the same order or will differ by a multiple of 360°; so IMs have the same fate as harmonics of the same order. We show this as follows. An nth-order IM may have frequency $[(n - q)f_1 + qf_2]$, where n and q are positive integers. The total phase shift will be n times the phase shift of the fundamental, θ. The case where $q = n$ or $q = 0$ is a harmonic. Other cases have the same phase shift, $(n - q)\theta + q\theta = n\theta$. Difference-frequency IMs have frequency $(n - q)f_1 - qf_2$ and phase shift $(n - q)\theta - q\theta = (n - 2q)\theta$. This is a change of $q \times 2\theta$ from the phase of the harmonic but, for $\theta = -180^\circ$, a change equal to a multiple of 2θ is ineffective.

6.1.2.4 Summary A composite amplifier using 180° hybrids at input and output ideally contains no even-order harmonics or IMs. These are all dissipated in the output load. Odd-order harmonics and IMs are not suppressed, nor are the input and output matches improved relative to the individual amplifiers.

6.1.3 Simple Push–Pull

A push–pull amplifier is shown in Fig. 6.5 (Hardy, 1979, pp. 301–302). Since other circuits that combine pairs of amplifiers are sometimes called push–pull, we will identify this form as "simple" push–pull. The circuit is similar to Fig. 6.4 except that the output combiner is not a hybrid, which would isolate the two amplifiers from each other, but is a transformer, which does not provide isolation. Difficulties associated with this lack of isolation may account for the restricted use of simple push–pull amplifiers in spite of other advantages, which will be instructive to consider. (Commonly, the 180° power division at the input would be accomplished using a transformer also.)

Efficiency can be improved by operating the individual amplifiers class B, where each amplifier is on during only half of the fundamental cycle. If this is done with a 180° hybrid combiner at the output (Fig. 6.4), the strong even-order harmonic content in the half cycles from each individual amplifier is routed to the output termination where it is dissipated, decreasing the amplifier's efficiency. With a transformer, whichever of Amp 1 or Amp 2 is conducting at any time drives the load. When an amplifier is not conducting, it sees the

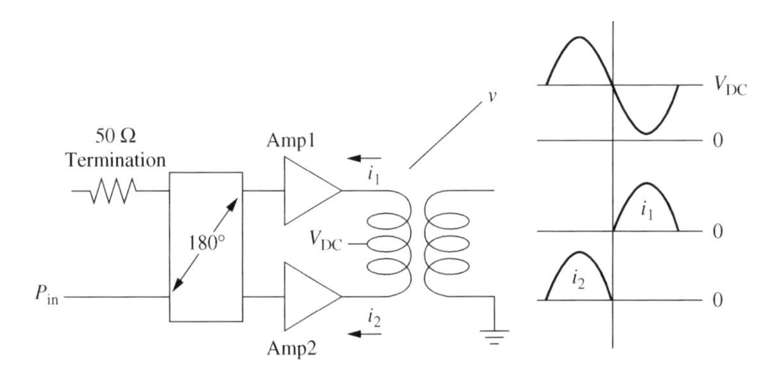

Fig. 6.5 Simple push–pull amplifier.

high-voltage swing generated by the other amplifier. The signals from the two amplifiers combine at the output. Ideally, all harmonics are even and cancel but are not dissipated. [Complementary devices (e.g., npn and pnp or n- and p-channel) are sometimes used to combine the two half cycles without requiring transformers.]

With the hybrid, the even-order harmonics are eliminated from each amplifier's output, leaving a sine wave that is added to the sine wave from the other amplifier. With class B operation of a simple push–pull, the two outputs are simply added and form a sinusoid as a result. In both cases balance is required for complete cancellation of even-order harmonics and odd-order harmonics are not canceled.

If the amplifiers should be operating class A (sinusoidal current from each amplifier), ideally the total current would add at the output for either type of 180° combiner. If one of the amplifiers should stop conducting, the power from the simple push–pull circuit would be halved whereas the output from the hybrid would drop to one quarter because, under those conditions of imbalance, half of the power would be dissipated in the hybrid's load. However, a damaged amplifier in a simple push–pull pair could affect the other amplifier, possibly destroying it, due to the lack of isolation.

6.1.4 Gain

If we remove the amplifiers from Fig. 6.2 or Fig. 6.4, we obtain the configuration shown in Fig. 6.6. It is apparent that the signals add at the output, since they arrive there in phase. Thus, for ideal hybrids, either 90° or 180°, the gain is one. The addition of amplifiers with gain g will increase the output by g, giving a module gain equal to the gain of each individual amplifier. This will be reduced by dissipation losses in the hybrids. Other deviations from ideal in hybrids (typical the magnitude and phase of the transfers vary some over the specified RF band) or differences in the two amplifiers will also cause losses. Amplifier input mismatches, which cause the input signal to be reflected into the input termination, are already accounted for in the way the transducer gains of the individual amplifiers are measured (presumably with the same standard impedance).

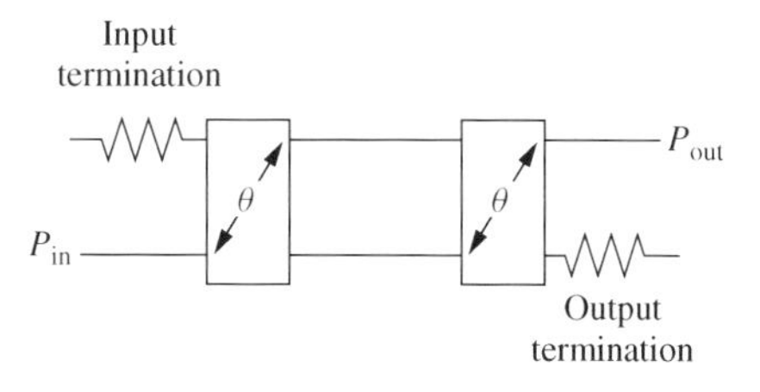

Fig. 6.6 Hybrids without amplifiers.

6.1.5 Noise Figure

When the composite amplifier is driven by the standard impedance R_0, the noise at the output of each individual amplifier will be $\overline{k}T_0Bf_{\rm Amp}g_{\rm Amp}$. The part of this noise originating in each amplifier is $\overline{k}T_0B(f_{\rm Amp}-1)g_{\rm Amp}$. Half of this goes to the combiner output and half goes to the output termination so the amplifier noise at the combiner output has the same level as the noise from one amplifier. The source noise is divided and amplified and recombines coherently at the combiner output along with the signal. Its power at the output is $\overline{k}T_0Bg_{\rm Amp}$. (The input termination noise combines coherently in the output termination at the same level.) Therefore the total output noise is $\overline{k}T_0Bf_{\rm Amp}g_{\rm Amp}$, which is $f_{\rm Amp}g_{\rm Amp}$ greater than the input noise to the module. Since the signal is greater by $g_{\rm Amp}$ at the output than at the input, the noise factor for the composite is the same as for the individual amplifiers:

$$f_{\rm module}=\frac{S_{\rm in}}{S_{\rm out}}\frac{N_{\rm out}}{N_{\rm in}}=\frac{1}{g_{\rm Amp}}f_{\rm Amp}g_{\rm Amp}=f_{\rm Amp}. \tag{6.13}$$

It is simple to account for loss in the input hybrid since it acts like an attenuator in front of the module and thus increases $f_{\rm module}$ by its attenuation. (Since noise factor for the individual amplifiers was presumably measured with a standard impedance source, reflections from the inputs of those amplifiers are again already accounted for.) Output attenuation, less one, will be divided by g before being added to f, so it will have less impact.

6.1.6 Combiner Trees

The amplifiers, shown in Fig. 6.2 or Fig. 6.4, might consist of modules that are again represented by either of these figures, thus combining four elementary amplifiers. Such a module might, in turn, serve as an amplifier for a higher level module, and so forth. Figure 6.7 shows three levels of power combining. Each level serves as an amplifier for the next higher level. Thus one can use the configuration in Fig. 6.2 or 6.4 repeatedly, increasing the number of devices combined and the maximum output power.

The power dividers and combiners can be 90° hybrids, 180° hybrids, or in-phase dividers and combiners. We might use combinations to gain the combined advantages of the different types. For example, we might use 90° hybrids in Level 1 for impedance matching and odd-harmonic suppression and 180° hybrids in Level 2 for even-harmonic suppression. We must be aware, however, that the hybrids may contain magnetic cores and so can produce harmonics and IMs themselves (Section 4.7).

Each level increases the total output power by 3 dB (assuming a fixed output power from each amplifier) less the loss in its output combiner, but the overall gain decreases by the losses in its input and output combiners, so amplifiers may be inserted in the input power division structure (or tree).

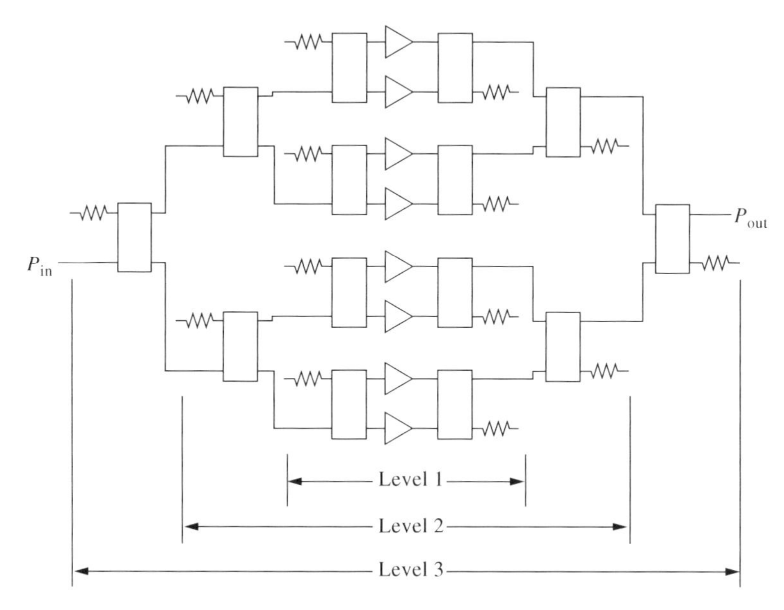

Fig. 6.7 Combiner tree.

6.1.7 Cascade Analysis of a Combiner Tree

We can analyze a combiner tree, such as is shown in Fig. 6.7, as a cascade by using total powers in all of the legs at each interface as the variables at that point in the cascade. Thus each power divider is represented as an attenuator with gain (in a matched circuit) of

$$g = \frac{p_{\text{out}}}{p_{\text{in}}}, \tag{6.14}$$

where p_{in} is the total power at all q inputs and p_{out} is the total power at all $2q$ outputs (e.g., $q = 1$ for the first divider). Ideally, the attenuation is 0 dB and $g = 1$.

The combined M amplifiers ($M = 8$ in Fig. 6.7) have M times the input power and M times the equivalent input noise of a single amplifier; so the combined noise figure is the same as that of a single amplifier. The combined output signal power and the combined output power at each intermod are all M times greater than for a single amplifier; so the intercept points for an nth-order nonlinearity are

$$p_{\text{IP}n,\text{combined}} = M \times p_{\text{IP}n,\text{amp}}. \tag{6.15}$$

Amplifiers that may appear at other levels can be treated similarly.

Each output power combiner also acts as an attenuator, and Eq. (6.14) applies again except that there are now $2q$ inputs and q outputs. However, if the combiner provides cancellation of an intermod, this must be accounted for by an increase in the system input IP occurring at that module. If the combiner is a 90° hybrid, there is an additional 3-dB reduction in second-order products [Eq. (6.6)], which corresponds to a 3-dB increase in the system IP2. Ideally 90° hybrids completely cancel third harmonics and some IM3s, so the system IP3 for those products would become infinite at that point. Realistically, the balance will be imperfect so a finite increase in IP3 should be used to represent the partial cancellation (1 dB increase for each 2 dB of cancellation). Similarly, a 180° hybrid theoretically provides infinite cancellation of second-order products, but we can represent actual performance by increasing the IP2 by an amount equal to the cancellation in dB.

Imperfections in power combining, caused by differences in the phase or amplitude of the two combined signals, lead to increased attenuation and decreased cancellation in the combiners. However, these errors are due not only to the combiners but also to imperfections in other components at that level. For example, an error of φ in the relative phases of the outputs from the power dividers at the front of Level 2 (Fig. 5.7) has the same effect as an error of φ at the inputs to the combiners at the other end of that level. Errors in the dividers might increase the attenuation in the combiners or they might tend to cancel errors in the combiners, thus decreasing their attenuations. The effective gain and phase errors at the combiners are the total path errors for the level. Likewise, differences in gains through supposedly identical devices within the level can contribute to losses in the combiners at the level output. A statistical analysis of the effects of variances in the various component parameters on the overall expected gain and gain variance can be important in some applications but is beyond the scope of this book.

6.2 FEEDBACK

Figure 6.8 shows an operational amplifier (op amp) circuit with negative feedback. We have seen this before in Fig. 3.18. The negative feedback in this circuit can cause the transfer function to be more a function of the passive components than of the active amplifier and, therefore, to be quite linear. Figure 6.9 shows a mathematical block diagram corresponding to Fig. 6.8. The standard equation for the closed-loop transfer function is

$$a = \frac{a_{\text{op}}}{1 + a_{\text{op}} a_{\text{FB}}}. \tag{6.16}$$

When the open-loop gain $|a_{\text{op}} a_{\text{FB}}|$ is much greater than one, this becomes

$$a \approx 1/a_{\text{FB}}, \tag{6.17}$$

and the circuit transfer function becomes dependent on the passive components that determine a_{FB}. [Note that the transfer function of the input block in Fig. 6.9, when multiplied by Eq. (6.17), produces the standard transfer function for this circuit, $R_{\text{FB}}/R_{\text{in}}$.]

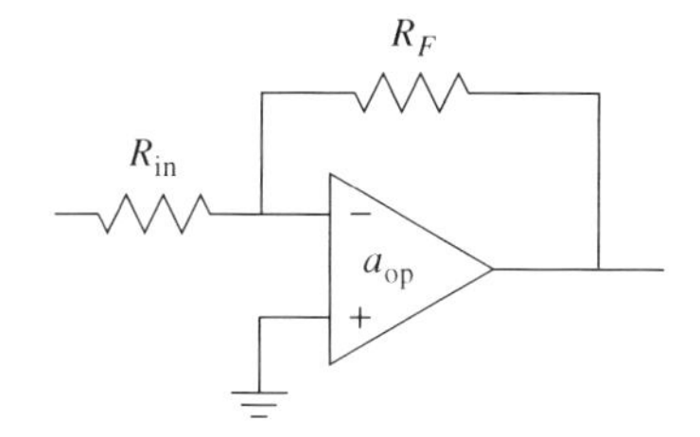

Fig. 6.8 Op amp.

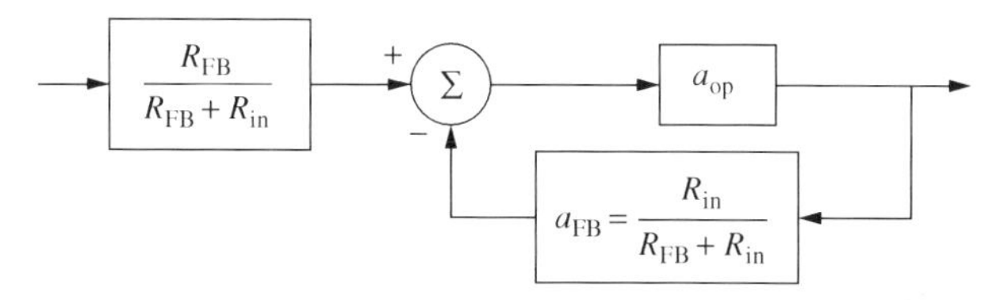

Fig. 6.9 Block diagram of op amp.

The main problem at higher frequencies is stability. For stability, the open-loop gain $|a_{op}a_{FB}|$ should be less than one by the time the open-loop excess phase $\angle a_{op}a_{FB}$ reaches $-180°$. For this reason, a single-pole roll-off is commonly incorporated into a_{op} to reduce the gain below unity by the time the unavoidable phase shift in the transfer function reaches $-90°$, which will add to the $-90°$ that accompanies the roll-off (Egan, 1998, pp. 49–54). As a result, the open-loop gain is often low at higher RF frequencies, limiting this method to the lower frequencies.

One method for overcoming this limitation feeds back the detected *amplitude* of the output for comparison to the detected *amplitude* of the input. When the modulation is sufficiently low in frequency, significant open-loop gain can be obtained in that loop to produce good modulation linearity. Phase can also be controlled this way in the case of quadrature amplitude modulation (QAM) signals where a coherent carrier signal is available to act as a reference for coherent detection. In that case, the signal can be separated into normal components and the AM of each can be controlled separately (Katz, 1999).

6.3 FEEDFORWARD[2]

In Fig. 6.10, a_1 is the linear voltage transfer function of the main amplifier and a_1' is the linear voltage transfer function of a secondary amplifier. Part of the input is sent to the main amplifier and part to the secondary amplifier. The output of the main amplifier is sampled in a directional coupler[3] and injected into the secondary line by another directional coupler (c_2 and c_3, respectively). The gains and delay τ_1 and phase shift φ_1 are ideally such that the versions of the input

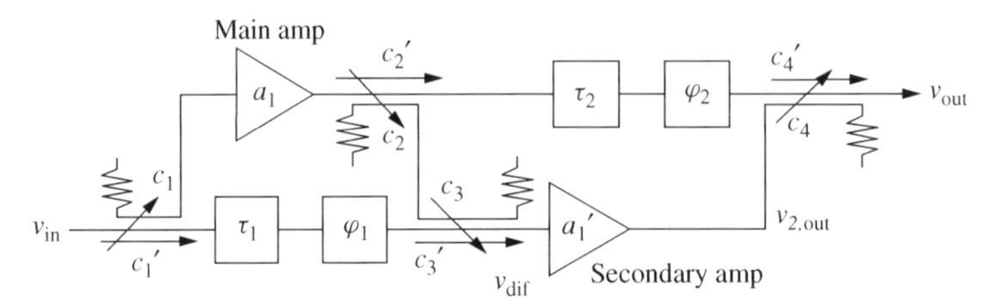

Fig. 6.10 Feedforward amplifier. Component amplifiers are represented by their linear voltage gains a; couplers by their coupling c and main-line gain c' (both voltage gains).

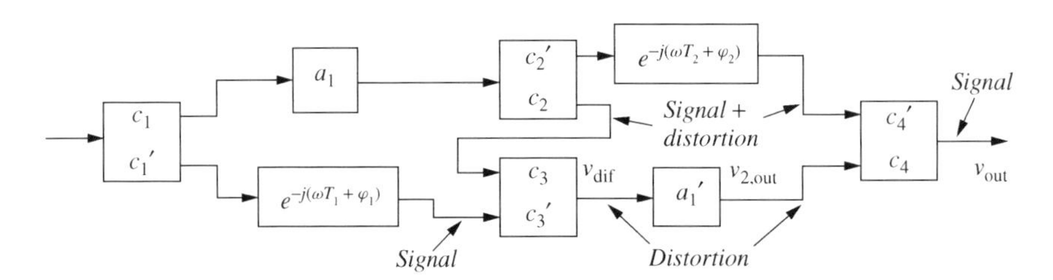

Fig. 6.11 Feedforward block diagram.

signal, arriving at the secondary amplifier by the two paths, cancel, leaving only the distortion that was generated in the main amplifier to enter the secondary amplifier. Since the secondary amplifier has only this small residual signal to amplify, it is presumably less subject to distortion than the main amplifier. The amplified distortion is subtracted from the main signal in the output coupler, canceling the distortion. Again, this cancellation requires proper values of gain and τ_2 and φ_2. A mathematical block diagram is shown in Fig. 6.11.

6.3.1 Intermods and Harmonics

Assuming all adjustments are correct, the signal entering the secondary amplifier can be written, from Eq. (4.1), as

$$v_{\text{dif}} = a_z(a_1 v_{\text{in}} + a_2 v_{\text{in}}^2 + a_3 v_{\text{in}}^3 + a_4 v_{\text{in}}^4 + a_5 v_{\text{in}}^5 + \cdots - a_1 v_{\text{in}}) \tag{6.18}$$

$$= a_z(a_2 v_{\text{in}}^2 + a_3 v_{\text{in}}^3 + a_4 v_{\text{in}}^4 + a_5 v_{\text{in}}^5 + \cdots), \tag{6.19}$$

where

$$a_z = c_1 c_2 c_3. \tag{6.20}$$

The output of the secondary amplifier is

$$v_{2,\text{out}} = a_z \left\{ \begin{array}{l} a_1'[a_2 v_{\text{in}}^2 + a_3 v_{\text{in}}^3 + a_4 v_{\text{in}}^4 + a_5 v_{\text{in}}^5 + \cdots] \\ \quad + a_2'[a_2 v_{\text{in}}^2 + \cdots]^2 + a_3'[a_2 v_{\text{in}}^2 + \cdots]^3 + \cdots \end{array} \right\}. \tag{6.21}$$

If this is subtracted from the output from the main amplifier, properly delayed and phase shifted, it will cancel the IMs and harmonics created in the main amplifier, producing

$$v_{\text{out}} = a_y \left\{ \begin{array}{l} a_1 v_{\text{in}} + a_2 v_{\text{in}}^2 + a_3 v_{\text{in}}^3 + a_4 v_{\text{in}}^4 + a_5 v_{\text{in}}^5 + \cdots \\ \quad -[a_2 v_{\text{in}}^2 + a_3 v_{\text{in}}^3 + a_4 v_{\text{in}}^4 + a_5 v_{\text{in}}^5 + \cdots] \\ \quad -a_2'[a_2 v_{\text{in}}^2 + \cdots]^2 - a_3'[a_2 v_{\text{in}}^2 + \cdots]^3 - \cdots \end{array} \right\} \tag{6.22}$$

$$= a_y \{a_1 v_{\text{in}} - a_2'[a_2 v_{\text{in}}^2 + \cdots]^2 - a_3'[a_2 v_{\text{in}}^2 + \cdots]^3 - \cdots\}, \tag{6.23}$$

where

$$a_y = c_1 c_2' e^{-j(\omega\tau_2 + \varphi_2)} c_4' \tag{6.24}$$

Here we have exchanged spurs (IMs and harmonics) produced by the secondary amplifier, which is amplifying only the relatively weak spurs from the main amplifier, for the spurs produced in the main amplifier, which is amplifying the relatively powerful main signal.

6.3.2 Bandwidth

The delay and phase shift in parallel with each amplifier are intended to duplicate the delay and phase shift within the amplifier and coupling devices and to add the 180° phase shift required for subtraction if that is not obtained in some other way. Only the phase shifter is necessary for this at any given frequency, but the delay is incorporated to try to match the phase shift in the other branch over a wide frequency range. Otherwise cancellation will occur at only one frequency. It is, of course, necessary that the various coupling factors c be adjusted to produce the same magnitude of gain in each path so some means of gain adjustment is desirable also.

A failure to match paths from input to output will result in incomplete cancellation of the IMs. A failure to match paths from input to the secondary amplifier will cause it to carry some of the main signal to the detriment of its linearity as well as a loss in overall gain due to unnecessary cancellation of the desired signal. The system tends to flatten the gain (i.e., to reduce ripple) since changes in a_1 from optimum cause error signals that are amplified by a_2 and used to cancel the change at the output.

6.3.3 Noise Figure

The noise figure of the overall amplifier is ideally (assuming perfect adjustment) that of the path from input to output through τ_1 and the secondary amplifier (Fig. 6.10).

There are three paths from input to output. In Fig. 6.11, let the upper path have a transfer function of a_u, the lower path have a transfer function of a_l, and the path that crosses from upper to lower at the couplers have a transfer function of a_x. We know that IMs in the crossing path cancel those in the upper path so

$$a_u = -a_x. \tag{6.25}$$

We also know that the crossing path and the lower path are the same after they join at the secondary amplifier input and that they cancel each other up to that point, so

$$a_l = -a_x. \tag{6.26}$$

Therefore, the net transfer function is

$$a = a_u + a_x + a_l = a_u = -a_x = a_l, \tag{6.27}$$

so we see amplified *input* noise, using the transfer function of any of the three paths. Noise generated in the common part of the upper and crossing paths cancels at the output. The rest of the upper path is just an attenuator at one port of the output coupler and is accounted for in that coupler's noise figure. Much as in the case of image noise when a mixer is driven by a diplexer (Section 3.9.1), the termination at that port is assumed in computing the noise figure of the coupler in the path through the other port. The remaining and uncanceled component noise is due to the lower path. Therefore, the lower path contains the input noise and all of the uncanceled component noise, including the effect of loss in the output coupler.

Since the noise figure is determined by the lower path, the best noise figure will occur when $c_1' \gg c_1$, which will require that a_1 be large for a given overall gain.

6.4 NONIDEAL PERFORMANCE

We have described how certain circuit configurations can ideally eliminate the effects of nonlinearities in some active components. Detailed discussion of how other imperfections in various parts of the configurations affect the results is beyond the scope of this book.

Feedforward and parallel configurations require accurate matching of paths to prevent loss of power and gain and to effectively cancel nonlinearities. Determining the effects of inaccurate transfer functions is an important part of design. It requires writing the detailed overall transfer function and introducing the various amplitude and phase perturbations that can be expected from components to determine their effects on the output.

The response of a feedback configuration ideally depends on only a few components, but the imperfections of the open-loop amplifier are attenuated by only a finite amount, and that amount depends on open-loop gain, which falls with increasing frequency. For example, an IM voltage v_{IM} that would appear at the amplifier output without feedback will be reduced to approximately $v_{\mathrm{IM}}/|a_L|$, where $|a_L|$ is the open-loop gain, as long as $|a_L| \gg 1$. This may practically eliminate the IM if it has a frequency well below the loop bandwidth but will have small effect if the frequency exceeds that bandwidth.

6.5 SUMMARY

- Modules that combine two identical amplifiers using 90° hybrids ideally have good input and output matches to the standard impedance.
- Third harmonics and third-order IMs that are near the harmonics (at frequency sums) generated in the two identical amplifiers are ideally eliminated when 90° hybrids are used to combine them. Third-order IMs near the fundamentals (at difference frequencies) are not reduced.
- Even-order harmonics and IMs generated in two identical amplifiers are ideally eliminated when 180° hybrids are used to combine them.
- Class B simple push–pull amplifiers are inherently more efficient than amplifiers combined using 180° hybrids.
- The gain of a module that combines two identical amplifiers using 90° or 180° hybrids ideally equals the gain of each individual amplifier.
- The noise factor of a module that combines two identical amplifiers using 90° or 180° hybrids ideally equals the noise factor of each individual amplifier.
- Multiple levels of combining modules can add the powers of many amplifiers.
- Combiner trees can be analyzed as cascades using the total powers at each interface.
- Hybrids that contain magnetic cores can cause harmonics and IMs.
- Feedback improves linearity but has stability problems at high frequencies.
- Feedforward techniques amplify the error and use it to cancel distortion.

ENDNOTES

[1]Tsui (1985, pp. 245–273), Vizmuller (1995, pp. 146–158), Anaren (2000), and MA-COM (2000).
[2]Arntz (2000), Huh et al. (2001), Myer (1994), Seidel (1971a, 1971b), and Seidel et al. (1968, pp. 675–711).
[3]A directional coupler couples part of a wave to another line. The direction of travel of the signal in the coupled (secondary) line depends on its direction of travel in the main line. The representation in Fig. 6.10 is for main- and secondary-line signals traveling in the same direction (e.g., left to right). The coupling factor is the ratio of the power of the coupled signal to the power of the signal entering the coupler. The directivity is the ratio of the signal power launched in a given direction in the secondary line with a given incident wave in the main line to the same power when the wave in the main line is reversed. Ideally, this is infinite, practically maybe 10–45 dB, depending on frequency and the bandwidth of the coupler.

CHAPTER 7

FREQUENCY CONVERSION

Nearly all traditional radio receivers,[1] as well as other electronic systems, employ frequency conversion. This is also called heterodyning and the radio architecture that uses it is called superheterodyne. Prior to the introduction of the superheterodyne system, selective radios required filters with many variable components, all changing synchronously to track the signal. With the superheterodyne system, the desired frequency is converted to a fixed frequency, and the primary filter can thus be fixed, a much easier and more effective design. Receivers are not the only applications that use heterodyning to change frequency.

7.1 BASICS

7.1.1 The Mixer

The device in which heterodyning occurs is called a mixer.[2] There are two inputs, the RF (radio frequency or radio-frequency signal) and the LO (local oscillator). The desired output is the IF (intermediate frequency or intermediate-frequency signal). This terminology corresponds well to the mixer's usage in a receiver, but we will so identify the mixer's ports and their signals in other frequency converters as well.

The mixer contains a device that multiplies the RF signal by the LO signal. The product of these two sinusoids can be decomposed into a sinusoid whose frequency is the sum of the RF and LO frequencies and another having the difference frequency. One of these is the desired frequency-shifted IF.

A simple mixer may consist of a single diode or some other electronic device (e.g., a field-effect transistor) that can be operated in such a way as to produce the required product. A general nonlinearity contains a squaring term that will produce the required product. (We will discuss the mathematics of this process in the following sections.). When a single diode is used, the RF, LO, and IF all occur at the same location and can only be separated by filtering. A singly balanced mixer can be created using two diodes whose inputs and outputs are phased and combined in such a way that one of the inputs (e.g., the LO) cancels at the IF output port. A doubly balanced mixer (DBM) (Fig. 7.1) can cancel the appearance of both inputs in the IF. Harmonics of the balanced signals are also canceled. (The degree of cancellation is finite in all cases.) The remainder of our discussion assumes a doubly balanced diode mixer but most of the material will be generally applicable (Egan, 2000, pp. 36–43, 64–67).

Usually the LO power is much greater than the RF power and, as a result, the mixer acts like a linear element to the through path (RF to IF), except for the frequency translation. To operate in this manner with large RF signals, the LO power may have to be increased, perhaps from 7 dBm for a low-level mixer to as much as 27 dBm for a high-level mixer. High-level mixers may have one or more additional diodes, or perhaps other passive elements, in series with each diode shown in Fig. 7.1, or they may combine two of these diode bridges.

Even more complex combinations of diodes and combiners can produce mixers with special advantages. For example, the IF at the sum frequency or at the difference frequency can be canceled, leaving a single-sideband mixer that produces an output at only the sum or the difference frequency. At the other extreme of complexity, LO and mixer are sometimes combined in one active device, called a converter.

Here are some of the parameters by which mixers are characterized:

Frequency ranges: the RF, LO, and IF ranges for which the mixer is designed.

LO power level: the design or maximum LO power.

Conversion loss: the ratio of IF to RF power, sometimes given as a function of LO power. This is also called *single-sideband conversion loss* because the output power of only one of the two converted signals (sum or difference frequency) is measured.

1-dB input compression level: the RF power at which the conversion loss increases by 1 dB over the low-level value.

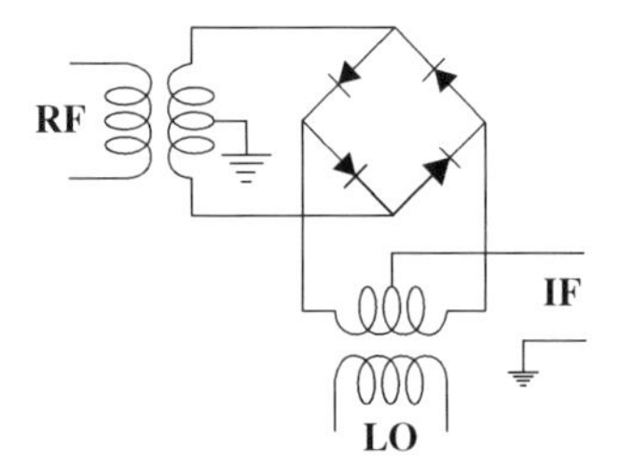

Fig. 7.1 Doubly balanced mixer. RF and LO ports shown are considered balanced but the IF port is unbalanced.

Noise figure: this is equal to or greater than the conversion loss.

Spurious levels: a list or table of the levels (usually typical) of various undesired products created in the nonlinearity. These are given for particular LO and RF power levels and generally are measured with broadband terminations on all ports. They are usually relative to the level of the desired IF signal.

IM intercept points: usually the $IIP3_{IM}$.

Isolation: between the various ports, LO, RF, and IF; for example, how much is the LO power attenuated in getting to the IF output.

Impedance and SWR: as for other active devices. The other characteristics depend on the impedance matches at the terminals.

7.1.2 Conversion in Receivers

Incoming RF signals are injected into a mixer, as is the stronger LO. The nonlinearity produces signals at the sum and difference of the LO and RF frequencies, and one of these becomes the IF, to which the IF filter is tuned. A radio is tuned by changing the frequency of the LO, and thus of the RF signal that will convert to the IF frequency. The range of incoming frequencies is restricted by a relatively broad filter, either fixed or tuned. This prevents the sum frequency from being received when the difference frequency is desired and visa versa. Among these two inputs, the undesired signal is called the image of the desired signal. The process is illustrated in Fig. 7.2.

The desired conversion process is indicated by Eq. (3.38) or (3.39), which can be combined to give the tuned frequency as

$$f_R = |f_L \pm f_I|. \tag{7.1}$$

Here the RF frequency that will pass through the IF filter after conversion is given as a function of the LO frequency. The sign in the equation is controlled by the

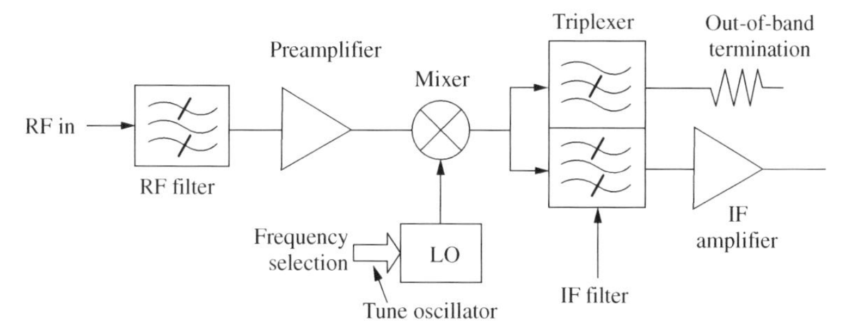

Fig. 7.2 Superheterodyne architecture. The out-of-band termination is good design practice but not essential. (The upper half of the triplexer is a bandstop filter; the lower half is a matching bandpass filter.)

RF filter, which should allow only one of these frequencies to pass — otherwise both can be received. The process is illustrated in Fig. 3.10. The bandwidths can be seen there from the width of the noise bands.

Since the sum or difference frequency is normally generated in a nonlinearity, spurious signals (spurs) at other frequencies are also generated, commonly at weaker levels. This is the same process that was described in Chapter 4, except that, here, one of the two significant inputs is the relatively large LO. We do not want to see either of the inputs in the IF. We are looking for one of the products of the RF and the LO, produced in the nonlinearity, and are trying to avoid other products of these two signals and of other, unavoidable, input signals, with the LO. This involves a more complex design process.

7.1.3 Spurs

When the LO is tuned to produce a signal at the IF frequency according to Eq. (7.1) with the intended sign, and a signal is produced in the IF, but by a process that gives a different relationship between the RF and IF frequencies, we say we have a spurious response, or spur. The spur appears to have been converted from the RF frequency that corresponds, by the equation for the desired response, to the LO setting; but it is, in fact, the response to some other signal. Spurious responses to the intended RF signal should be rejected by the IF filter while the RF filter limits the range of RF frequencies that might otherwise produce spurs. A designer may say that there is a spur at some frequency, referring either to the frequency of an IF signal resulting from a spurious response or to the frequency of an RF signal that causes a spurious response in the IF. The former might be produced by the desired signal; the latter by what can be termed an interferer since it can cause interference with the desired signal.

Spurs that only occur when a certain RF frequency, or range of frequencies, is received, are called single-frequency spurs — IMs require two RF signals. Spurs that occur without an RF signal are called internal spurs. They are produced by contaminating signals elsewhere in the receiver.

Single-frequency spurs are described by

$$f_{\mathrm{IF}} = m f_{\mathrm{LO}} + n f_{\mathrm{RF}}. \tag{7.2}$$

These are called m-by-n spurs or $|m|$-by-$|n|$ spurs. For example, if $m = -2$ and $n = 3$, the spur may be called minus-two-by-three or two-by-three (or -2×3 or 2×3). If no sign is given, it is probably safer to assume it has been left out rather than to assume that both signs are positive. If we want to specify $m = 2$ and $n = 3$, we can say plus-two-by-plus-three. We will put the LO multiplier m first; sometimes it is done the other way.[3]

Figure 7.3 is a chart that gives the expected level of various spurious responses. It is organized as an $|n| \times |m|$ matrix of spur levels relative to the level of the desired 1×1 signal. This particular chart is unusual in that it gives information for three different mixers at two RF power levels and in the large number of spurs for which it gives values.

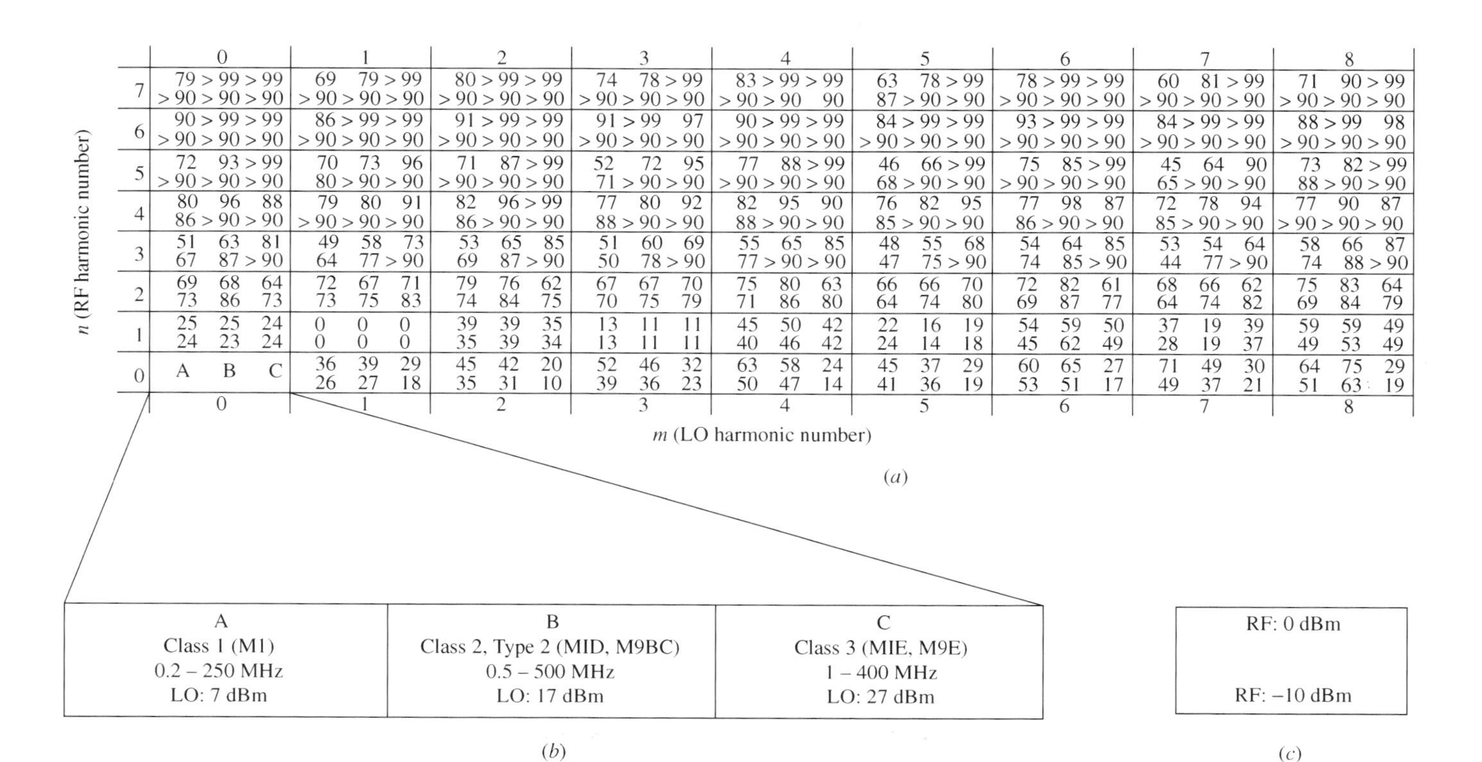

n (RF harmonic number)	0	1	2	3	4	5	6	7	8
7	79 >99 >99 >90 >90 >90	69 79 >99 >90 >90 >90	80 >99 >99 >90 >90 >90	74 78 >99 >90 >90 >90	83 >99 >99 >90 >90 90	63 78 >99 87 >90 >90	78 >99 >99 >90 >90 >90	60 81 >99 >90 >90 >90	71 90 >99 >90 >90 >90
6	90 >99 >99 >90 >90 >90	86 >99 >99 >90 >90 >90	91 >99 >99 >90 >90 >90	91 >99 97 >90 >90 >90	90 >99 >99 >90 >90 >90	84 >99 >99 >90 >90 >90	93 >99 >99 >90 >90 >90	84 >99 >99 >90 >90 >90	88 >99 98 >90 >90 >90
5	72 93 >99 >90 >90 >90	70 73 96 80 >90 >90	71 87 >99 >90 >90 >90	52 72 95 71 >90 >90	77 88 >99 >90 >90 >90	46 66 >99 68 >90 >90	75 85 >99 >90 >90 >90	45 64 90 65 >90 >90	73 82 >99 88 >90 >90
4	80 96 88 86 >90 >90	79 80 91 >90 >90 >90	82 96 >99 86 >90 >90	77 80 92 88 >90 >90	82 95 90 88 >90 >90	76 82 95 85 >90 >90	77 98 87 86 >90 >90	72 78 94 85 >90 >90	77 90 87 >90 >90 >90
3	51 63 81 67 87 >90	49 58 73 64 77 >90	53 65 85 69 87 >90	51 60 69 50 78 >90	55 65 85 77 >90 >90	48 55 68 47 75 >90	54 64 85 74 85 >90	53 54 64 44 77 >90	58 66 87 74 88 >90
2	69 68 64 73 86 73	72 67 71 73 75 83	79 76 62 74 84 75	67 67 70 70 75 79	75 80 63 71 86 80	66 66 70 64 74 80	72 82 61 69 87 77	68 66 62 64 74 82	75 83 64 69 84 79
1	25 25 24 24 23 24	0 0 0 0 0 0	39 39 35 35 39 34	13 11 11 13 11 11	45 50 42 40 46 42	22 16 19 24 14 18	54 59 50 45 62 49	37 19 39 28 19 37	59 59 49 49 53 49
0	A B C	36 39 29 26 27 18	45 42 20 35 31 10	52 46 32 39 36 23	63 58 24 50 47 14	45 37 29 41 36 19	60 65 27 53 51 17	71 49 30 49 37 21	64 75 29 51 63 19
	0	1	2	3	4	5	6	7	8

Fig. 7.3 Spur-level chart for three doubly balanced mixer classes and two signal levels. Relative spur levels are shown at (*a*). Each rectangle contains three columns, one for each of the mixer classes shown at (*b*). Each rectangle contains two rows, one for each of the RF levels shown at (*c*). The LO frequency is 50 MHz and the RF frequency is 49 MHz (Cheadle, 1993, p. 485). The higher mixer classes (Henderson, 1993c, p. 481) have another diode or other passive components in series with the diode in each leg and are designed for increasingly higher LO power levels. A minus is understood for all of the relative spur levels.

Spurs that are produced at the desired IF frequency by the desired RF frequency are called crossover spurs. Here an RF signal is converted to the same IF frequency by each of two processes, the intended conversion and the spurious response. Even if we should find no harm from superimposing two copies of the same signal, any slight detuning from the LO frequency that produces the crossover spur produces two copies of the signal separated by some finite frequency. Crossover spurs are particularly troublesome because they cannot be preventing by filtering since the desired signal must be passed. Appendix X contains a list of crossover spurs.

Design involves consideration of all possible RF input signals, whether desired or undesired, and the choice of the LO frequency range and filtering to minimize interference due to spurious responses. Sometimes the RF filter becomes a preselector, which is tuned or broken into selectable segments. Sometimes the conversion is done in more than one step to avoid undesired responses. Mixers can be selected for desirable spurious performance and balanced (Egan, 1998, pp. 36–43) to reduce the appearance of the LO and input signals and their harmonics in the IF.

7.1.4 Conversion in Synthesizers and Exciters

Another use for heterodyning is in frequency synthesis. This can be represented in a manner similar to Fig. 7.2, but the RF and LO are fixed or synthesized frequencies, and the object is to combine them to produce a new synthesized frequency at the IF.[4] Here we have control over the signals existing in the RF, rather than being subject to whatever is picked up by an antenna, so no RF filter is required. We also have control of signal levels. Now the spurious responses of interest *are* IF signals, produced by the intended, actual, RF, that are passed by the IF filter. We must prevent these undesired signals in the IF, and the acceptable level of such signals in the output is often much lower than for the receiver. Heterodyning in exciters, which provide signals for transmitters, is similar to that in synthesizers.

In upconverters, the mixer port that is labeled "IF" may be used as the input port because its designated frequency range is lower than the port labeled "RF." This is generally acceptable, but we may need a different spur level chart (Fig. 7.3) for this usage. Regardless of its label on the physical device, we will still call the input port the RF port in our discussions.

7.1.5 Calculators

Appendix C describes two calculators that can be helpful in computing frequency ranges in receiver and synthesizer conversions.

7.1.6 Design Methods

The design method for frequency conversion that we will discuss uses a two-dimensional picture of the spurious products in the frequency regions of interest.

On this we superimpose a representation of the passband, the range of frequencies that our design must pass. An important feature of this representation is that it allows us to picture the entire design at once, rather than observing the results of stepping one or more parameters through its range of interest.

However, there are, in general, three frequencies of importance, the LO, the RF, and IF. The application of the two-dimensional representation is straightforward if one of these frequencies is fixed. Otherwise we must reduce a three-dimensional problem to a two-dimensional representation for visualization. We can do this by normalizing two of the frequencies to the third. This complicates the interpretation of the picture somewhat (although this can be mitigated by a computer aid) but still allows us to visualize the whole design.

Software that simulates testing of a converter design (e.g., Kyle, 1999; Wood, 2001b), perhaps permitting the specification of filter responses and mixer characteristics, may be initially easier to comprehend; it is closer to the designer's experience. However, its realism can be its downfall. Actual testing of converter performance, especially in the common situation where both RF and LO vary, can be a time-consuming process. (It is not unusual for designers to use spurious frequencies that are computed during design to guide their search during testing, at least initially.) Simulation can be faster than actual testing, but we still must investigate all of the combinations of these two variables, requiring that each be stepped in acceptably small increments. The method that we will use requires no stepping of variables; the variables are continuous. The entire design is visualized at once. More importantly, this allows us to more easily visualize alternatives.

Perhaps all design of this complexity involves trial and error, where a particular design is analyzed and then changed until the results of analysis are acceptable. Commonly the designer's imagination is involved in selecting alternatives to analyze, looking for the most satisfactory solution. The method that we will use seems better suited to this process than does simulation. We may find simulation satisfying as a check on the final design and for optimizing parameters (e.g., filter characteristics), particularly for multiple (series) conversions. Even there, we must deal with the fact that a simulation employs one set of frequencies at a time.

7.1.7 Example

Appendix E gives an example of a frequency conversion with its desired and spurious responses and illustrates the method used for analysis and visualization. The reader can refer to it at any point to clarify the processes.

7.2 SPURIOUS LEVELS[5]

We will first look at the levels to be expected from undesired signals and then at their frequencies.

7.2.1 Dependence on Signal Strength

We have seen that the DC term in Eq. (4.8) results in frequencies associated with a nonlinearity of order k being produced by all of the terms of order equal to k, or higher than k by some multiple of 2. Thus a spur of frequency

$$f = nf_a + mf_b, \tag{7.3}$$

where

$$|n| + |m| = k, \tag{7.4}$$

can be produced by the nonlinearity of order $k + 2i$, where i is zero or any positive integer. [Equation (7.3) is the same relationship that is expressed by Eq. (7.2).] The spur amplitude produced by that nonlinearity would be proportional to

$$A^{|n|}B^{|m|}(A^2 + B^2)^i. \tag{7.5}$$

In the case where f_b is the LO frequency, the LO amplitude B is much greater than the RF amplitude A. Therefore,

$$A^2 + B^2 \approx B^2, \tag{7.6}$$

and the amplitude from Eq. (7.5) becomes

$$A^{|n|}B^{|m|+2i}. \tag{7.7}$$

Thus the general form of a spur is

$$v_{|n||m|} = \left(\sum_{i=1}^{\infty} c_{|n||m|i}A^{|n|}B^{|m|+2i}\right)\cos[n\varphi_a(t) + m\varphi_b(t)] \tag{7.8}$$

$$= d_{|n||m|}A^{|n|}\cos[n\varphi_a(t) + m\varphi_b(t)], \tag{7.9}$$

where $d_{|n||m|}$ is a constant for a given spur and LO level.

Because $A^2 \ll B^2$, there is only one power of A in this equation, but there are many powers of B, and B cannot be said to be small, so we are left to simply write that sum of powers (each multiplied by the appropriate value of c) as a constant, $d_{|n||m|}$. While this tells us nothing of the relationship between the strength of the m-by-n spur and the LO amplitude, it does tell us that the spur's amplitude is proportional to the $|n|$th power of the RF amplitude.

These equations apply to each diode in a balanced mixer. The signals in each diode differ in sign; in a doubly balanced mixer all four possible combinations of signs on the two signals (LO and RF) appear in the four diodes. The four diode signals are combined in such a manner that the RF and LO inputs are canceled at the output. In addition, all spurious responses, except those for odd m and n, are theoretically canceled. This trend can be seen in Fig. 7.3, especially for $n = 1$. Since the mixer spur levels are a sum of diode voltages such as in Eq. (7.9), they will have the same form.

7.2.2 Estimating Levels

We will find it convenient to consider the amplitude of the spur $v_{|m||n|}$ relative to the amplitude of the desired signal v_{11}, since this ratio $R_{|m||n|}$ does not change in linear components once the spur has been created (assuming flat frequency response and no other spurs created at the same frequency). Moreover, this is also the equivalent ratio of the spur-to-signal amplitudes preceding the mixer, that is, this is the amplitude of an equivalent spurious input relative to the desired signal. Since the level of the signal at the output of the mixer is related to its level at the input by conversion loss, $1/g_{\text{mixer}}$, we can write, based on Eq. (7.9),

$$R_{|m||n|} \triangleq \frac{|v_{|m||n|}|}{|v_{11}|} \sim \frac{A^{|n|}}{|v_{11}|} = \frac{A^{|n|}}{g_{\text{mixer}} A} \sim A^{|n|-1}. \tag{7.10}$$

We will use this proportionality to predict the ratio of spur-to-signal amplitude at a given signal level from the ratio at some other signal level.

While we have established no theoretical basis for the dependence of spur amplitude on LO amplitude, Henderson (1993a) has found that the spur-to-signal amplitude ratio $R_{|m||n|}$, in doubly balanced diode mixers, tends to be given by[6]

$$R_{|m||n|} \sim (A/B)^{|n|-1}. \tag{7.11}$$

Note that the value of m does not enter into this expression. We can express this relationship in dB as

$$(\Delta R_{|m||n|})_{\text{dB}} = (|n| - 1)[(\Delta A)_{\text{dB}} - (\Delta \text{B})_{\text{dB}}], \tag{7.12}$$

where $(\Delta R_{|m||n|})_{\text{dB}}$ is the change in spur-to-signal-level ratio resulting from a change in signal level $(\Delta A)_{\text{dB}}$ and a change in LO level $(\Delta B)_{\text{dB}}$, all in dB.

Thus we can predict changes in the spur-to-signal ratio as a function of signal amplitude for small enough signals based on theory, and we can estimate the effect of a change in LO strength based on observation. We would like the basic data to be as close to design values as practical in both amplitude and frequency. This is especially true for the LO signal strength since we lack a theoretical basis for predicting its effect. Fortunately, we have control over the LO levels, whereas the RF levels often vary over a wide range.

Figure 7.4 shows a spreadsheet that predicts the changes in spur levels based on this relationship. Data for mixer A in Fig. 7.3 has been entered in the upper table along with the LO and RF levels that occurred during their measurement. LO and RF levels in our system are entered in the lower part. Based on all of that information, relative (to signal) spur levels are displayed in the bottom part of the figure. A minus is understood for all of the relative spur levels and $>x$ means that the spur is at least x below the signal and is, therefore, at a relative level of $< -x$. This dependence of spur levels on signal and LO levels influences the choice of mixers and of LO power and the distribution of gain in a cascade.

Spur levels vary from unit to unit, so design margins are required. They vary with terminations, so broadband terminations at the design impedance are usually

Given Data

RF: -10 dBm
LO: 7 dBm

n (RF mult.) \ m (LO multiple)	0	1	2	3	4	5	6	7	8
0		26	35	39	50	41	53	49	51
1	24	0	35	13	40	24	45	28	49
2	73	73	74	70	71	64	69	64	69
3	67	64	69	50	77	47	74	44	74
4	86	> 90	86	88	88	85	86	85	> 90
5	> 90	80	> 90	71	> 90	68	> 90	65	88
6	> 90	> 90	> 90	> 90	> 90	> 90	> 90	> 90	> 90
7	> 90	> 90	> 90	> 90	> 90	87	> 90	> 90	> 90
8	?	?	?	?	?	?	?	?	?

Derived

RF: -20 dBm
LO: 10 dBm

n (RF mult.) \ m (LO multiple)	0	1	2	3	4	5	6	7	8
0		13	22	26	37	28	40	36	38
1	24	0	35	13	40	24	45	28	49
2	86	86	87	83	84	77	82	77	82
3	93	90	95	76	103	73	100	70	100
4	125	> 129	125	127	127	124	125	124	> 129
5	> 142	132	> 142	123	> 142	120	> 142	117	140
6	> 155	> 155	> 155	> 155	> 155	> 155	> 155	> 155	> 155
7	> 168	> 168	> 168	> 168	> 168	165	> 168	> 168	> 168
8	?	?	?	?	?	?	?	?	?

Fig. 7.4 Levels of spurs relative to signal (minus understood) for given LO and RF levels. The upper table is measured data and the lower table estimates values with the RF and LO levels given there.

important to reproducing results obtained during characterization. They can also vary with frequency so we should try to obtain characterizations at frequencies close to those in the intended operations. Further, as we shall see, the predicted dependence on RF level can be inaccurate if the signal is too strong.

Broadband terminations are important because the mixer performance is influenced by impedances seen by spurious responses as well as by the desired responses. Maas (1993, pp. 188–189) indicates that reactive out-of-band terminations at the IF port of a DBM (Fig. 7.1) can change spur and IM levels by as much as ±20 dB, while such mismatches on the LO port can account for ±10 dB. Only a 1- or 2-dB effect is expected from such mismatches at the RF input port.

Even-order terms in the signal or signals that are balanced tend to cancel (Henderson, 1993c, pp. 482–483). In a DBM we therefore expect spurs with m or n even to be small compared to odd spurs and spurs with both m and n even to be even smaller. This is commonly observed to be true (McClaning and Vito, 2000, p. 306). The trend can be seen in Fig. 7.3 along with the decrease in level at higher orders and the particularly high level of $m \times 1$ spurs. Since the unbalanced IF port in a DBM is usually rated lower in frequency than the other two ports, it is sometimes used as an input port for upconversion (unlike the configuration of Fig. 7.1). This can change the spur levels. Lacking a separate chart for this configuration, Henderson (1993a) recommends increasing by 10 dB the estimated levels of spurs that are both of odd order in the low-frequency signal that enters the IF port and of even order in the other input.

7.2.3 Strategy for Using Levels

Our goal will be to limit the maximum spur level that is produced for a given range of possible input signal levels. This range will include the maximum levels of undesired signals and possibly of the desired signal, if its spurs can be a problem. The maximum spur level in a synthesizer is set by spectral purity requirements. In a receiver, it may be set below the minimum desired signal by some required signal-to-interference ratio or, if we are concerned about misidentifying received signals, it might be related to a detection threshold or the noise level. As noted above, it is helpful to deal with *relative* spur levels, how far the spurs are below the desired 1×1 product. Relative spur levels can be improved by reducing signal strength as long as n exceeds 1. The greater the value of n, the faster the spur level changes with signal strength. Thus, if we use operating regions where n is large, we can more effectively control the relative spur level by the strength of the RF signal at the mixer input. However, noise figure is degraded when the signal strength is lowered at the input to a mixer, so compromise is required.

Example 7.1 Spur Levels The strongest signal to be received is -15 dBm, and the weakest desired signal will be -80 dBm. We require a 10-dB signal-to-spur ratio so the strongest allowed spur, referred to the input, is -90 dBm — 10 dB below the weak signal and 75 dB below the strong signal. Therefore we require the relative spur amplitude to be -75 dB with an RF level of -15 dBm. We consider an operating region in which an $|m| \times |n| = 2 \times 3$ spur is present, and the upper table in Fig. 7.4 applies to our mixer. (Therefore, the -15-dBm received input must have been amplified by 5 dB before the mixer so its level can be -10 dBm, for which the table applies, at the mixer input.) The relative level of the 2×3, according to the table, is -69 dBc, 6 dB larger than allowed. We know it will decrease by $(n - 1 =)$ 2 dB for each dB decrease in the signal strength, so the signal at the mixer input must be reduced by (6 dB/2 =) 3 dB relative to the -10 dBm for which the table was made, giving -13 dBm maximum input to the mixer. (For clarity, we are not including design margins here.) This means we are only allowed 2 dB of net gain preceding the mixer, and the gain to the mixer output will be a loss, not good for noise figure. We might seek a more spur-free operating region or one where the spurs are weaker or we might find another mixer with better performance for the spur of concern. We might also find a mixer designed for a higher LO power. If the spur had $n = 1$, we could not have improved its relative level by changing the signal strength.

7.3 TWO-SIGNAL IMs

In Chapter 4 we studied the production of the intermodulation products (including harmonics) of two signals in a module, and we have just studied the special case where one of these signals, the LO, was much larger than the other. Now we look at what might be considered a combination of these two cases, the production of IMs in a mixer (Cheadle, 1993, pp. 489–494). To a large degree, the mixer acts like other modules except that it changes the frequency of the signals that pass through it. As in the case of other modules, it needs to be characterized for IMs so we can determine what spurious products will be generated from the interaction of two signals that pass through it. These are not products that are created by interaction between the LO and the signals—we intend to control those products so they do not create significant problems. Here we are concerned with the interaction between two converted signals. In the absence of specific characterization for IMs, we can make use of a theoretical relationship between the mixer spur products and these IMs, which is due to the fact that they are all based on the same nonlinear coefficients. The disadvantage of using spur-level tables to find IM levels is due to the possible frequency dependence of these products, which can cause spurs and IMs that are based on the same nonlinearity to not be related as expected when their frequencies are significantly different. Nonetheless, in the absence of more specific data, it is worth understanding what information about IM levels is contained in the spur-level table. We show, in Appendix P, that the ratio r of the amplitude of the largest nth-order IM, resulting from two signals of equal amplitude, to the amplitude of either $1 \times n$ spur (which has order $n+1$) is given by

$$r = c[n, \text{int}(n/2)], \tag{7.13}$$

where c is the binomial coefficient and int(x) is the integer part of x. For $n = 2$, these are IMs c and e in Fig. 4.2 and, for $n = 3$, they are IMs c, d, f, and g in Fig. 4.6, while the harmonics in these figures correspond to (single-frequency)

TABLE 7.1 Ratio (r) of Largest IM to Mixer Spur

IM order n	n for spur	IM-to-spur ratio, r	
2	2	2	6.0 dB
3	3	3	9.5 dB
4	4	6	15.6 dB
5	5	10	20.0 dB
6	6	20	26.0 dB
7	7	35	30.9 dB
8	8	70	36.9 dB
9	9	126	42.0 dB
10	10	252	48.0 dB

mixer spurs. In Fig. 4.6, the typically large separation between the important IMs, at c or d, and the harmonics, at e or h, illustrates the danger that frequency response will alter the theoretical relationship between the two. The values for r in Eq. (7.13) are shown in Table 7.1.

Intercept points can be computed, as in Chapter 4, once the IM levels have been determined for a given signal level.

7.4 POWER RANGE FOR PREDICTABLE LEVELS

Figure 7.5 shows output IM3 levels plotted against input power in each of two equal tones. Curves are plotted for the Class 1 and the Class 3 mixer types of Fig. 7.3. If we base the IP3 on some output level P_x taken in the nonlinear regions, all predicted IM levels in the linear region (i.e., where the IM power is proportional to input power in dB) will be in error by the vertical offset between P_x and the linear extension from the low-power region. For example, the data point for the Class 3 mixer at +10 dBm input power would lead to estimated low-level IMs that are 13 dB low.

The maximum input levels for which the theoretical relationship holds have been given as −20, −10, and 0 dBm for Class 1, Class 2, and Class 3 mixers, respectively (Cheadle, 1993, p. 490).

Since the IM level is closely related to a corresponding spur level, we would expect that the 1 × 3 spur level would not follow the theoretical relationship to input power above these levels either. One way to gain confidence that we are in the linear range is to compare measured spur levels at one RF input level to those predicted from measurements at another level. This is done in Fig. 7.6. Note the large errors for the Class 1 mixer especially,[7] not surprising in light of the top of the linear range for the third-order IMs given above. We will usually want to use the spur level for the lower of the two RF levels unless the IMs are only measurable at the higher level (or if the higher level is closer to the design value). As we progress in our design and narrow down the mixer that will be used, measurements on a number of mixers of those types may be warranted. This would provide an opportunity for using the expected frequency ranges and terminations also.

Example 7.2 Mixer IM We will compare the reported IP3 for three mixers to the levels that we compute from their 1 × 3 spurs, which are shown in Fig. 7.3. We begin with the M9E Class 3 mixer with 27 dBm LO power.

With 0 dBm RF input level, the relative level of the 1 × 3 spur from Fig. 7.3 is −73 dBc. With two input signals at 0 dBm, each would produce this spur level, but they would also produce close-in third-order IMs at a level 9.5 dB higher, according to Table 7.1. These IMs will appear near the converted signals at a relative level of (9.5 − 73 =) −63.5 dBc. The IIP3 will be higher than the signal level by (63.5/2 =) 31.8 dBc [Eq. (4.24) or Appendix H, Eq. (32)], so the input intercept point will be (0 dBm + 31.8 dBc =) **31.8 dBm**. The measured value is 32.5 dBm (Stellex Catalog, 1997, p. 467), within 0.7 dB of the estimated value.

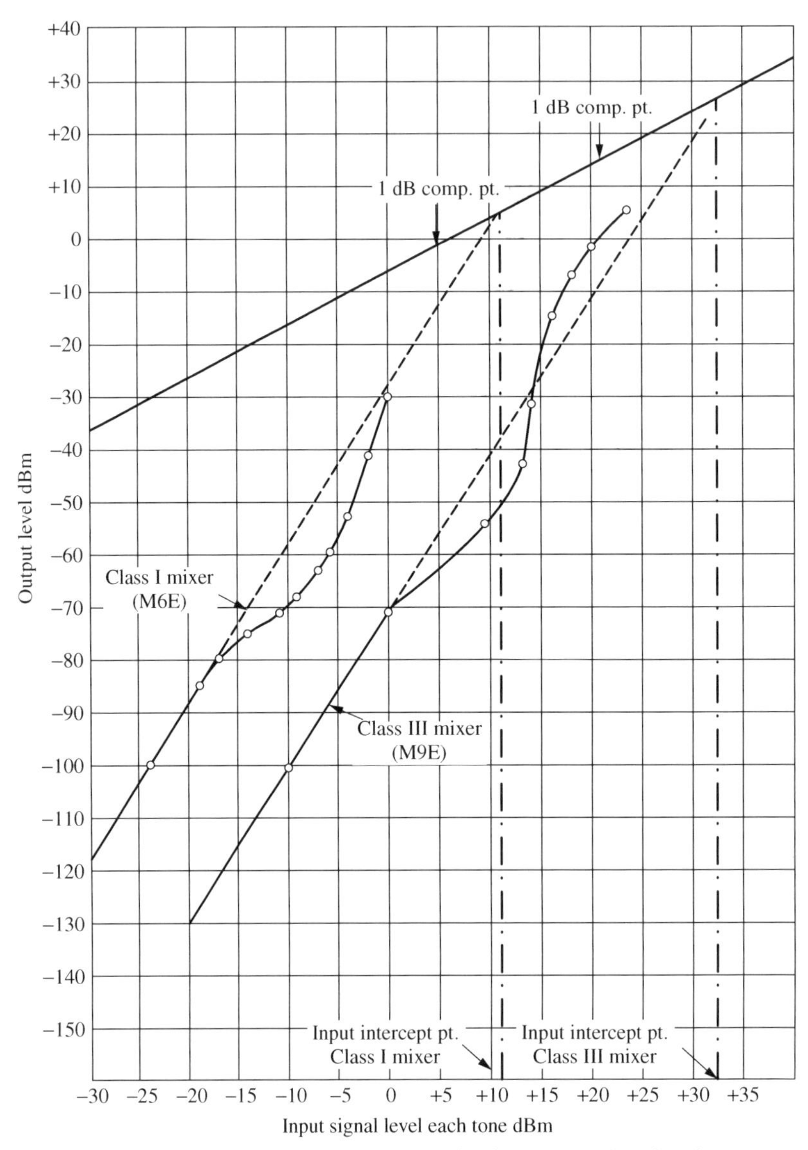

Fig. 7.5 IM3 output level for Class 1 and Class 3 mixers plotted against input power in each of two tones (Cheadle, 1993, p. 490).

Next we look at the M9BC Class 2 mixer with +17 dBm LO power. With −10 dBm RF input level, the relative level of the 1 × 3 spur is given (Fig. 7.3) as −77 dBc. With two input signals at −10 dBm, each would produce this level of 1 × 3 spurs plus close-in third-order IMs at a level 9.5 dB higher. These IMs will appear near the converted signals at a relative level of (9.5 − 77 =) −67.5 dBc. The intercept point will be higher than the signal level by (67.5/2 =) 33.8 dBc,

	0	1	2	3	4	5	6	7	8
7	139 > 159 > 159 > 90 > 90 > 90	129 139 > 159 > 90 > 90 > 90	140 > 159 > 159 > 90 > 90 > 90	134 138 > 159 > 90 > 90 > 90	143 > 159 > 159 > 90 > 90 > 90	123 138 > 159 87 > 90 > 90	138 > 159 > 159 > 90 > 90 > 90	120 141 > 159 > 90 > 90 > 90	131 150 > 159 > 90 > 90 > 90
6	140 > 149 > 149 > 90 > 90 > 90	136 > 149 > 149 > 90 > 90 > 90	141 > 149 > 149 > 90 > 90 > 90	141 > 149 147 > 90 > 90 > 90	140 > 149 > 149 > 90 > 90 > 90	134 > 149 > 149 > 90 > 90 > 90	143 > 149 > 149 > 90 > 90 > 90	134 > 149 > 149 > 90 > 90 > 90	138 > 149 > 148 > 90 > 90 > 90
5	112 133 > 139 > 90 > 90 > 90	110 113 136 30 80 > 90 > 90	111 127 > 139 > 90 > 90 > 90	92 112 135 21 71 > 90 > 90	117 128 >139 > 90 > 90 > 90	86 106 > 139 18 68 > 90 > 90	115 125 > 139 > 90 > 90 > 90	85 104 130 20 65 > 90 > 90	113 122 > 139 25 88 > 90 > 90
4	110 126 118 24 86 > 90 > 90	109 110 121 > 90 > 90 > 90	112 126 > 129 26 86 > 90 > 90	107 110 122 19 88 > 90 > 90	112 125 120 24 88 > 90 > 90	106 112 125 21 85 > 90 > 90	107 128 117 21 86 > 90 > 90	102 108 124 17 85 > 90 > 90	107 120 117 17 > 90 > 90 > 90
3	71 83 101 4 −4 67 87 > 90	65 78 93 5 1 64 77 > 90	73 85 105 4 −2 69 87 > 90	71 80 89 21 2 50 78 > 90	75 85 105 −2 77 > 90 > 90	68 75 88 21 0 47 75 > 90	74 84 105 0 −1 74 85 > 90	73 74 84 29 −3 44 77 > 90	78 86 107 4 −2 74 88 > 90
2	79 78 74 6 −8 1 73 86 73	82 77 81 9 2 −2 73 75 83	89 86 72 15 2 −3 74 84 75	77 77 80 7 2 1 70 75 79	85 90 73 14 4 −7 71 86 80	76 76 80 12 2 0 64 74 80	82 92 71 13 5 −6 69 87 77	78 76 72 14 2 −10 64 74 82	85 93 74 16 9 −5 69 84 79
1	25 25 24 1 2 0 24 23 24	0 0 0 0 0 0 0 0 0	39 39 35 4 0 1 35 39 34	13 11 11 0 0 0 13 11 11	45 50 42 5 4 0 40 46 42	22 16 19 −2 2 1 24 14 18	54 59 50 9 −3 1 45 62 49	37 19 39 9 0 2 28 19 37	59 59 49 10 6 0 49 53 49
0	A B C	26 29 19 0 2 1 26 27 18	35 32 10 0 1 0 35 31 10	42 36 22 3 0 −1 39 36 23	53 48 14 3 1 0 50 47 14	35 27 19 −6 −9 0 41 36 19	50 55 17 −3 4 0 53 51 17	61 39 20 12 2 −1 49 37 21	54 65 19 3 2 0 51 63 19
	0	1	2	3	4	5	6	7	8

Predicted IM at −10 dBm RF
Difference = Predicted - Measured
Measured IM at −10 dBm RF

Fig. 7.6 Predicted and measured spur levels. The data in Fig. 7.3 at 0 dBm RF level is used to predict the level at −10 dBm by reducing it by $(n-1)$ 10 dB. This is shown in the upper row in each rectangle. The measured data at −10 dBm is shown in the bottom row (the same as in Fig. 7.3) and the difference between predicted and measured values is shown in the middle row. No values are shown where the predicted level is below the measurement limit (indicated by >).

so the input intercept point will be (−10 dBm + 33.8 dBc =) **23.8 dBm**. Repeating this process for a 0-dBm input level, for which the 1 × 3 spur is given as −58 dBc, we obtain an IIP3 of **24.3 dBm**. The measured IM3 level at 50 MHz for this mixer is −70 dBc with a −10 dBm RF input (Stellex Catalog, 1997, p. 467). The corresponding IIP3 would be (−10 dBm + 70 dB/2 =) **25 dBm**, within about 1 dB of the estimate from the spur levels.

However, we compute an IIP3 of **17.3 dBm** for the M1 Class 1 mixer, using spur data for −10 dBm RF input, whereas the IP3 given for that mixer is only **11.5 dBm** (Watkins-Johnson Catalog, 1993, p. 449), and the value implied from data for low-level mixers, such as this, in general is **15.5 dBm** (Stellex Catalog, 1997, p. 467). The disagreement is even greater if we use spur data for 0-dBm RF input. This should not be too surprising since the RF levels exceed the −20 dBm maximum given for linear IM response for Class 1 mixers (although the error is in the opposite direction of that implied by Fig. 7.5).

7.5 SPUR PLOT, LO REFERENCE

We would like a plot that shows all of the spurious frequencies so we can superimpose a representation of our passbands and see if the spurs fall within them. Spurious frequencies occur when a frequency implied by Eq. (7.9),

$$f_I = m f_L + n f_R, \tag{7.2}$$

is in the IF band. Here f_I is the IF, f_R is the RF contained in $\varphi_a(t)$, and f_L is the LO frequency in $\varphi_b(t)$. We want a plot of Eq. (7.2) for the various combinations of m and n, but there are too many variables for a two-dimensional plot; we must eliminate one of them. One possibility is to fix f_L. This will be particularly useful for conversions where the LO *is* fixed, nontunable frequency-band converters. In this case we can plot f_I against f_R for a fixed f_L and various m and n. Alternately, we can normalize to f_L, plotting f_I/f_L versus f_R/f_L for various m and n:

$$f_I/f_L = m + n f_R/f_L. \tag{7.14}$$

This normalized version is most useful for making a plot that can be used for different projects. We could carefully plot these curves, label each with m and n, and use a copy of the plot for any project. We can also create a spreadsheet to give this plot, as illustrated by Fig. 7.7, which represents data on an associated spreadsheet.

7.5.1 Spreadsheet Plot Description

In Fig. 7.7, the LO has the value 5.5. We can use this to represent 5.5 GHz or 5.5 kHz. The units are arbitrary, but the same units apply to all of the numbers,

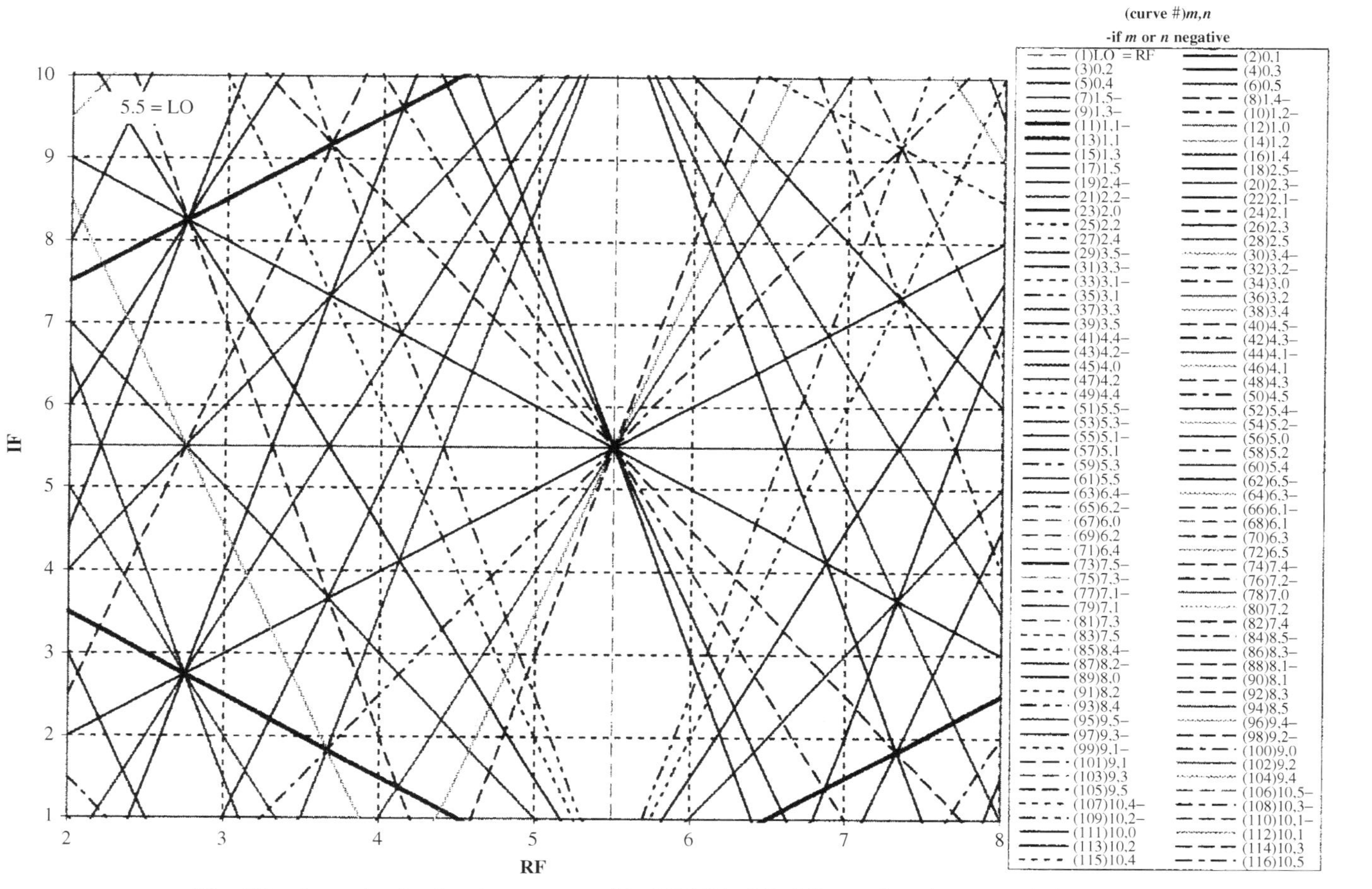

Fig. 7.7 Spur plot for band converter with 5.5 MHz LO. Minus after the curve designation indicates that either m or n is negative.

LO, RF, and IF. This spreadsheet is done for $0 \le m \le 10$ and $0 \le n \le 5$. Some spur plots and their accompanying spreadsheets are designed to provide 116 curves (Fig. 7.7), while others provide only 61. The spreadsheet is designed so a high maximum m can be easily exchanged for high maximum n within these limits. While 61 curves can provide a clearer presentation, the larger number may be needed in practice because, as can be seen from Fig. 7.3, spur levels do not fall very fast with m.

The spurs are listed in the legend to the right in Fig. 7.7, each spur having its curve number in parentheses and its values of $|m|$, $|n|$. Curves are color coded in the operating spreadsheet, and touching a line with the cursor causes the legend information for that curve to be displayed. Clicking on a line causes the line equation, written in terms of cell coordinates and ending in the curve number, to appear at the top of the window.

The heavy lines are $|m| \times |n| = 1 \times 1$ products. One of them normally represents the desired IF. The upper 1×1 represents upconversion, where the IF is the sum of the RF and LO frequencies. The lower-right heavy curve represents low-side downconversion, where the LO is below the RF. The lower-left heavy curve represent high-side downconversion where the LO is above the RF and the IF; here $n = -1$ in Eq. (7.2), causing spectral inversion. By this we mean that increasing RF frequencies cause decreasing IF frequencies. Thus, if signal a has a higher frequency than signal b at the RF port, a will have a lower frequency than b at the IF port.

Crossovers, where spur curves cross these heavy curves, are listed in Appendix X. The frequency ratios, labeled as RF/LO, there can be multiplied by the LO frequency to give the RF at these crossovers. (We will sometimes use R, L, and I to represent the three mixer ports and sometimes use RF, LO, and IF.)

7.5.2 Example of a Band Conversion

Example 7.3 Let us represent a high-side downconversion from an RF band extending from 4 to 4.5 MHz using this plot. (The LO frequency is still 5.5 MHz.) The representation is shown in Fig. 7.8, where we have changed the RF range on the spreadsheet and the display limits on the graph to concentrate around this area. We have drawn a "rectangle," extending from 4 to 4.5 MHz on the RF axis, with corners on the 1×-1 curve. This represents the minimum RF and minimum IF band to accomplish the desired conversion, which can be seen to be a conversion to an IF band from 1 to 1.5 MHz. This, of course, also corresponds to Eq. (7.1). Now we see, by touching the lines that go through the conversion region represented by the rectangle, that the spurs that will occur in band are, from left to right at the top of the rectangle, numbers 40, 20, and 30. From the legend (or the display by the cursor), these are ($m \times n =$) 4×-5, -2×3, and -3×4 spurs. (However, the legend and cursor display do not indicate to which of the two numbers the minus sign belongs. We have assigned it to the number that results in IF > 0.) If the mixer should have the characteristics of the

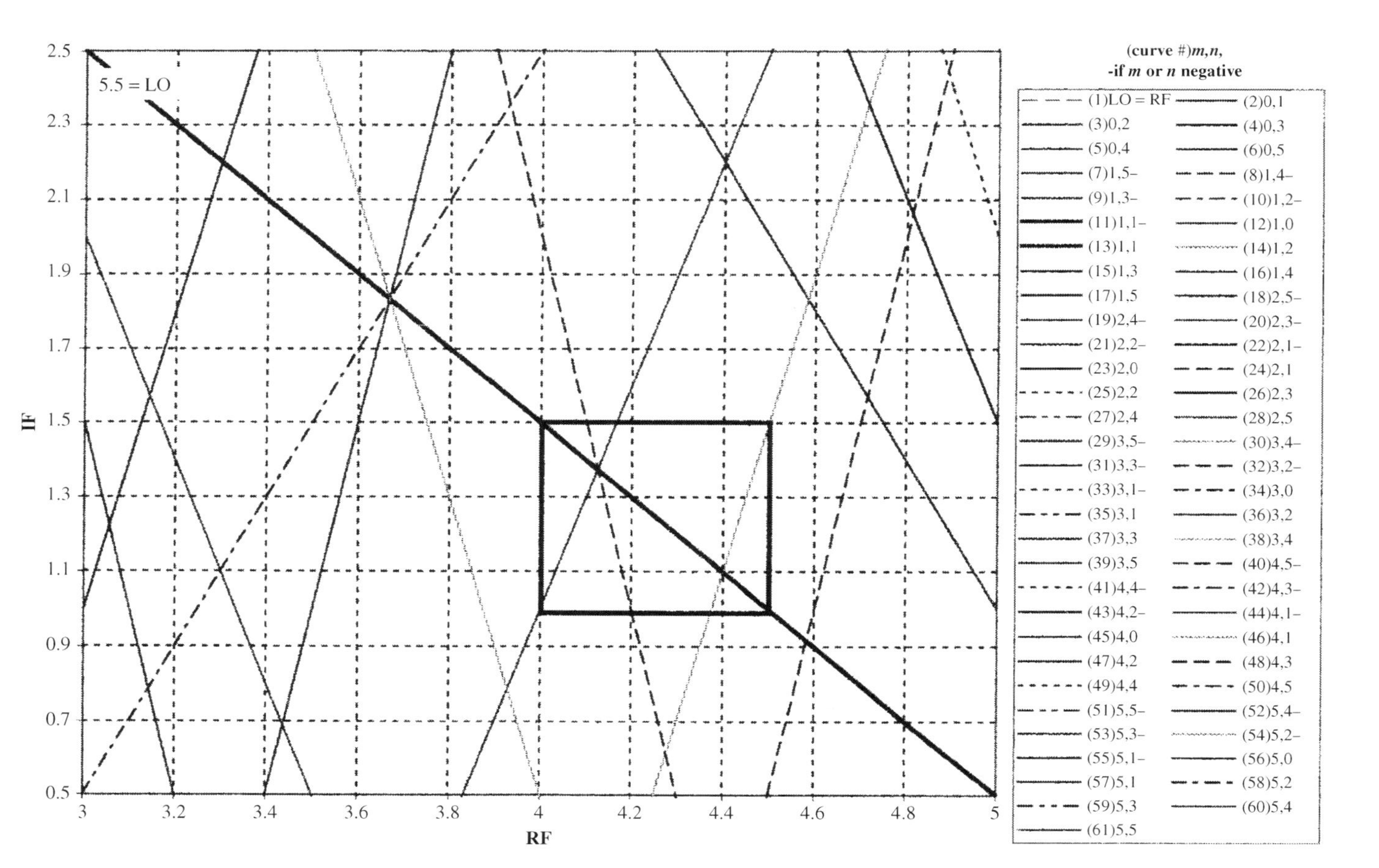

Fig. 7.8 Spur plot for band converter, fixed LO.

mixer represented by Fig. 7.4, and if the LO and RF levels should be as given in the upper table there, the spur-to-signal ratios for these would be < -90 dB, -69 dB, and -88 dB, respectively. Most of the nearby out-of-band spurs have the same orders, which becomes apparent when they are selected (and viewed in color). The closest new spur (i.e., not with the same m and n as an in-band spur) is at RF equal to 4.75 when the IF is 1.5. This is 0.5 from the RF band center. Since the RF bandwidth also equals 0.5, the RF filter shape factor at that point is

$$\mathrm{SF} = \mathrm{BW}_{\mathrm{spur}}/\mathrm{BW}_{\mathrm{pass}} = (2 \times 0.5)/0.5 = 2. \tag{7.15}$$

Here $\mathrm{BW}_{\mathrm{spur}}$ is twice the separation of the spur from the filter center and $\mathrm{BW}_{\mathrm{pass}}$ is the filter passband width. Whatever attenuation is required from the filter would be required at that SF. However, this is curve 21, a 2×2 spur, which Fig. 7.4 shows to be 74 dB below the signal, lower than one of the in-band spurs, so we will not improve the worst-case signal-to-spur ratio by reducing it.

7.5.3 Other Information on the Plot

The vertical dashed line in Fig. 7.7, where the RF equals the LO (equals 5.5), is not a spur in the same sense as the others. It represents potential LO leakage out the RF port and through the RF filter. This can be a significant problem in some designs so the line provides a warning if it is in or near the conversion rectangle. The horizontal line at IF = 5.5 is curve 12, representing leakage of the LO into the IF, another strong signal to be avoided in or near the operating region. Its level equals the LO power reduced by the LO-to-IF isolation. This gives an IF power level (dBm), not a level relative to the signal (dB).

Example 7.4 Relative Level of LO Leakage For a mixer, LO-to-IF isolation is 30 dB. Conversion loss is 8 dB. LO level is +7 dBm and signal level is −20 dBm. The LO strength in the IF is

$$P_{\mathrm{LO\text{-}in\text{-}IF}} = +7\ \mathrm{dBm} - 30\ \mathrm{dB} = -23\ \mathrm{dBm}. \tag{7.16}$$

The signal level there is

$$P_{\mathrm{signal\text{-}in\text{-}IF}} = -20\ \mathrm{dBm} - 8\ \mathrm{dB} = -28\ \mathrm{dBm}. \tag{7.17}$$

The relative level of the undesired product is

$$R = P_{\mathrm{LO\text{-}in\text{-}IF}} - P_{\mathrm{signal\text{-}in\text{-}IF}} = -23\ \mathrm{dBm} + 28\ \mathrm{dBm} = 5\ \mathrm{dB}. \tag{7.18}$$

So the LO provides a very strong undesired signal. Good designs usually make this relatively easy to filter.

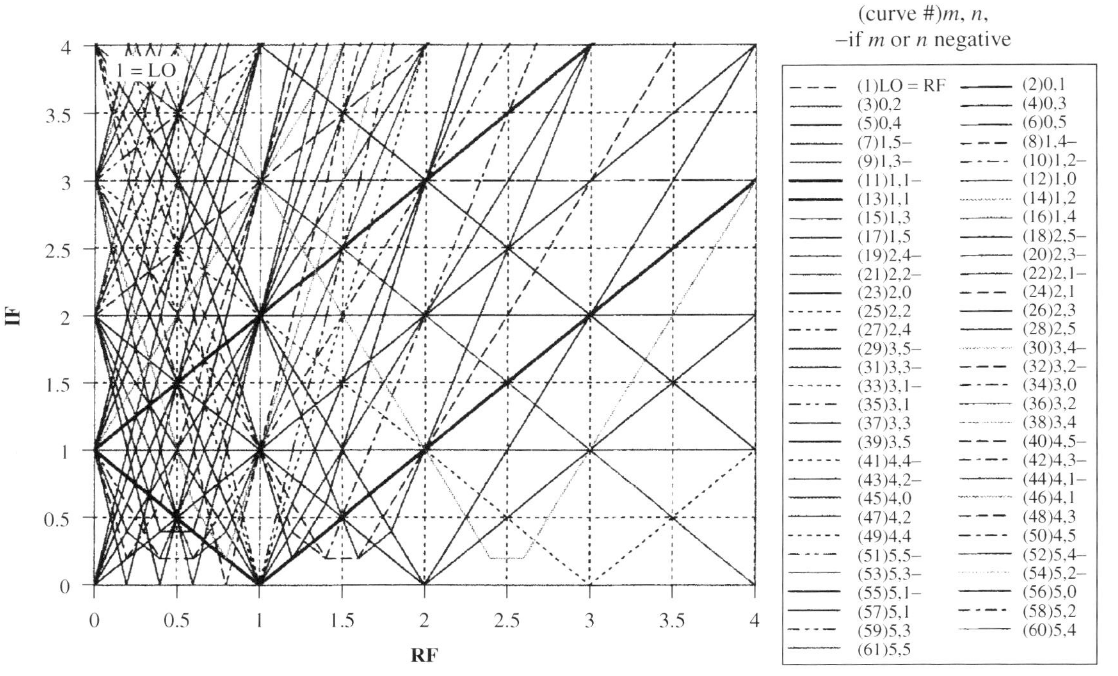

Fig. 7.9 Linear spur plot normalized to LO. Curves are distorted below IF = 0.25 because of the limited number of plotted points.

If we were preparing a plot for general use, we would write the spur orders (m and n) on the curves and normalize to an LO frequency of 1, which we can easily do by selecting that value in this spreadsheet. Figure 7.9 shows a normalized linear plot. It also illustrates a spreadsheet problem in the region below IF = 0.5 (for this particular plot). Because no point happens to be plotted where IF = 0 for some curves, they become distorted at low values of IF; points either side of the true minimum are connected without going through the minimum. As used here, the plotted points were automatically distributed evenly between the minimum and maximum specified values on the spreadsheet. The spacing is 0.2, so points at multiples of 0.5 are missed. The problem will be reduced if smaller regions of RF are plotted. The required points can also be entered into the spreadsheet or more points can be used. The use of this graph is not restricted to fixed LOs. We can represent an LO range on the normalized graph. We will treat this topic in the next section.

7.6 SPUR PLOT, IF REFERENCE

From here we will use a spur plot for a fixed IF (rather than a fixed LO), possibly normalized to the IF. Such plots are shown in Figs. 7.10 and 7.11, the latter being a logarithmic plot. (These are 61-curve plots, but 116-curve plots are available in the workbook that contains these plots.) The version of Eq. (7.2) that we plot now is

$$f_L = (f_I - nf_R)/m \tag{7.19}$$

with f_I fixed. The version normalized to f_I is obtained by dividing by f_I:

$$f_L/f_I = (1 - nf_R/f_I)/m, \tag{7.20}$$

but that plot can also be obtained by setting $f_I = 1$. Then the axes are understood to be f_L/f_I and f_R/f_I. Note that the heavy curve with the negative slope (part of curve 8) represents upconversion, $f_I = f_R + f_L$. The rest of that curve, with the positive slope at the lower right, represents low-side downconversion, $f_I = f_R - f_L$. Heavy curve 6, with the positive slope at the top, represents high-side downconversion, $f_I = f_L - f_R$. The ratios R/I, from Appendix X, can be multiplied by the IF to find RFs at the crossovers.

Example 7.5 Conversion to a Single IF Suppose we wish to convert a band from 4.8 to 5.6 GHz to a narrow band at 2 GHz. We will approximate the IF bandwidth as zero. This problem fits well our fixed IF value. Figure 7.12 shows the *normalized* plot for such a condition; RF (4.8–5.6 GHz) and LO (6.8–7.6 GHz) frequencies are divided by IF = 2 GHz. Figure 7.13 shows essentially the same plot with spurs and their levels, from the lower table in Fig. 7.4, labeled. Looking at Fig. 7.4, we can see that spur levels do not fall off with increasing m as they do with increasing n. For that reason, we are interested in higher LO multiples, even though no spur-level information is available for $m > 8$. Fortunately, we

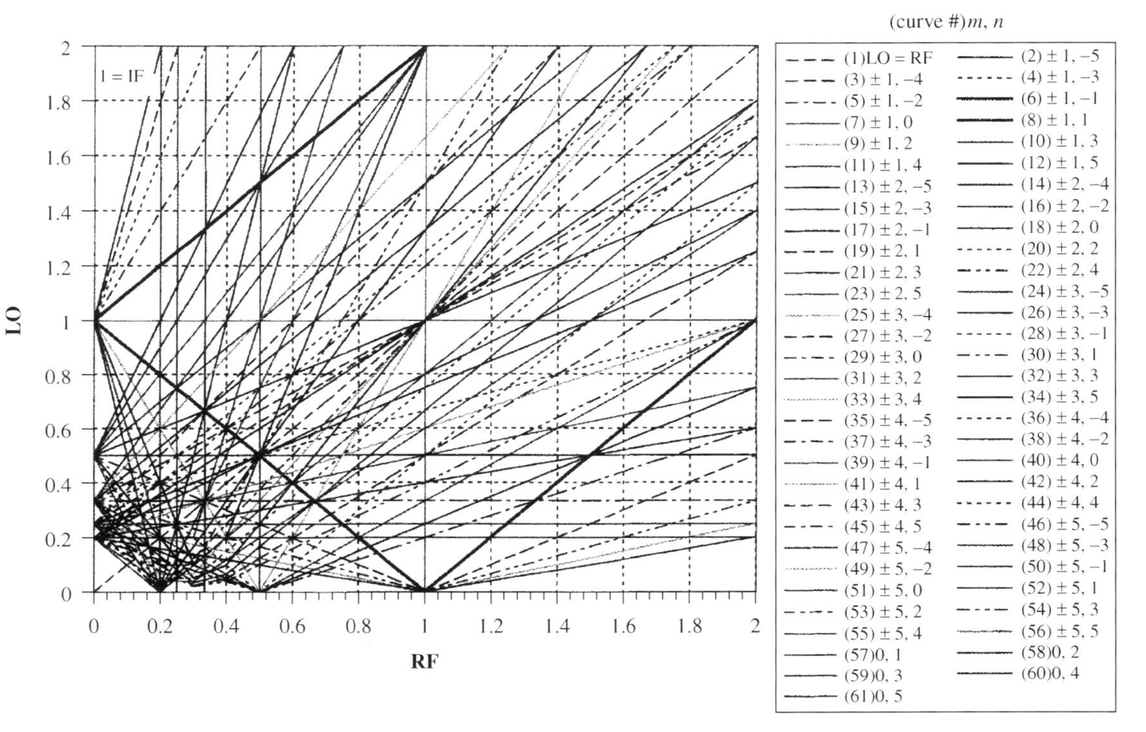

Fig. 7.10 Linear spur plot normalized to IF.

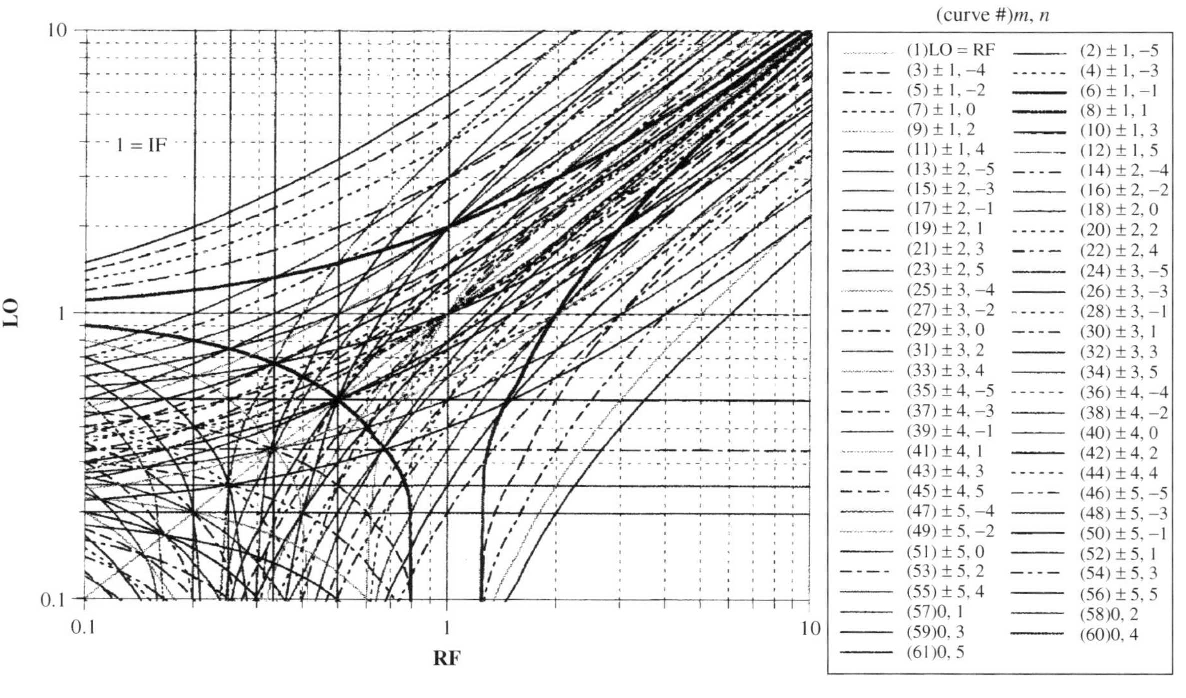

Fig. 7.11 Log spur plot normalized to IF.

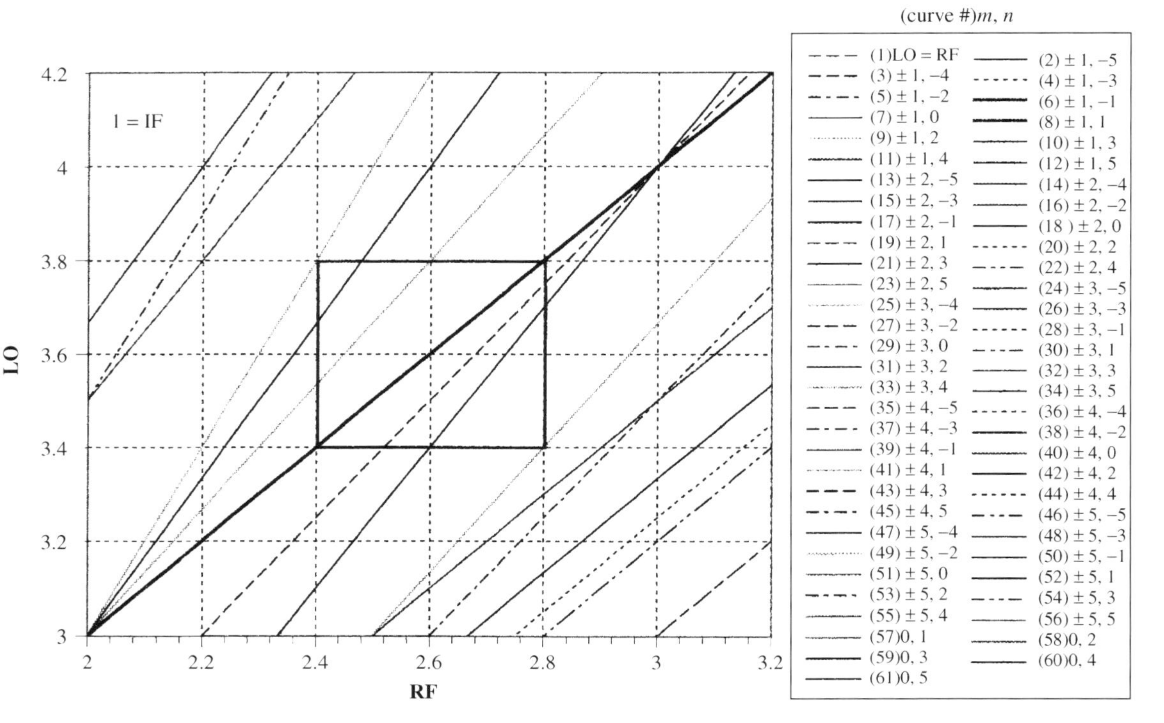

Fig. 7.12 Conversion of 4.8–5.6 GHz to 2 GHz, high-side LO.

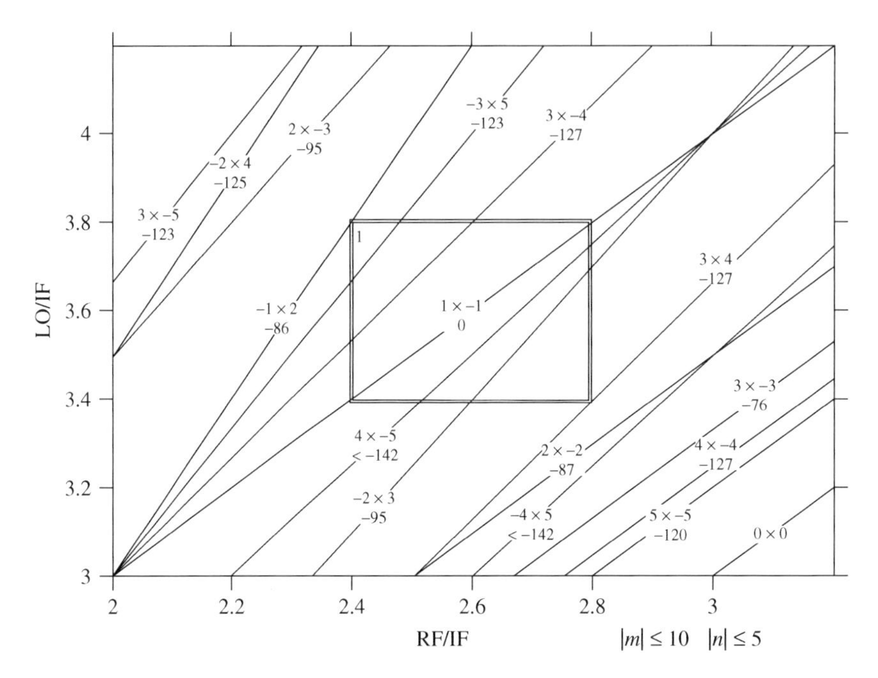

Fig. 7.13 Conversion with spur levels labeled.

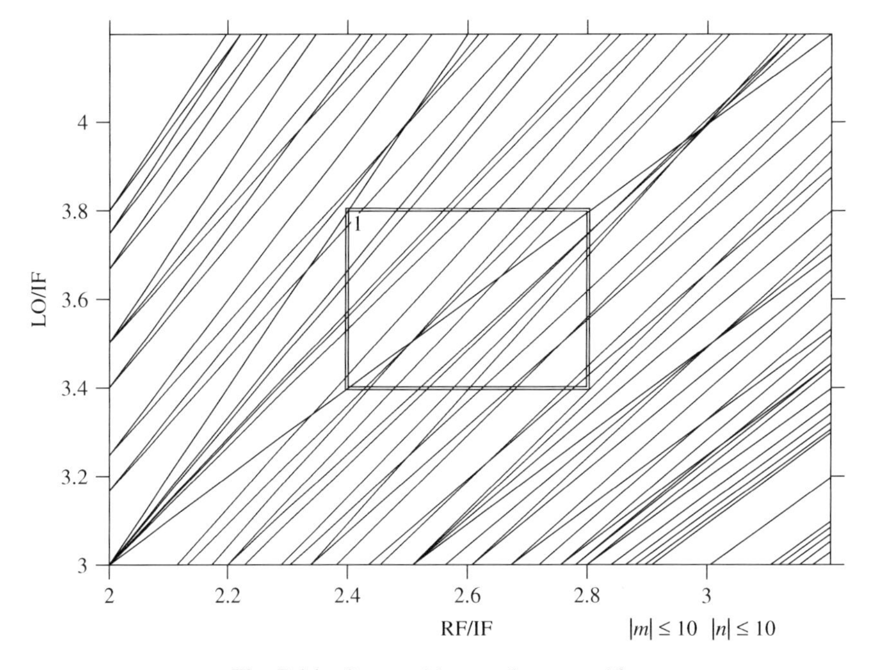

Fig. 7.14 Spurs with m and n up to 10.

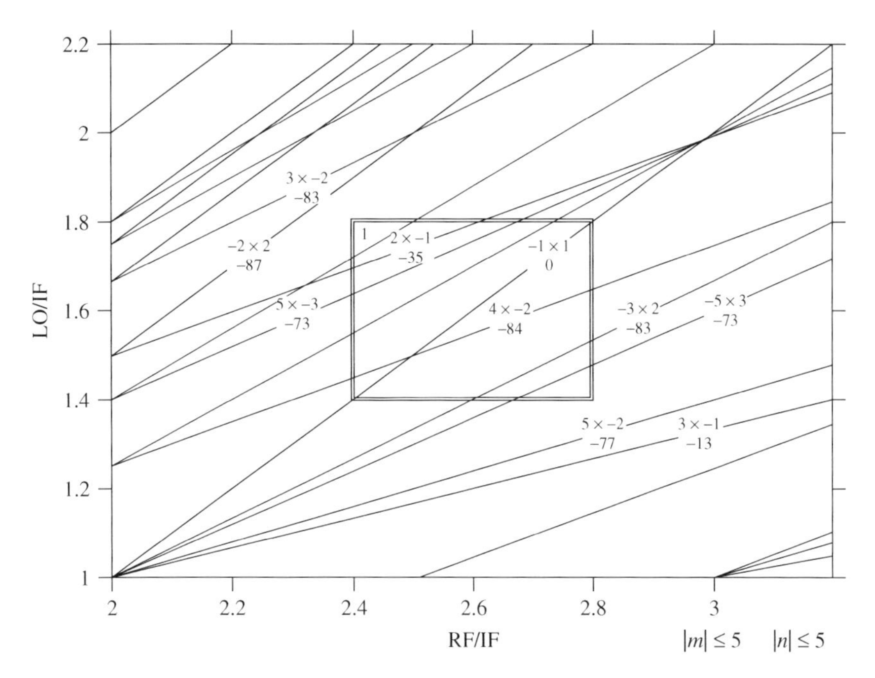

Fig. 7.15 Low-side downconversion.

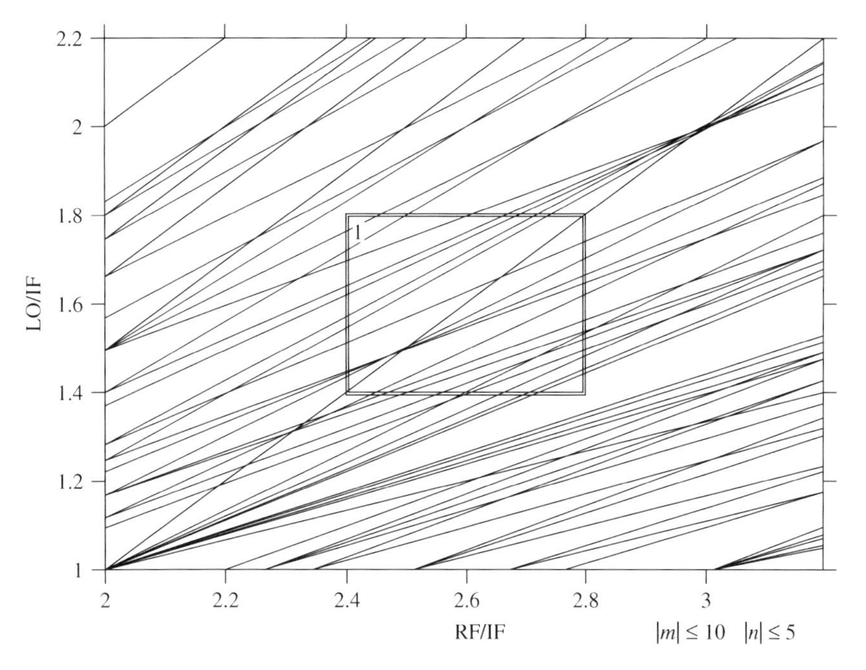

Fig. 7.16 Low-side downconversion with m up to 10 but n only up to 5.

find that no spurs with $m > 5$ appear in Fig. 7.13. Increasing both m and n to 10 does produce additional spurs, as is evident in Fig. 7.14—apparently spurs will not occur in this region if there is too much difference between the values of m and n—but we know, from Fig. 7.4, that the higher levels of n tend to produce weak spurs.

High-side downconversion (LO > RF > IF) is usually preferable to low-side downconversion (RF > LO > IF). Let us look at the graph for the latter to see if the reason might be apparent. Figure 7.15 shows the same RF-to-IF conversion using a low LO. The spurs are generally larger, especially the very large 2×1 that appears in band. Moreover, if we look at m up to 10 with n still only as high as 5, we get Fig. 7.16, so we can expect many higher-order spurs with low values of n, and therefore at high levels. The advantages of high-side over low-side downconversion are discussed further in Section 7.9.3.

If the IF varies, in a plot that is normalized to the IF, the conversion rectangle will move diagonally because both axes are normalized to the IF.

Example 7.6 Conversion to an IF Range Figure 7.17 shows the same LO range as in Fig. 7.13, but the 2-GHz IF has been changed into a range from 1.9 to 2.1 GHz. The conversion rectangles at the ends of this range are shown in the figure, where they are interconnected to form a conversion "polygon" that shows the path along which the rectangle moves as the IF changes. (These lines meet at the origin since both coordinates are divided by an infinite IF at that extreme.) The RF bands have been widened by ±0.1 GHz (to 4.7–5.7 GHz)

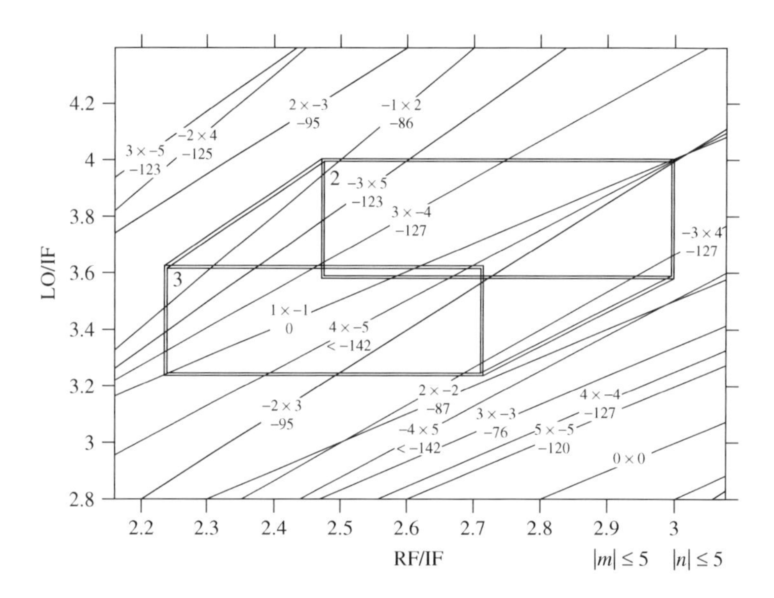

Fig. 7.17 Finite IF band, linear plot.

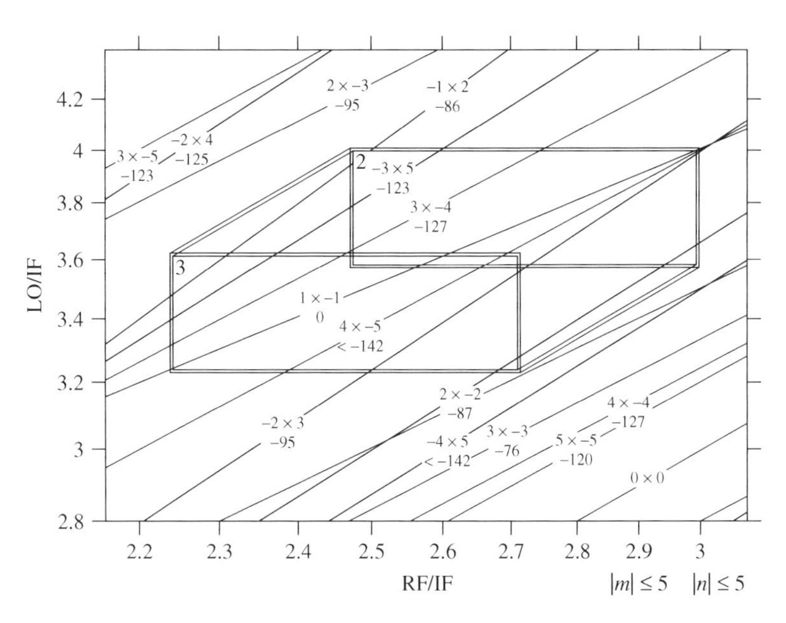

Fig. 7.18 Finite IF band, log plot.

also, to accommodate wider incoming signal bandwidths corresponding to the IF bandwidth. In a log plot (Fig. 7.18), the rectangle maintains its size as it moves with changing IF and the diagonal sides of the polygon are parallel.

While we found that the LO-referenced spreadsheet was particularly suited for band conversions, in which the LO is fixed, they can also be represented in a normalized IF-referenced plot.

Example 7.7 Band Converters Figure 7.19 shows what happens if we fix the LO in the center of the range it had in Fig. 7.18, at 7.2 GHz. The two rectangles have shrunk to single lines since the LO has only one value (the normalized LO has many values but that is a result of the changing IFs). Now we are only converting a 200-MHz band to the IF, however, whereas we had been able to receive a 1-GHz-wide band. (The 1×-1 curve extends from RF $= 5.1$, IF $= 2.1$, at the bottom of the polygon, to RF $= 5.3$, IF $= 1.9$, at the top.) To again receive the wider band with a fixed LO we must widen the IF (to 1.5–2.5 GHz). The result of the wider IF is illustrated in Fig. 7.20. The 1×-1 desired curve now goes corner to corner, indicating that the entire IF band is being used.

Appendix B summarizes the various shapes used to represent passbands with the IF-referenced spur plot and considers the representation of passbands and rejection bands in greater depth.

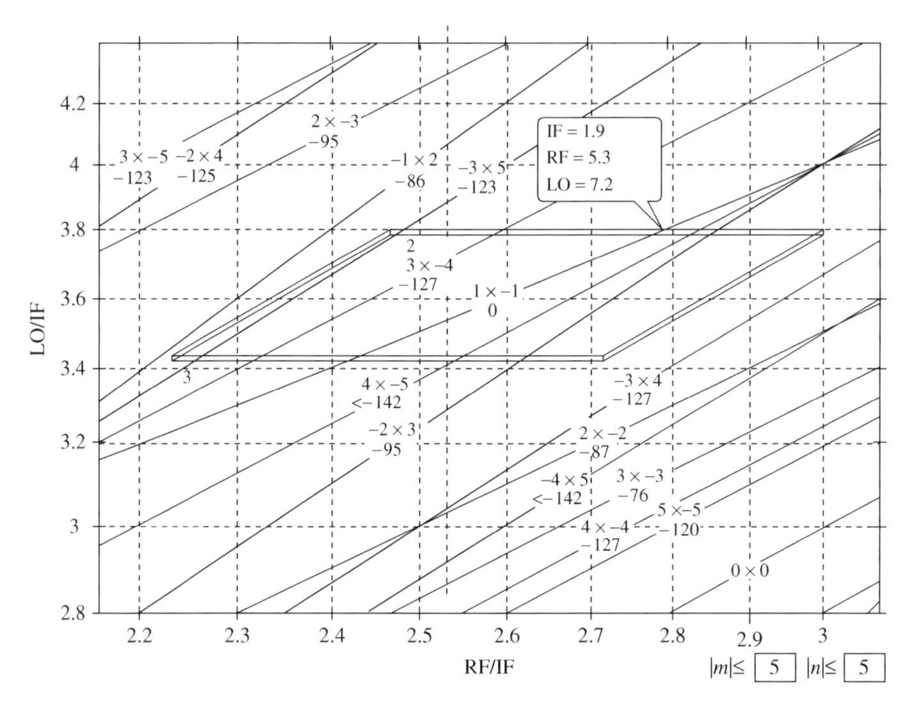

Fig. 7.19 Conversion of Fig. 7.18 with LO fixed in midrange.

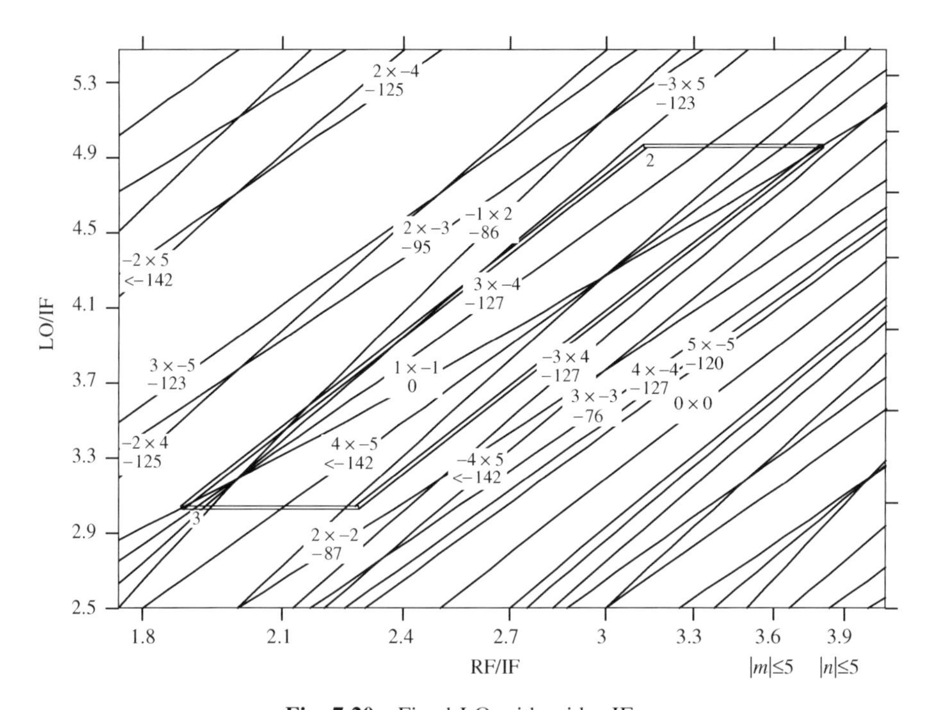

Fig. 7.20 Fixed LO with wider IF.

Sometimes the LO frequency is placed in the middle of the RF passband, converting the center of the incoming spectrum to zero, or near zero, frequency (baseband) (Mashhour et al., 2001). Such systems may be called *direct-conversion, homodyne*, or *zero-IF*. Typically this is done in two parallel paths, the LO signals at the two mixers being in quadrature to each other. In this case the mixer might be considered a phase or amplitude detector, but we can still analyze the spurs.

Example 7.8 Zero IF Let us look at a 2.4-GHz ±22 MHz RF input that is converted (detected) in this manner with a baseband that extends from near zero to 22 MHz. We will use the spur levels from the bottom table in Fig. 7.4. Figure 7.21 shows the spur plot at the maximum IF. If we look at lower IFs (up and to the right on the graph) we find that the 1×-1 and -1×1 desired products and the spur curves come closer together, heading for zero separation at zero IF. All of the in-band spurs are $k \times k$, essentially the kth harmonic of the IF. The nearest out-of-band spur, for m, $n \leq 10$, is a 10×9 at a shape factor of about 12 (not visible in Fig. 7.21). There is probably an 11×10 closer, but these high-n spurs should be small.

In cases where the baseband extends to zero frequency, DC generated in the mixer can be a problem. Imperfect balance can allow some of the detected signal [the DC terms in the even-order responses (Section 4.2)] to appear at the output. Detection can occur between the LO at the LO port and any of the LO signal

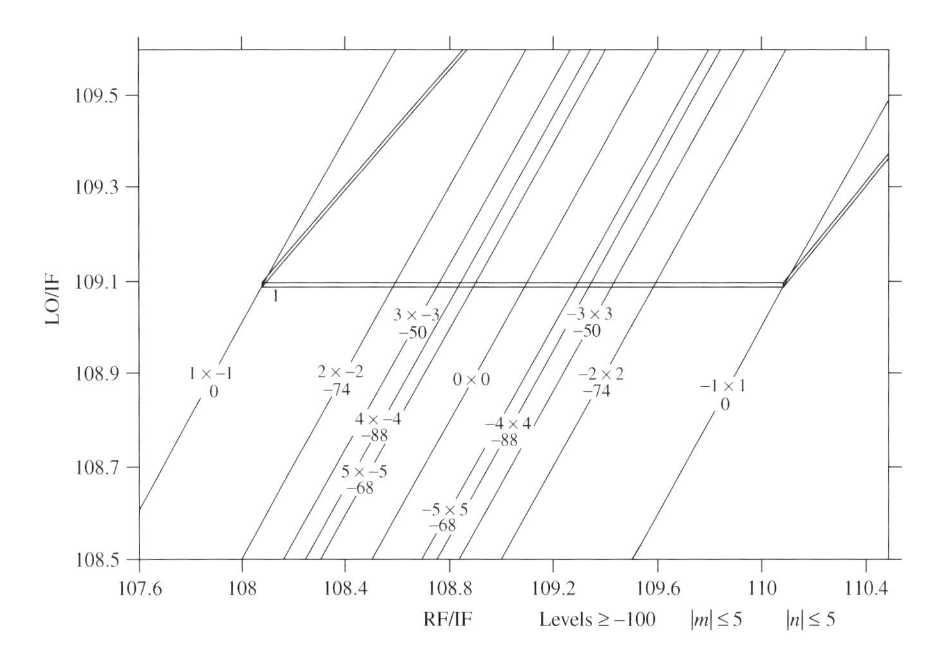

Fig. 7.21 Conversion to baseband.

that leaks into the RF port. That DC is not reduced by balance. It is the intended output of a mixer acting as a phase detector or a coherent amplitude detector, and its magnitude depends on the phase relationship between the signals at the two ports. The problem is exacerbated if LO leakage is amplified before entering the RF port as a result of leaking into the preceding RF cascade. Flicker noise, whose power is inversely proportional to frequency, can also cause problems at low baseband frequencies.

7.7 SHAPE FACTORS

The closer is the spur curve to the required passband, the more difficult it is to attenuate the spur by filtering. We use the *shape factor* parameter to indicate the degree of difficulty in rejecting a spur due to this fact. See Fig. 7.22*a*. The shape factor is the ratio of the required rejection bandwidth to the required pass bandwidth. The rejection bandwidth is twice the difference between the filter center frequency and the frequency of the spur under consideration.

The degree of difficulty also depends on the required rejection level. Thus, as the spur becomes stronger, we want a larger shape factor so the required high value of rejection will be more easily obtained. Note, however, that a filter that attenuates the RF by 1 dB attenuates the resulting mixer product by n dB. Thus, for example, if a 1×3 spur is too large by 6 dB, the RF filter need provide only 2 dB more attenuation. Here we are assuming that the filter provides attenuation at frequencies that produce the spur, but not at frequencies where the desired signal occurs. On the other hand, a reduction of 6 dB in the spur, relative to the signal, could also be obtained by a 3-dB attenuation of the RF. In that case the spur falls 9 dB but, unlike the reduction caused by filtering, the desired signal will also drop, by 3 dB, producing again a relative improvement of 6 dB.

The shape factor does not, by itself, define the required filtering, but it is one of two necessary parameters, the other being the required attenuation at that shape factor. It can also be important to know whether the spur frequency is

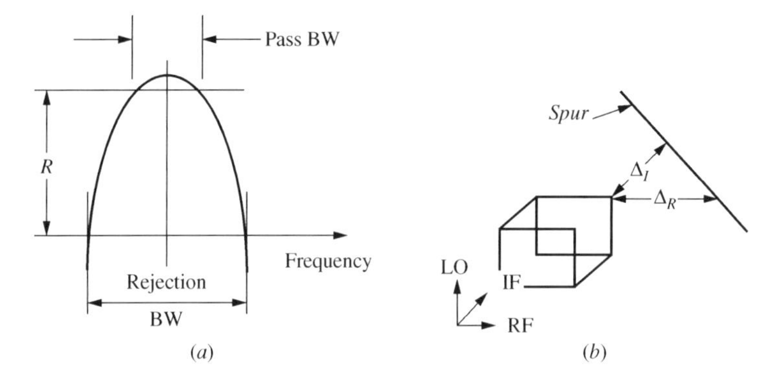

Fig. 7.22 Shape factor definitions.

above or below the filter passband since many filters do not possess arithmetic symmetry; this is most noticeable in filters having high percentage bandwidths.

7.7.1 Definitions

Both RF and IF shape factors are defined. The latter is important if, for example, the detected power in the defined IF is being measured because spurious responses that fall out of the IF passband might then be measured as signals. If the IF is to be further converted, the IF filter may be the RF filter for the next conversion or it might be supplanted in importance by such a filter. If the IF is to be further analyzed, say by passing through multiple contiguous filters, an IF shape factor may not be significant.

The shape factor (SF) is the separation from band center divided by half of the bandwidth. For RF this is

$$\mathrm{SF}_R = (\Delta_R + B_R/2)/(B_R/2) \tag{7.21}$$

$$= 2(\Delta_R/B_R) + 1, \tag{7.22}$$

and for IF it is

$$\mathrm{SF}_I = (\Delta_I + B_I/2)/(B_I/2) = 2(\Delta_I/B_I) + 1, \tag{7.23}$$

where Δ_R and Δ_I are shown in Fig. 7.22*b*, and B_R and B_I are the RF and IF bandwidths. Note that Δ_R and Δ_I are negative for in-band spurs, allowing shape factors as small as zero for spurs that go through band center, in our treatment.

The point from which we measure Δ_R (at an extreme of the IF) implies that we are concerned only with frequencies that are converted into the designated IF passband; there the attenuation is provided only by the RF filter. Similarly, the point from which we measure Δ_I (at an extreme of the RF) implies that only frequencies converted from the designated RF passband, for which attenuation is only due to the IF filter, are of interest. This need not be true always. For example, it is conceivable that a spur that is slightly out of the IF passband might receive less attenuation from the combined RF and IF filters than one that is within the IF passband. (See Appendix B, Section B.2.) Nevertheless, the computed shape factors are of great value in initial design and probably are close to the requirements obtained from final design calculations in most cases.

The attenuation obtained at a given shape factor from a given filter often depends upon whether the rejection frequency is above or below the passband (see Section B.3).

7.7.2 RF Filter Requirements

Example 7.9 Filter Requirements Table Figure 7.23 shows a table containing data from the spur plot in Fig. 7.24. The third column gives the spur amplitude relative to the signal amplitude. The fourth column gives the shape factor, to which a sign has been appended to indicate whether the spur is below or above the

Spurs are shown up to $m = 10$ by $n = 5$, where $\pm m \times \text{LO} \pm n \times \text{RF} = \text{IF}$.
Shape Factors (SF) less than 15.00 are shown.
Mixer File is "High Level"
LO Level: 13 dBm. RF Level: –5 dBm.
IF Levels below –100 dBc excluded.

Rectangle dimensions, spur levels, filter shape factors follow.
*is input change from given RF Level to give output at
–75 dB relative to given RF Level.

1 IF: 2.0 E + 00
(x : 6.100 – 6.400) RF: 1.220 E + 01 – 1.280 E + 01
(y : 7.400 – 7.100) LO: 1.420 E + 01 – 1.480 E + 01

m	n	dBc	SF	*	Use for plot	
–2	3	–69	–6.56	–2	*–2*	*–100*
3	–4	–77	–6.33	0.5	*0.5*	*–100*
4	–5	–92	–3.53	3.4	*3.4*	*–100*
–3	4	–77	–3	0.5	*0.5*	*–100*
–4	5	–92	–0.87	3.4	*3.4*	*–100*
1	–1	0	0	--	*–80*	*–80*
2	–2	–47	2.33	–14	*–14*	*–100*
3	–3	–57	3.44	–6	*–6*	*–100*
4	–4	–81	4	1.5	*1.5*	*–100*
5	–5	?	4.33	--	*–80*	*–80*
0	0	?	5.67	--	*–80*	*–80*
–5	5	?	7	--	*–80*	*–80*
–4	4	–81	7.33	1.5	*1.5*	*–100*
–3	3	–57	7.89	–6	*–6*	*–100*
–2	2	–47	9	–14	*–14*	*–100*
–1	1	0	12.33	–75	*–75*	*–100*
6	–5	?	13.8	--	*–80*	*–80*

Plot unknowns with * at: –80
For knowns, plot * off graph at: –100

Fig. 7.23 Table showing spur levels, RF shape factors, and required changes in input levels.

center of the RF passband. Even in-band spurs are given shape factors here. The shape factors can be obtained from measurements on a plot or, for greater accuracy, by solving Eq. (7.2) along the appropriate line. For example, in Fig. 7.13, if we wanted the shape factor for the 2×-2 spur, we would set f_L/f_I to 3.4, its value at the bottom of the rectangle, with $m = 2$ and $n = -2$, in a normalized version (we could also use true frequencies) of Eq. (7.2), and obtain

$$1 = nf_R/f_I + mf_L/f_I = -2f_R/f_I + 2(3.4), \tag{7.24}$$

$$f_R/f_I = 2.9. \tag{7.25}$$

Knowing that the filter is centered at $f_R/f_I = 2.6$ and has a normalized width of 0.4, we compute the RF shape factor, using Eq. (7.21), as

$$\text{SF}_R = (2.9 - 2.6)/(0.4/2) = 1.5. \tag{7.26}$$

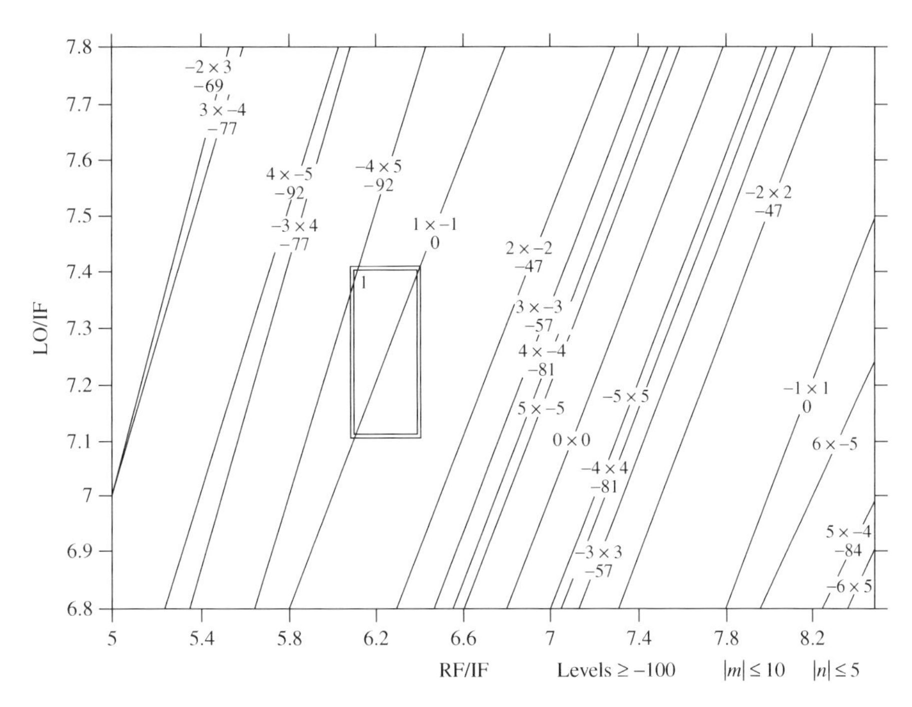

Fig. 7.24 Spur plot for Fig. 7.23.

The $*$ column in Fig. 7.23 contains the change in input level that would cause the spur to go to the specified level (−75 dB) relative to the given "RF Level" (−5 dBm at the input). For the case shown, that allowed spur level is (−5 dBm −75 dB =) − 80 dBm. Looking at this in a little more detail, the third column tells the relative amplitude of spurs caused by an input at the specified RF level. The $*$ column shows what increase or decrease in the specified RF level would be necessary to adjust a spur to its required level (−80 dBm in this case). This is n times the required change in the spur, which implies that the reference level is not changing (otherwise the ratio would be $n - 1$). For example, the first row in the spur list shows that the -2×3 spur is at −69 dBc, 6 dB above the specified level. The $*$ column in that row indicates that 2 dB of attenuation is required at RF. That would produce the required 6 dB reduction, since $|n| = 3$.

The RF Level is set equal to the maximum level of the desired signal. Filtering does not change that level because the desired signal is within the passband, but filtering must change the level of the undesired signal that produces the spur if it is excessive. If the maximum level of the interferer should be the same as that of the desired signal, any attenuation indicated in the $*$ column would have to be provided by filtering. If the level of the interferer should be greater than the RF Level, an additional attenuation equal to that excess would also be required and visa versa for weaker interferers.

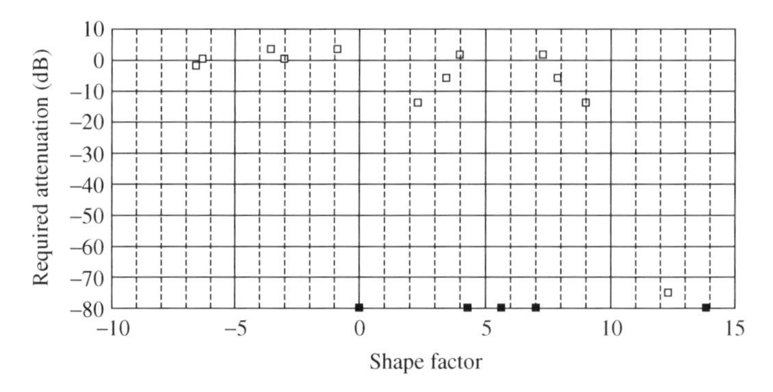

Fig. 7.25 Points giving required filter shape.

If there is significant ripple in the passband of the filter, we can use the minimum loss in obtaining the signal strength. Then required attenuation will be relative to that loss.

Example 7.10 Plotting the Filter Requirement The $*$ levels from Fig. 7.23 have been plotted in Fig. 7.25. Unknown levels have been placed at the bottom of the graph, as has the desired signal, which was identified by 0-dBc level and zero shape factor (meaning it has gone through the center of the RF band). Asterisks have been placed on these symbols to indicate that their levels do not have the same meaning as the others. Perhaps we will want to measure the amplitude of these spurs, whose levels are unknown. The filter response curve should be below all of the other points. Thus an attenuation of 14 dB is required at a shape factor of 2.33, and an attenuation of 75 dB is necessary to attenuate the image frequency at a shape factor of 12.55, both on the high side. The low side requires only 2 dB at a shape factor of -6.56.

The spur designated as 0×0 is also shown at the bottom of Fig. 7.25. It is the LO-to-RF leakage. We must take into account the allowed "reradiation" and the mixer and RF amplifier isolations to determine how much reduction the filter must provide this signal at a shape factor of 5.67.

Figures 7.24 and 7.13 both represent converters with zero IF bandwidths. In general, the RF filter computations must be made at the extremes of the IF band (e.g., for rectangles 2 and 3 in Fig. 7.18), unless the spur plot shows that one end dominates. It is conceivable that an in-band spur could cross through a conversion polygon and be out of band at both IF extremes, but such cases seem unlikely in practice and would be revealed by the spur plot or by a difference in sign of the shape factor at the two ends.

7.7.3 IF Filter Requirements

If the IF is to be detected, that is, if its power is to be measured or if modulation is to be extracted from it, all spurious signals should be below some threshold.

Then the total of IF and RF filtering must be adequate to reduce all spurs to that allowed level everywhere, and required IF shape factors can be computed in the same manner as were the RF shape factors.

Example 7.11 IF Filter Figure 7.26 shows a list of IF shape factors for the design illustrated in Fig. 7.17. The required attenuation is determined by comparison of the dBc column to the allowed spur level. In this example, the 3×-3 spur exceeds the allowed −85-dBc level by 9 dB so this much attenuation is required at the indicated shape factor of 13. The value is shown in the right column, which differs from the $*$ column in Fig. 7.23 in that the level difference is here not divided by n since the attenuation comes after the nonlinear process. As before, the RF level represents the desired signal, and, presumably, the largest in-band RF signal, which is the signal producing the spurs. Shape factors are measured along lines corresponding to changes in IF with the RF and LO fixed. These extend from the corners of the rectangles toward (0,0) in

Spurs are shown up to $m = 5$ by $n = 5$, where $\pm m \times \text{LO} \pm n \times \text{RF} = \text{IF}$.
IF Shape Factors (SF) less than 30 are shown.
Mixer File is "Figure 7.4"
LO Level: 10 dBm. RF Level: −20 dBm.
Rectangle dimensions, spur levels, filter shape factors follow.
"Allowed Gain" is minus required IF Filter attenuation.
−85 dB relative to given RF Level.

2 — IF: 1.9
(x : 2.473 − 3) — RF: 4.699 − 5.699
(y : 4 − 3.578) — LO: 6.799 − 7.6

3 — IF: 2.1
(x : 2.238 − 2.714) — RF: 4.699 − 5.699
(y : 3.619 − 3.238) — LO: 6.799 − 7.6

IF Shape Factors for Connected Pairs

For rectangle pair 3,2:

m	n	dBc	SF	Allowed Gain (dB)
−4	5	?	−7.00	--
−3	4	−127	0.00	42
−3	5	−123	0.00	38
−2	3	−95	0.00	10
−2	4	−125	16.00	40
−1	2	−86	0.00	1
1	−1	0	0.00	--
2	−3	−95	−9.00	10
2	−2	−87	2.00	2
3	−4	−127	0.00	42
3	−3	−76	13.00	−9
4	−5	?	0.00	--
4	−4	−127	24.00	42

Fig. 7.26 IF shape factor data for Fig. 7.17.

linear plots and are along lines parallel to the diagonal edges of the polygons in log plots.

If the IF is to be further processed in a manner that provides additional filtering, that requirement may be relieved or replaced by requirements peculiar to the process.

In addition to their role in preventing adverse effects from spurs, another requirement usually accommodated by an IF filter is selectivity. Selectivity is the ability of a receiver to prevent interference from adjacent signals. A selectivity specification may just give the attenuation required of a signal separated from the signal to which the receiver is tuned by a given frequency. It can be applied directly to the IF filter.

7.8 DOUBLE CONVERSION

What is the design process when multiple conversions are employed? There can be various combinations of fixed and tunable conversions. We will probably design each conversion separately, ensuring that reasonable RF and IF filters can be used successfully. Eventually, optimizing the design will cause us to consider the interaction between the specifications of the stages.

Let us consider double conversions. That should also provide a guide for even greater numbers of serial conversions.

A simple and conservative approach in designing the first converter stage is to provide filters that cause all spurs to be below the ultimately required level. Then, even if they are converted linearly in the second stage, they will not be a problem. A more optimum approach is to allow the parameters of the first-IF filter to be determined by design of the second converter stage, where it becomes effectively the RF filter.

The RF filter requirements for the second stage can be determined to meet spur requirements in the same manner as for the first stage. Once determined, the required second-stage RF filter attenuation at a given frequency can be reduced by subtracting the minimum loss for the same signal in the first-stage RF filter since the two are added to determine the strength of received signals as seen at the input to the second mixer. The minimum loss is determined at a given IF by considering the attenuations that occur as the LO is tuned (Fig. 7.27).

This relief applies to signals but not to spurs in the first IF since they do not pass through the first-stage RF filter. However, the spurs are weaker than the converted signal, so they do not need as much attenuation. If we allow relief that exceeds the relative level of the signal to the spur at the output of the first mixer, there is a possibility that a spur from the first conversion might cause an excessively large spur in the second conversion.

For example, if 50 dB of attenuation is required for the second-stage RF filtering at some frequency and the signal at that frequency will receive at least

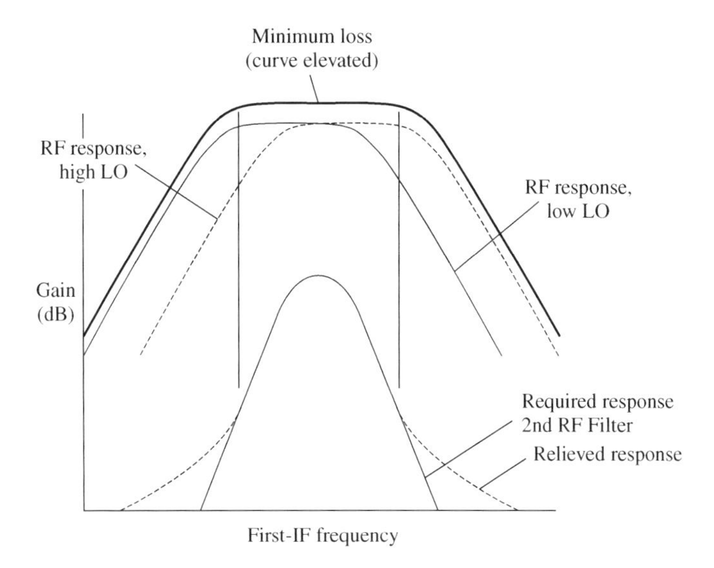

Fig. 7.27 Second RF (first IF) filter requirement relieved by contribution of first RF filter.

30 dB attenuation from the first RF filter, only 20 dB is required from the second-stage RF filter (which is also the first-stage IF filter). That puts the signal at −50 dB at the input to the second mixer. However, a −25-dB spur at the output of the first mixer is stronger than the input signal attenuated by 30 dB, and reducing it by the relieved attenuation of 20 dB would still leave it at −45 dB, 5 dB higher than allowed. The problem occurs because the relief, 30 dB, exceeds the signal-to-spur ratio, 25 dB.

7.9 OPERATING REGIONS

Here we consider the properties of the various regions of the spur plot that affect their usefulness as operating regions.

7.9.1 Advantageous Regions

Figure 7.28 is a log plot showing some spurs that are particularly important. LO reradiation ($f_{LO} = f_{RF}$) is also shown. RF passbands that include this line allow the strong LO signal to pass through the RF filter after reduction by the LO-to-RF isolation. This can be a problem; "reradiation" refers to the possibility that the LO might radiate from a receiver's antenna.

The 0×1 RF feedthrough is also shown. Along this line, the RF could leak to the IF without conversion, attenuated by the RF-to-IF isolation. This is often called the IF response since it appears when an RF input occurs at the IF frequency. It is often required to be as small as the image (the undesired $\pm 1 \times \pm 1$), which response may be specified separately.

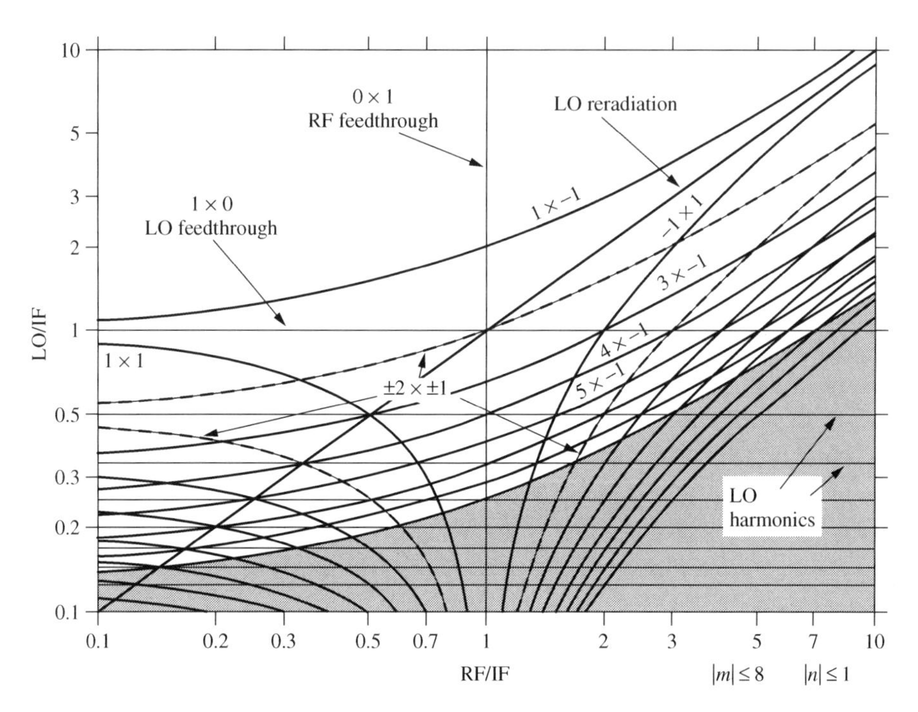

Fig. 7.28 Spurs of primary concern.

The horizontal lines represent harmonics of the LO, occurring where the LO frequency is a fraction of the IF. These are "internal spurs," meaning that they occur even in the absence of an input signal. They are relatively large and do not decrease with weaker input levels. In fact their *relative* levels get worse with weaker signals.

The other curves are $m \times \pm 1$ spurs. These are of particular concern because they can be strong, as can be seen from Fig. 7.4, and their relative strength cannot be decreased by decreasing the RF level. Included are the three sections of the $|m| \times |n| = 1 \times 1$ curve, one of which will be used to give the desired response. The $\pm 2 \times 1$ curves have the same shape but are shifted downward so they come to LO/IF $= \frac{1}{2}$, rather than to 1, as RF/IF $\Rightarrow 0$. The same pattern can be seen for the other $\pm m \times 1$ curves. They have three segments with the same form as the desired 1×1, but they are at lower levels of LO/IF, approaching $1/m$ as RF/IF $\Rightarrow 0$. They will fill the shaded area as m increases (reaching the lower right corner at $m = 90$).

While small regions that are free of large spurs can be found along the $\pm 1 \times \pm 1$ curves throughout the plot, we tend to pick regions that appear clear in Fig. 7.28 to minimize interference by large spurs, especially for wide bands. We will now consider three regions that are identified in Fig. 7.29. We can use Fig. 7.30, which has spurs up to $n = 3$, to see some other spurs in these regions.

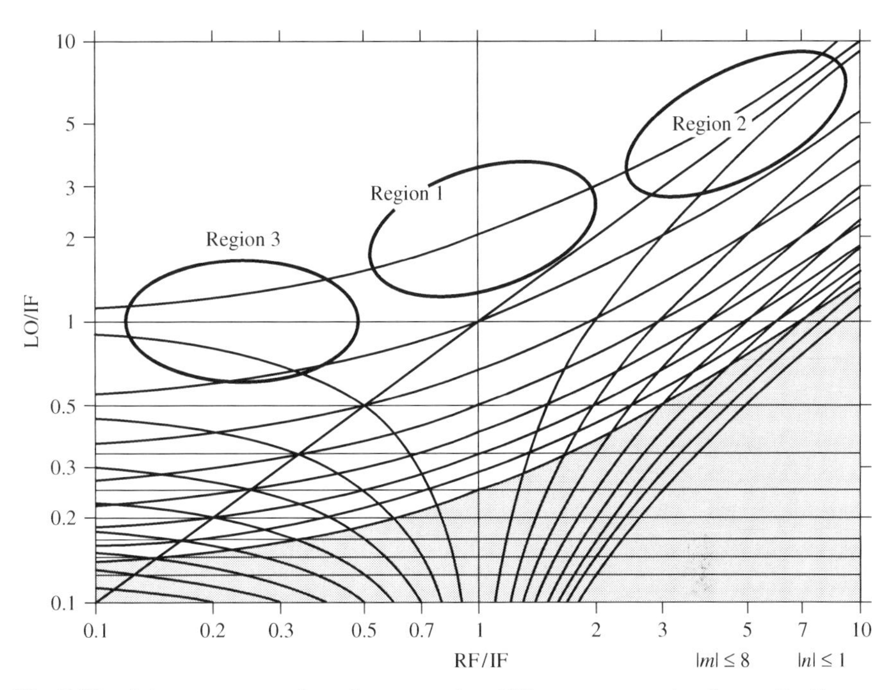

Fig. 7.29 Advantageous regions for conversion. Ellipses are meant to focus attention on certain areas of the plot rather than to define boundaries.

7.9.1.1 Region 1: $f_{LO} > f_{RF}, f_{IF}$

This is where the IF and RF are relatively close and the LO frequency is their sum. It is along the 1×-1 curve and is relatively clear of spurs excepting possibly the lower harmonics of the RF (e.g., the second at RF/IF = 0.5, visible in Fig. 7.30) and the -1×3. LO reradiation ($f_{LO} = f_{RF}$) and the IF response (vertical line at RF/IF = 1) can be problems in this region. The latter can sometimes force the use of two serial conversions (double conversion), neither of which is in Region 1, to translate a frequency band by a relatively small amount.

7.9.1.2 Region 2: $f_{LO}, f_{RF} \gg f_{IF}$

This region in the upper-right corner is the region for significant downconversion, commonly in a receiver. It includes both the 1×-1 and the -1×1 desired conversion curves and is clear for narrow bandwidths. The primary problems are the image and LO reradiation ($f_{LO} = f_{RF}$). Where spurs must be very low, $\pm 2 \times \pm 2$ spurs (see below) can be a significant problem, discouraging large ratios between the RF and IF frequencies.

7.9.1.3 Region 3: $f_{LO} \approx f_{IF} > f_{RF}$

This is the region of significant upconversion, commonly in an exciter or possibly a spectrum analyzer (McClaning and Vito, 2000, p. 715). It includes both the 1×1 and 1×-1 desired conversions. Here the main problems are LO feedthrough ($f_{LO} = f_{IF}$), $\pm 1 \times \pm 2$ spurs, and possibly harmonics of the RF ($f_{RF}/f_{IF} = 1/n$).

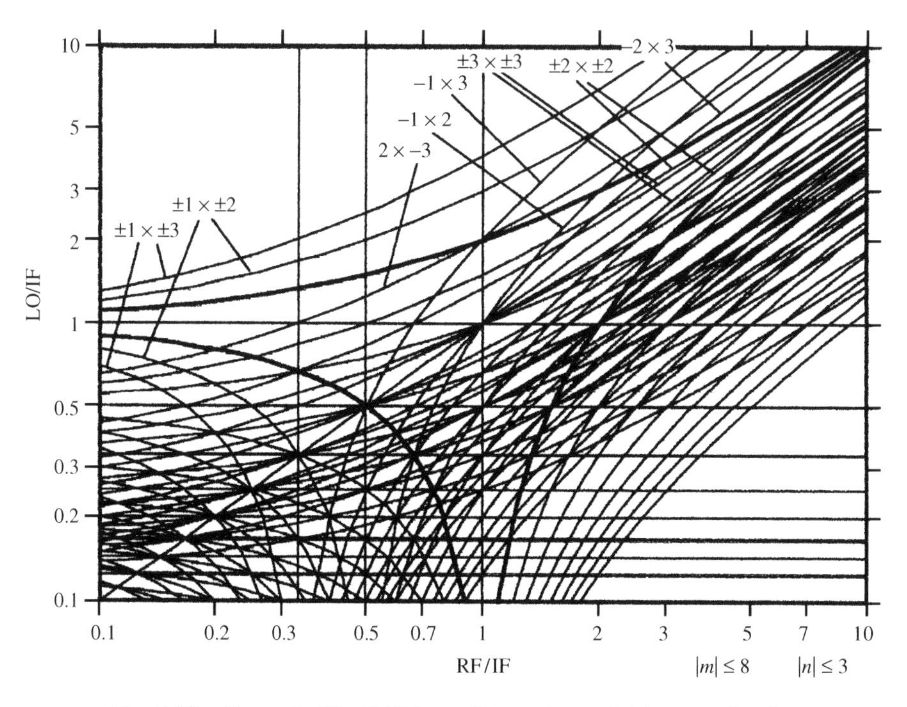

Fig. 7.30 Plot as in Fig. 7.29 but with maximum $|n|$ increased to 3.

7.9.2 Limitation on Downconversion, Two-by-Twos

In this section, we see that the relative IF bandwidth (percent bandwidth) in Region 2 is limited by the 2 × 2 spur and that the limitation is more severe for wider RF bandwidths or tuning ranges. This can prevent the desired downconversion from being accomplished in a single conversion.

We can see from Fig. 7.30 how the $\pm 2 \times \pm 2$ response parallels the desired $\pm 1 \times \pm 1$ response in Region 2 (this region is also shown expanded in Fig. 7.21). It is apparent that the $\pm 2 \times \pm 2$ curves can be the limitation here. These spurs occur in the IF at the second harmonic of the IF frequency:

$$f_{\text{IF}2\times 2} = \pm 2 f_{\text{RF}} \pm 2 f_{\text{LO}} = 2(\pm f_{\text{RF}} \pm f_{\text{LO}}) = 2 f_{\text{IF}1\times 1}. \qquad (7.27)$$

They can occur whenever the IF band is an octave wide but, more generally, whenever a signal can be converted to half of an IF frequency. That is, if the RF filter rejects any signal that would not be converted into the IF passband, the problem occurs when the IF bandwidth reaches one octave. However, if the RF filter is wider such that it allows signals to be converted to frequencies below the IF band, even though those converted signals are rejected by the IF filter their second harmonics may pass through it.

Example 7.12 Limitation Due to 2 × 2 Spurs Figure 7.31 shows the spur plot for a one-octave IF (5–10 MHz) with a fixed LO (85 MHz) and a 1×-1

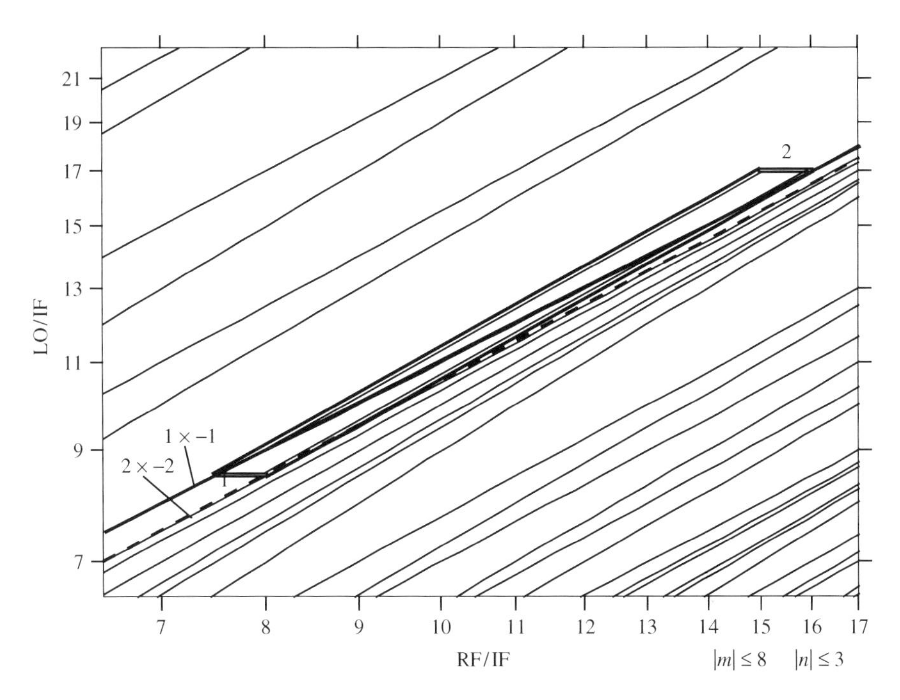

Fig. 7.31 2 × 2 spur (curve is marked with dashes for emphasis).

conversion ($f_{\text{IF}} = f_{\text{LO}} - f_{\text{RF}}$). The signal at the low end of the RF band (75 MHz) is converted to the high end of the IF band (10 MHz, at 1) by the desired process. The spur is converted to the same frequency when its RF is at the high end of the band (80 MHz, at 1):

$$2 \times 85 \text{ MHz} - 2 \times 80 \text{ MHz} = 10 \text{ MHz}. \tag{7.28}$$

When the RF increases to the top of the band, at 80 MHz, the resulting IF is 5 MHz, one octave lower (at 2). If the IF were not an octave wide, the RF bandwidth could have been smaller and the RF band would not have enclosed both the 1×-1 and the 2×-2 at 1.

Even if the RF signal that would cause this spur is outside the RF bandwidth, it is worth considering the shape factor that is imposed on the RF filter by this spur. Figure 7.32 shows the converted RF band (i.e., plotted against the corresponding IF frequencies) and the IF band. The highest frequency of the IF is $f_{I,\max}$, and the highest IF signal whose second harmonic can be in the IF band is half this frequency. The shape factor for rejecting it will be minimum when the converted RF band is lowest in IF frequency. This minimum converted RF will be B_R below $f_{I,\max}$; to tune lower would be to eliminate part of the IF band from use.

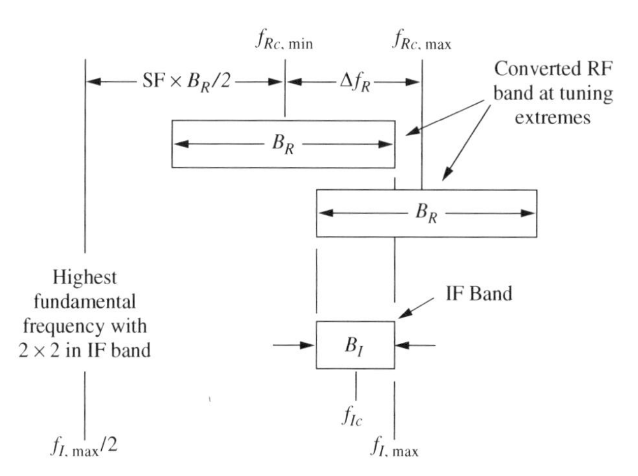

Fig. 7.32 Geometry for calculating 2 × 2 filtering requirement. (f_{Ic} and f_{Rc} are center frequencies.)

From this geometry, if we measure frequency changes from the upper edge of the IF band, we can write

$$\mathrm{SF} \times B_R/2 + B_R/2 = f_{I,\max} - f_{I,\max}/2 = f_{I,\max}/2, \tag{7.29}$$

$$\mathrm{SF} = \frac{f_{I,\max}}{B_R} - 1. \tag{7.30}$$

We can write this in terms of the IF center frequency f_{Ic} and bandwidth B_I as

$$\mathrm{SF} = \frac{f_{Ic} + B_I/2}{B_R} - 1 = \frac{f_{Ic}/B_I + 0.5}{B_R/B_I} - 1. \tag{7.31}$$

Since the tuning range is $\Delta f_R = B_R - B_I$, this can also be written as

$$\mathrm{SF} = \frac{f_{Ic}/B_I + 0.5}{1 + \Delta f_R/B_I} - 1. \tag{7.32}$$

If the tuning range is zero (band conversion), this is

$$\mathrm{SF} = f_{Ic}/B_I - 0.5. \tag{7.33}$$

The results are represented in Fig. 7.33*a*, where we can see that higher shape factors, required for ease of filtering, result in smaller percentage IF bandwidths, especially when the RF bandwidth is large compared to the IF bandwidth. The latter situation corresponds to relatively wide tuning ranges. The same data is shown in a different form in Fig. 7.33*b*.

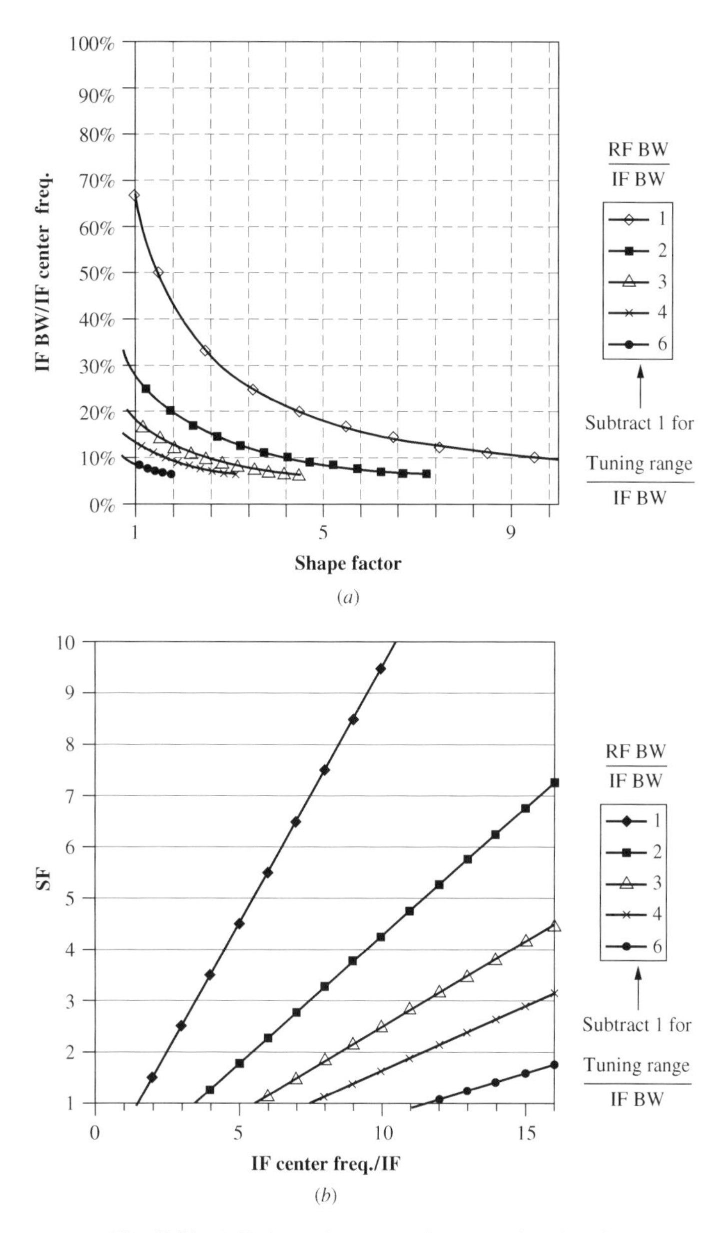

Fig. 7.33 RF shape factor required to reject 2 × 2.

7.9.3 Higher Values of *m*

We have seen that spur levels do not drop rapidly with increasing LO multiples ($|m|$ values), as they typically do with increasing RF multiples ($|n|$ values), and there will always be a limit to the number of spur curves we will draw. It is therefore important to know whether or not we might be impacted by LO multiples that are higher than we have explicitly considered. To this end, we will consider where spur curves appear as $|m|$ increases, restricting our study to the lower values of $|n|$, where large spur levels are more likely. (Review Fig. 7.4 to see the reason for these choices.) We have already identified a region that is covered as $|m|$ increases, with $|n| = 1$, by the shaded area in Fig. 7.29, but here we will look at these curves more explicitly, and at those with higher $|n|$ values.

Figure 7.34 shows curves with $|n| \leq 3$ and $|m| = 10$, except where those curves are off the plot, in which case the curve with the highest $|m|$ that occurs on the plot is shown. For any value of $|m|$ (e.g., 10), there are 3 curves for each $|n|$, corresponding to the various perturbations of signs of m and n. In addition to these, sections of curves for the two next lower values of m are also shown to enable us to see the separation and to estimate how curves with higher values of m would be spread out. In the lower left corner, entire curves with the highest three values of m are shown, since they are too short to justify showing only segments—the main reason for showing just segments of some curves is to reduce the clutter of the plot.

We can see that, as the values of $|m|$ increase, the curves move lower on the plot, tending to fill in the region at the bottom near RF/IF = 1. While this is

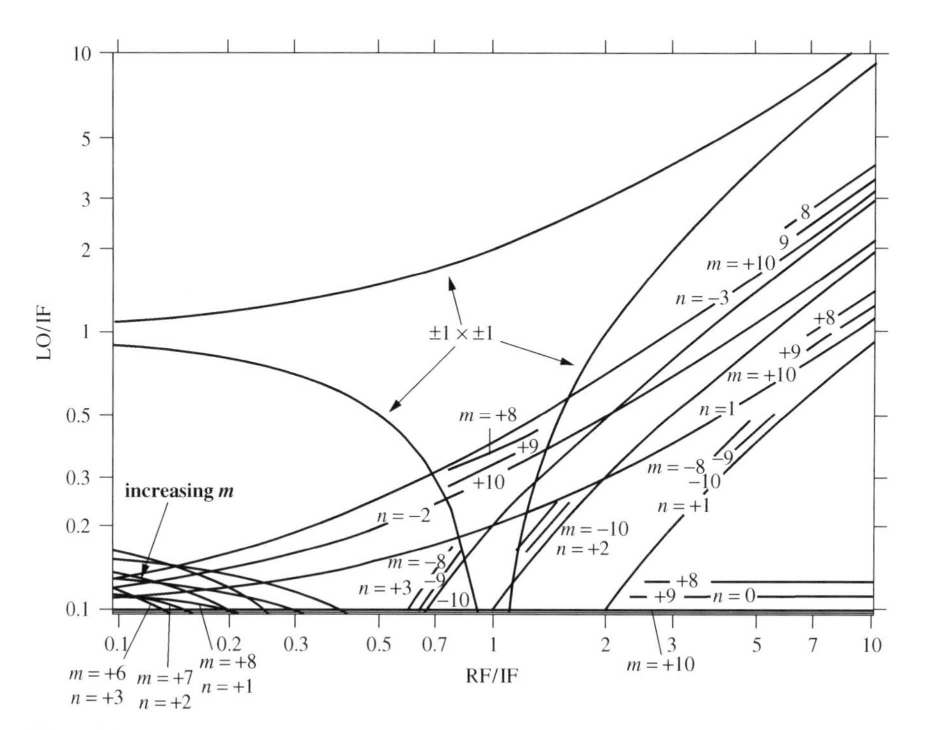

Fig. 7.34 Spurs at high values of $|m|$. Curves are shown for $0 < |n| \leq 3$. Segments are shown for two lower values of $|m|$ to indicate spacing.

an alternative to Region 1 for producing relatively small changes in frequency, it has not been recommended in spite of some problems that we have described for Region 1. (Here the LO frequency is the difference between the LO and RF frequencies rather than being their sum as in Region 1.)

7.10 EXAMPLES

Here are some examples of use of the spur plots.

Example 7.13 AM Radio We use the following specification for this example:

Frequency range: 535–1605 kHz
IF: 455 kHz
IF bandwidth: ±5 kHz
Image rejection: 60 dB

Here we use again the mixer spur level table of Fig. 7.4 but with the signal level set at −40 dBm. In typical AM radios, the mixer will usually be much less elaborate, possibly being part of a converter stage, one that combines oscillator and mixer.

First we try a fixed RF filter covering the whole RF band (Fig. 7.35). However, we find that the image, the -1×1 curve, which must be attenuated 60 dB, is in-band. At the LO frequencies where this occurs, we could simultaneously receive two different stations with equal ease. The LO reradiation (0×0) might also be a problem as well as the IF response (0×1). Reradiation can be attenuated by isolation from RF amplifiers. To reduce the 0×1 IF response with the filter shown would require an impractical shape factor of 1.14.

For these reasons, we go to the traditional AM receiver design with a tuned RF filter (Fig. 7.36), choosing a 20% bandwidth (so we need not tune very accurately). Now the -1×1 image can be filtered with a shape factor of 5.6 and the 0×0 LO reradiation with a shape factor of 2.8. The 0×1 IF response might still be a problem, requiring a shape factor of only 1.4 (or a separate filter) to reduce signals received at the IF frequency (from mobile marine radios) when the receiver is tuned to its lowest frequencies.

With an FM radio, we do not have the 0×1 IF response problem because the 10.7 MHz IF is far removed from the 88- to 108-MHz RF band. However, the image would again force us to tune the RF filter.

Example 7.14 Switched Preselector Another possible (if unusual) solution for the AM radio is to use a switched preselector, choosing different fixed RF filters for different parts of the RF band. Figure 7.37 shows such a realization. Because the channels are well defined, no overlap is shown, but, in practice, there would be overlap at the maximum allowed bandwidths due to finite tolerances. Now the 0×1 IF response can be filtered with a minimum shape factor of 1.75, and the minimum shape factor to attenuate the image is 3.18, the same for

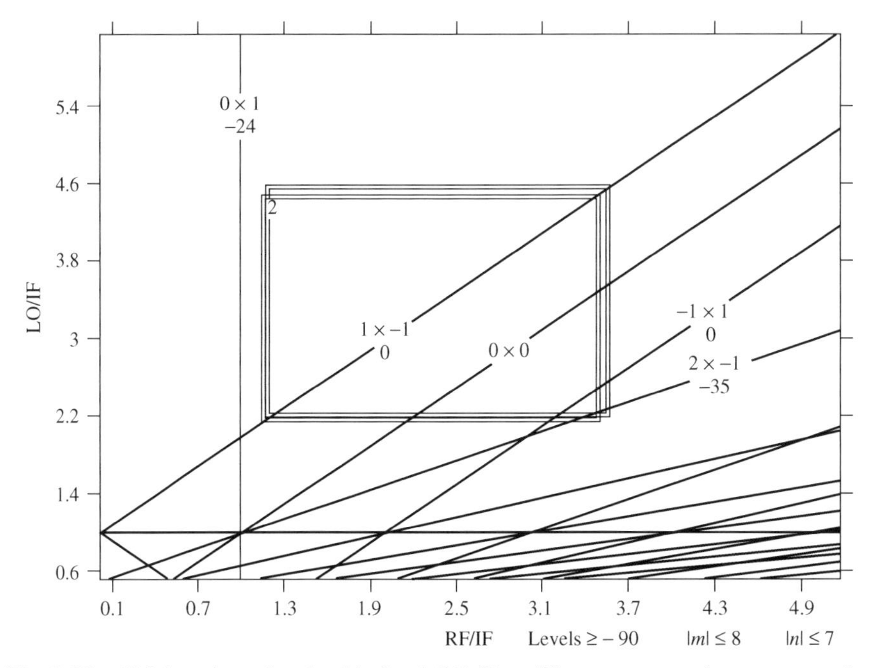

Fig. 7.35 AM broadcast band with fixed RF filter. The two rectangles represent the extremes of the IF band.

both upper bands. We have chosen not to attenuate the 0×0 LO reradiation by filtering—otherwise more and smaller preselector bands would be used.

Example 7.15 Multiband Downconverter A multiband downconverter is a special case of the switched preselector converter in which the LO has only one frequency per band. In the example shown in Fig. 7.38 we are downconverting 10–13 GHz to 3–4 GHz in 1-GHz-wide bands, using a high-side LO at three fixed frequencies (14, 15, and 16 GHz). The unnormalized and normalized frequencies are shown in Table 7.2. We are using the mixer table of Fig. 7.4 again but with the signal level set to −12 dBm. We have an in-band spur at −73 dB and must reduce the 3×3 response at least 19 dB to keep it lower than that in-band spur (not to imply that this is how spurious requirements are ordinarily set). This will require a little more than (19 dB/3 $\approx$) 6-dB attenuation by the RF filter at a shape factor of 4.33 for each filter, a relatively easy requirement. Some band overlap would be needed in practice to permit signals with finite widths to be received at the band breaks, if for no other reason (e.g., frequency drift).

Example 7.16 Design Aid for Switched Preselectors Figure 7.39 shows a spreadsheet aid for use in the design of converters with switched preselectors. As an example, the first four segments have been used to plan a multiband downconverter (Fig. 7.40) in which the frequency range of a 30- to 165-MHz

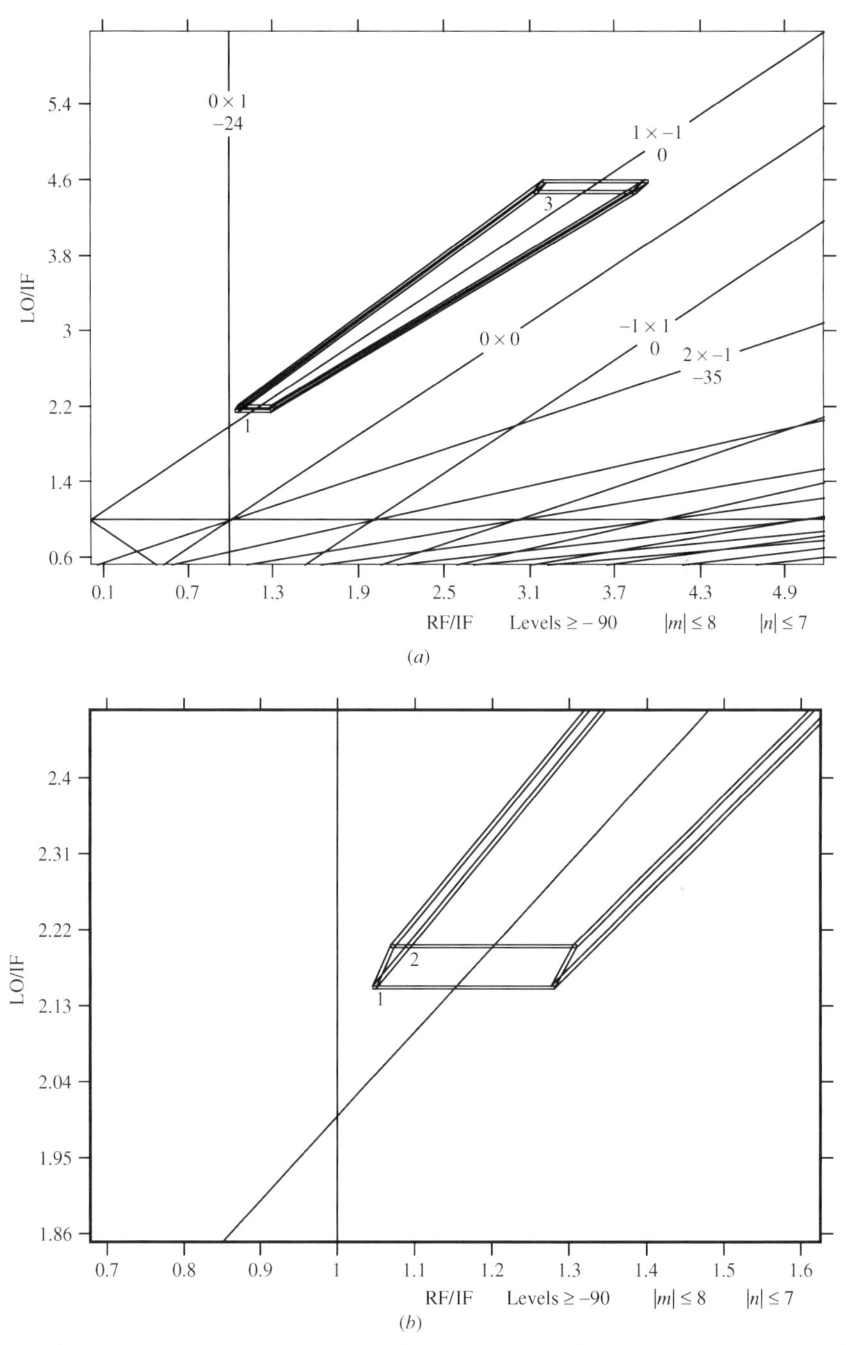

Fig. 7.36 AM broadcast band with 20% bandwidth RF filter. Bottom of polygon is expanded at (*b*). Parallelogram 1–2 represents a 106-MHz-wide RF band at the low end of the RF range and a fixed LO frequency of 990 MHz with a 10-MHz-wide IF band. It is similar to Fig. 7.19. This progresses toward 3 as the LO and the RF filters are tuned to the high end of the band.

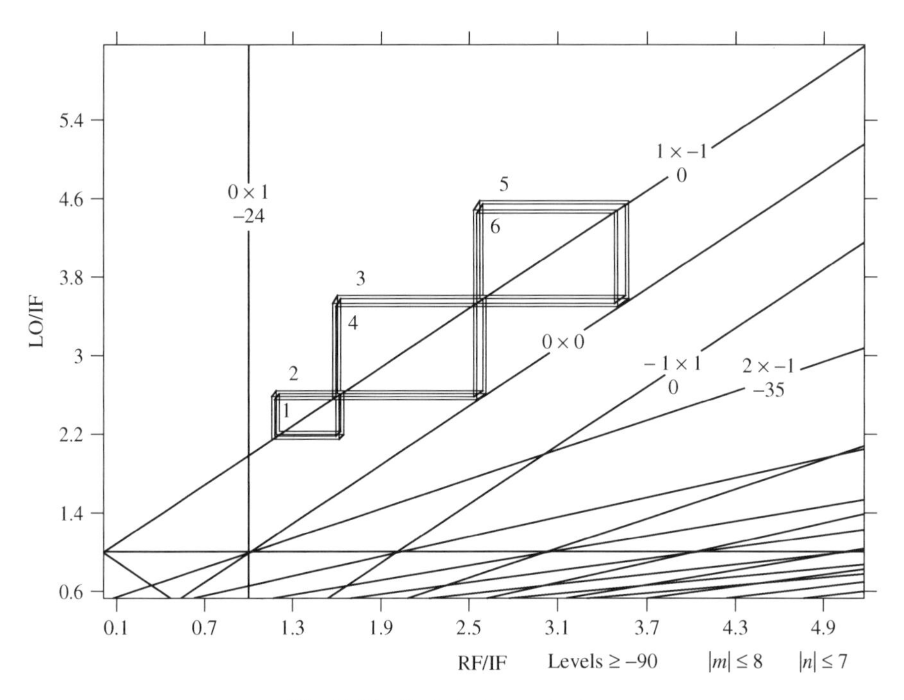

Fig. 7.37 Switched preselector.

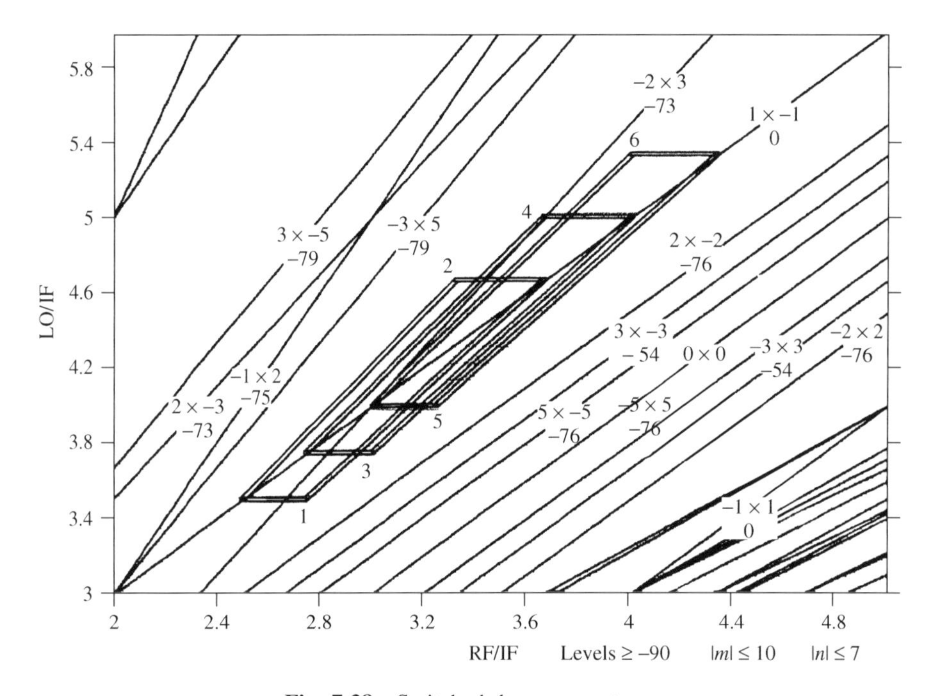

Fig. 7.38 Switched downconverter.

TABLE 7.2 Values for Fig. 7.38

	RF		LO	IF	
	min	max		max	min
	10	11	14	4	3
	11	12	15	4	3
	12	13	16	4	3
Rectangle	Normalized Values at max IF = 4				
1	2.50	2.75	3.50		
3	2.75	3.00	3.75		
5	3.00	3.25	4.00		
Rectangle	Normalized Values at min IF = 3				
2	3.33	3.67	4.67		
4	3.67	4.00	5.00		
6	4.00	4.33	5.33		

	A	B	C	D	E	F	G	H	I	J	K	L	M	N	O	P
1	12	9	2000	@	13	40	15	**Spur Spreadsheet**								
2	A	B	C	D	E	F	G	H	I	J	K	L	M	N	O	
3		Information about:			ER1	more			goal SF:	*2.80*	*1.90*	*1.70*	*1.50*	*1.50*	*1.90*	
4					ER2						m→	2	5	7		
5	ER'R	Seg.	IF1	RF1	LO1	normalized coordinates					n→	−1	3	3	LO	BW
6	#	#	IF2	RF2	LO2	R/IF2	L/IF2	R/IF1	L/IF1	Image	IF	Spur1	Spur2	Spur3	Rerad	CF = Fo
7	*Title:*		DESIGN EXAMPLE (extend 30-165 to 500)							Shape Factors (SF) over				*safety factors*		
8	ER2	1	151.67	500	348.33	10.27	7.157	3.297	2.297	2.89	5.76	1.87	1.58	6.09	1.95	103
9	ER2		48.67	397	348.33	8.157	7.157	2.618	2.297	*1.03*	*3.03*	*1.10*	*1.05*	*4.06*	*1.02*	448.5
10	ER2	2	120.33	397	276.67	10.36	7.218	3.299	2.299	2.87	5.75	1.88	1.59	6.08	1.93	82
11	ER2		38.33	315	276.67	8.218	7.218	2.618	2.299	*1.02*	*3.03*	*1.10*	*1.06*	*4.06*	*1.02*	356
12	ER2	3	95	315	220	10.5	7.333	3.316	2.316	2.85	5.77	1.92	1.62	6.13	1.92	65
13	ER2		30	250	220	8.333	7.333	2.632	2.316	*1.02*	*3.04*	*1.13*	*1.08*	*4.09*	*1.01*	282.5
14	ER2	4	40	250	290	2	2.32	6.25	7.25	2.88	1.94	5.82	5.51	10.06	1.94	85
15	ER2		125	165	290	1.32	2.32	4.125	7.25	*1.03*	*1.02*	*3.43*	*3.67*	*6.71*	*1.02*	207.5
16	##	5				#####	#####	#####	#####	#####	#####	#####	#####	#####	InBand	

	A	B	C	D	E	F	G	H	I	J	K	L	M	N	O	P
27	##		*			#####	#####	#####	#####							
28		minimum safety factor for each spur→								*1.02*	*1.02*	*1.10*	*1.05*	*4.06*	*1.01*	
29		LO→	max:	348	min:	220	1.58	:1 ratio		BW:	min:	65.00	max:	103.00	1.585	:1 ratio
30			*	**⇐ used to disable unused rows. Otherwise minimums safety factors will not be correct.**												

Fig. 7.39 Part of spreadsheet used as aid in developing converters with switched preselectors.

(or kHz or GHz) device is extended to 30–500 MHz (or kHz or GHz). The user manipulates RF2 and LO1 to find an optimum division of bands. Other variables that can be derived from these, such as the LO2, the IFs, and RF1 for the next band, are computed automatically. Overlap is not included but could be built into the equations. Shape factors for certain spurs, including those with m and n values that the user has predetermined to be important (e.g., by looking at the spur graph), are automatically computed. Shape factor goals, something attainable with the required attenuation and the technology to be used, have been entered and the safety factors of the computed shape factors relative to those goals are

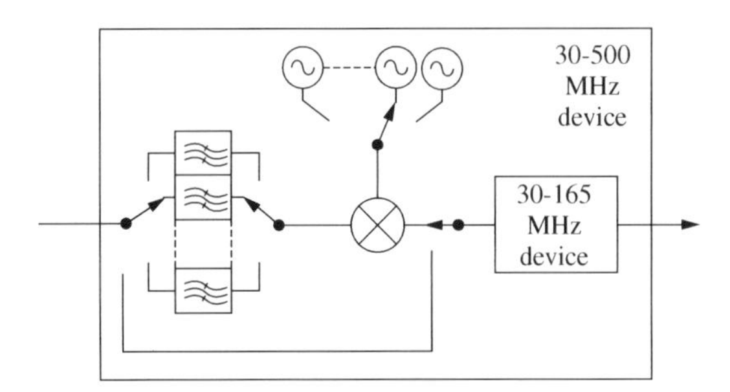

Fig. 7.40 Multiband design example.

also computed and displayed. Summary information for the design is displayed at the bottom. More information, which is not shown in this figure, is presented to the right. This spreadsheet can be used or modified for other designs. ER2, shown in column 1, is just a warning that refers to a note concerning the proper interpretation of certain negative numbers that appear on that line. ER1 would be displayed if the IFs, RFs, and LOs in a segment did not relate in a correct manner. The user modifies cells D9, D11, and so forth to change band breaks and changes cells E8, E10, and so forth to change LO frequencies and resulting IFs, observing cells J28–O28 for filter safety factors. Sometimes such tools are worth using for optimization; sometimes they are overkill.

7.11 NOTE ON SPUR PLOTS USED IN THIS CHAPTER

Spur plots were used long before the advent of computer-aided engineering. A careful plot of the normalized spur equation, either Eq. (7.14) or (7.20), could be copied and used for many projects by many engineers. Today the task can be aided by the use of a spreadsheet, as in Figs. 7.7–7.12.

Starting with Fig. 7.13, the spur plots were generated using specialized software, which allowed the displayed region and the maximum values of m and n to be easily chosen, curves for spurs below a specified level to be deleted, spur information to be printed on the display, and polygons representing the passbands to be displayed. This allowed more efficient use of space in this book and aided in the explanation of the concepts, its main purpose.

The specialized software would also be helpful for design but, at the time of this printing, it is not generally available. However, some of its features are incorporated in available software, for example the "Mixer Spur Chart Calculator" (Roetter and Belliveau, 1997) and *Spur Finder* (Wood, 2001a).

7.12 SUMMARY

- Heterodyning, or frequency mixing, is used in most radio receivers and many frequency synthesizers and transmitter exciters.

- The desired IF is the sum or difference between the RF and a stronger LO frequency.
- Spurious products are also created in mixing.
- A spur-level table describes the levels of spurious products for a particular mixer.
- An $m \times n$ spur occurs at a frequency $f_{IF} = mf_{LO} + nf_{RF}$.
- The ratio of spur amplitude to the desired $(\pm 1 \times \pm 1)$ IF amplitude decreases $(n - 1)$ dB for each dB reduction in the RF signal (with fixed LO power).
- A study of doubly balanced mixers has found that this ratio tends to remain unchanged when the LO power changes proportionally to the signal power (equal changes in dB).
- Two-signal IM levels depend on the same nonlinearities that produce spurs.
- IMs are produced effectively at the mixer input with amplitudes related to the $1 \times n$ spurs as if those spurs were nth harmonics.
- Plots of spurious responses can be drawn for constant LO frequency or for constant IF frequency and can be normalized to either.
- Advantageous regions for conversion can be identified from spur plots.
- Rectangles and polygons can be drawn on the spur plots to represent conversion regions. The spur curves pass through them where in-band spurs occur.
- On a spur plot that is normalized to the IF, a rectangle represents a given LO range and RF band. As the IF changes, the rectangle moves diagonally, producing polygons that represent also the IF range.
- Shape factors required to filter spurs can also be found from these plots.

ENDNOTES

[1]Bullock (1995), Rohde and Bucher (1988), and Tsui (1985).

[2]Henderson (1989, 1993a, 1993b, 1993c), Cheadle (1993), Maas (1993), and Egan (1998, pp. 36–43, 64–66).

[3]A harmonic of the LO is sometimes used to generate the IF ($f_{IF} = mf_{LO} \pm f_{IF}$), particularly at very high frequencies. We will not treat this case here but there would be many similarities.

[4]In some other cases, the frequency conversion occurs in the feedback path of a phase-locked loop (Egan, 2000, pp. 344–348). Considerations are similar to those described here but can be complicated by sampling effects (Egan, 2000, pp. 87–94).

[5]See also Egan (2000, pp. 8–10).

[6]Henderson (1993a) does develop equations for spur levels in doubly balanced mixers as functions of various balance parameters (see also Roetter and Belliveau, 1997). While these show dependence on LO power as well as RF power, Henderson indicates that the latter dependence is more reliable in practice.

[7]We use data for a Class 1 mixer at −10 dBm in Fig. 7.4 because there are more measured levels for this mixer. The fact that these levels may not be as accurate for predicting spur levels as we would like in some actual mixer does not detract from its usefulness in explaining the theory. If we desire, we can assume that Fig. 7.4 represents a mixer that does perform in accordance with theory over the range of signal powers at which we use it, or we can just recognize the errors possible in the approximation.

CHAPTER 8

CONTAMINATING SIGNALS IN SEVERE NONLINEARITIES

It is debatable whether a source of undesired, or contaminating, power should be called a signal, but that is what we are referring to here, for lack of a better term. A contaminating signal in a linear module or cascade can be treated in the manner we have studied, like a desired signal. There are times, however, when an undesired signal is sent through a severe nonlinearity, and we must understand what happens at the output of that nonlinearity. When we use a nonlinear module, it is usually characterized so we can tell how it responds to a single driving signal, but how does it respond to an accompanying contaminating signal?

There are three characteristics, which commonly apply to such a process, that give us a handle on the analysis. First, there is a desired signal driving the nonlinearity; otherwise we would just shut down the path to isolate the contaminant. Second, the nonlinearity is severe so it generally limits the output amplitude of the desired signal. Third, the contaminant is much smaller than the desired signal.

In looking for an alternative to a complicated nonlinear analysis, we take advantage of these relationships to find an easier way to analyze the effect of a contaminant that accompanies a large signal as they both pass through a device that provides amplitude limiting. Our process (Egan, 1981) will be to characterize the contaminant in the presence of the desired signal as a single sideband on that signal and to decompose that sideband into an equivalent combination of sidebands that represent AM and that represent FM (or phase modulation, PM — they are basically the same). We do this because we can often determine the responses to the AM and FM separately.

8.1 DECOMPOSITION

Figure 8.1 shows the sum of a strong and a weak sinusoid in three representations.[1] The waveforms are shown as a function of time at (*a*), in a Fourier frequency domain representation (Bracewell, 1965, pp. 79–80) at (*b*), and in a phasor representation at (*c*). The right part of the representations at (*b*) and (*c*) are straightforward equivalents to the signals at (*a*). They are also, however, entirely equivalent to the representations to their left, where certain components add and others cancel. The advantage of the representations on the left is that they give us another way to look at the two signals. They represent a sum of AM and FM on a carrier, the latter being the strong signal.

The nature of the representations on the left can be verified by referring to the representations of AM and FM shown in Figs. 8.2 and 8.3, respectively. Consequently, we can represent a small contaminating signal plus a strong desired signal as simultaneous AM and FM of the strong signal.

Each member of each pair, whether AM or FM, is half (−6 dB) the amplitude of the original small signal and is offset in frequency from the large signal by the

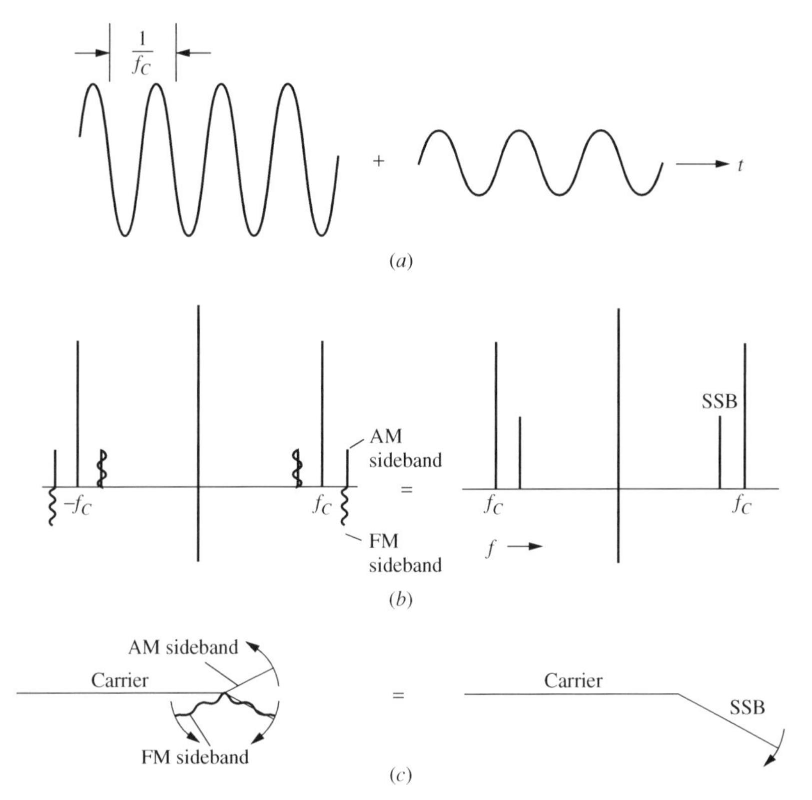

Fig. 8.1 Decomposition of SSB into AM and FM: (*a*) time, (*b*) Fourier, and (*c*) phasor representations. (From Egan, 2000.)

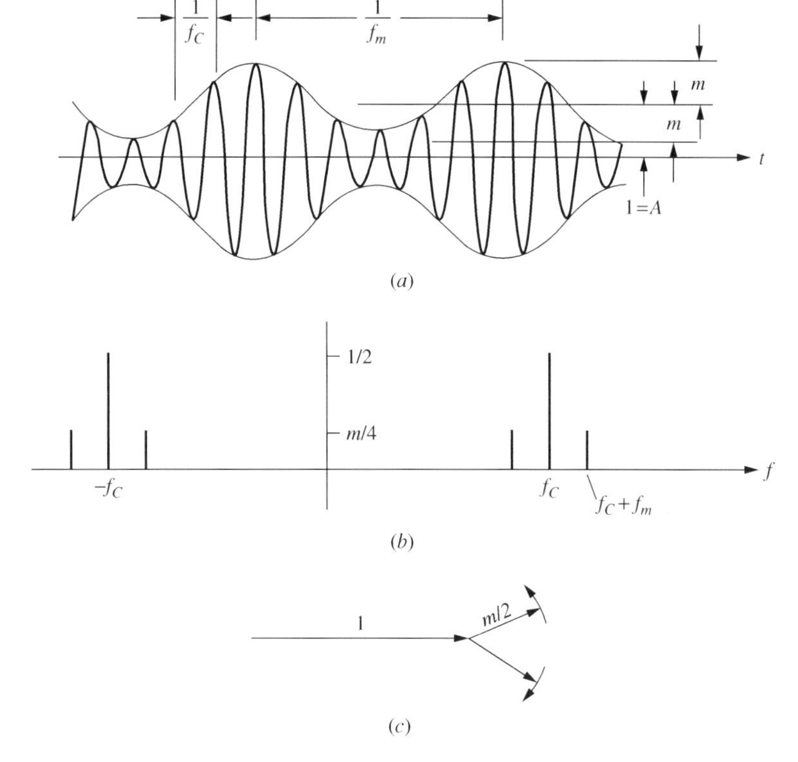

Fig. 8.2 AM: (*a*) time, (*b*) Fourier, and (*c*) phasor representations. (From Egan, 2000.)

separation between the original signals, which is f_m, the modulation frequency of the AM and the FM. The amplitudes of these sidebands relative to the strong signal (their carrier) equal $m/2$, where m is the modulation index. For FM or PM, this is the peak phase deviation (in radians):

$$m = \frac{\Delta f}{f_m}, \tag{8.1}$$

where Δf is the peak frequency deviation and f_m is the modulation frequency (both in the same units). For AM, it is

$$m = \frac{\Delta a}{a}, \tag{8.2}$$

where a is the average amplitude of the sinusoid and Δa is the peak change in amplitude. Therefore, a small contaminant that has amplitude relative to the large signal of

$$m = \frac{a_{\text{weak}}}{a_{\text{strong}}} \tag{8.3}$$

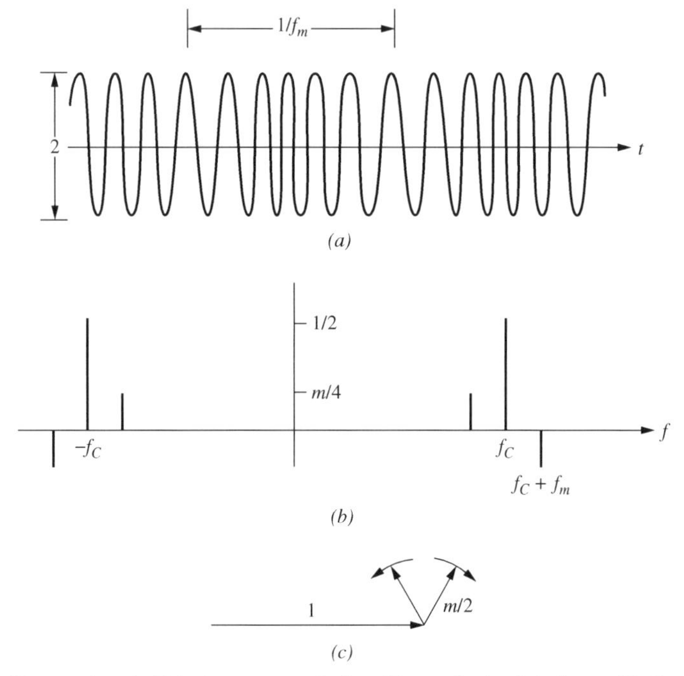

Fig. 8.3 Narrow-band FM (exaggerated for illustration): (*a*) time, (*b*) Fourier, and (*c*) phasor representations. (From Egan, 2000.)

is equivalent to a pair of sidebands representing AM with a modulation index of m plus a pair representing FM, also with a modulation index of m. The phases are such that two of the sidebands from each pair add to produce the original small signal, and the other two cancel each other.

This representation is only valid for a contaminant that is much weaker than the strong signal because Eq. (8.1) is only valid for small m. FM sidebands occur at all offsets from the carrier that are multiples of f_m, and we are able to use this representation only because, for small enough m, the other sidebands can be neglected. The carrier has amplitude $AJ_0(m) \approx A$, and the first sideband at $\pm f_m$ has amplitude $AJ_1(m) \approx m/2$. For example, for $m = 0.1$, the error in approximation for the carrier amplitude is only 0.25% and it is 0.12% for the first sideband while the second (unrepresented) sideband has an amplitude that is only about 2.4% of the represented sideband.

Parts (*b*) of Figs. 8.1, 8.2, and 8.3 represent amplitude modulation by a cosine and frequency modulation by a negative cosine (or phase modulation by a negative sine) of a cosine carrier and also show the equivalence of that to the addition of a small cosine. When the absolute or relative phases of the two original signals change, the decomposition still works and produces the same AM and FM deviations as given above, but the phases of the component signals change. For example, 180° phase shift in the frequency modulation to give a positive cosine would produce the upper sideband rather than the lower.

8.2 HARD LIMITING

A hard limiter (Egan, 1998, p. 398) produces a rectangular waveform of fixed amplitude with transitions synchronized to those of the input signal. The fundamental component of this output has the same frequency as the input signal. The harmonics may be removed by filtering.

What do we see at the output of a hard limiter if a strong signal, which becomes hard-limited, and a weak contaminant enter the limiter? We can decompose the pair entering the limiter into the strong signal plus AM and FM sidebands. The AM will be eliminated by hard limiting, leaving only the FM. The limiting level will determine the amplitude of the strong output signal. The phase deviation of the output and of its fundamental component will not be altered by limiting so the FM sidebands on the fundamental will retain the same level, relative to the desired signal, at the limiter output that they had at its input. Thus there are sidebands on the fundamental output at $\pm f_m$ that have voltage amplitudes, relative at the carrier (strong signal), of $m/2$, half (−6 dB) of the relative amplitude of the single-sideband contaminant. This is illustrated in Fig. 8.4. (We do not show the harmonics that are also produced by this process but which may have been filtered out. Relative sideband levels on those harmonics will be higher in proportion to the harmonic number, as explained for frequency multipliers in Section 8.6.)

While a hard limiter does not pass AM, AM can be converted to FM if the input level at which the output transition occurs is not centered on the input waveform. In that case, a change in input amplitude will change the time at which the transition occurs.

8.3 SOFT LIMITING

Figure 8.5 shows the amplitude response characteristic for typical saturating nonlinearities. The operating point is often close to the flat region of maximum output. We can write the slope S of the curve in Fig. 8.5 as

$$\frac{dP_o}{dP_i} = \frac{d(10\text{ dB}\log_{10} p_o)}{d(10\text{ dB}\log_{10} p_i)} = \frac{d\ln(p_o)}{d\ln(p_i)} = \frac{dp_o/p_o}{dp_i/p_i}. \tag{8.4}$$

For small changes, this is also the ratio of the relative change in output amplitude to the relative change in input amplitude, that is, the ratio of AM modulation

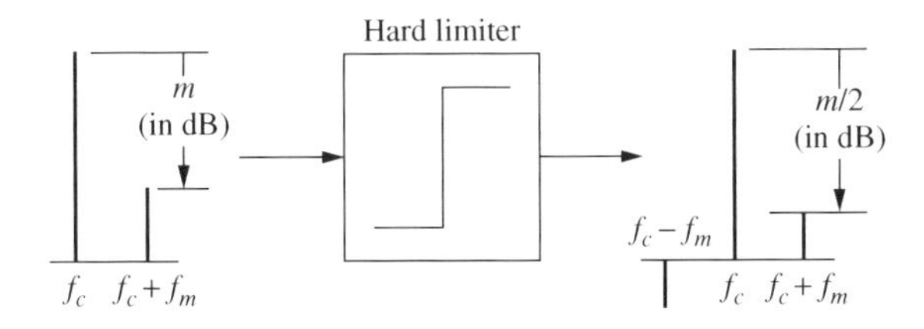

Fig. 8.4 Contaminant passing through hard limiter.

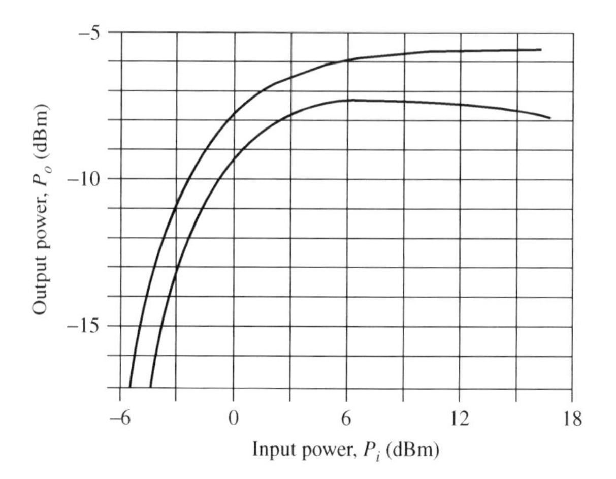

Fig. 8.5 Saturating nonlinearities.

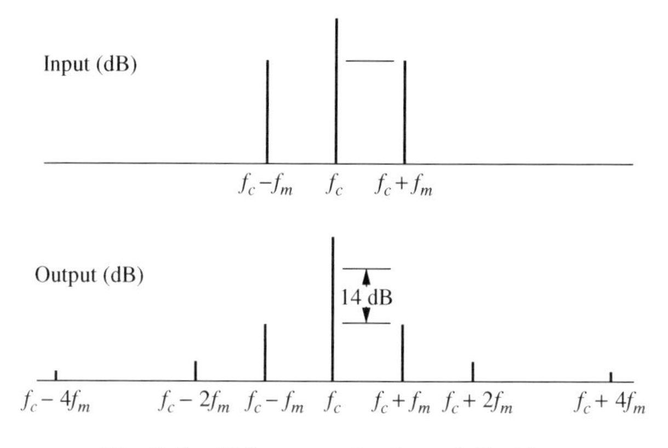

Fig. 8.6 AM suppression by soft limiting.

indexes. To see this, we write the right side of Eq. (8.4) in terms of voltages:

$$\frac{dP_o}{dP_i} = \frac{[(v_o + dv_o)^2 - v_o^2]/v_o^2}{[(v_i + dv_i)^2 - v_i^2]/v_i^2} = \frac{[2v_o dv_o - d^2 v_o]/v_o^2}{[2v_i dv_i - d^2 v_i]/v_i^2}. \tag{8.5}$$

Neglecting the relatively small squared differentials, this is

$$\frac{dP_o}{dP_i} = \frac{dv_o/v_o}{dv_i/v_i} \approx \frac{m_{\mathrm{AM}o}}{m_{\mathrm{AM}i}}. \tag{8.6}$$

Thus the transfer gain for the AM modulation index is the slope of the transfer characteristic in dB. If the slope is 0.2 dB out per dB in, m_{AM} will be multiplied by 0.2 (−14 dB) in passing through the nonlinearity (Fig. 8.6).

The modulation would generally become distorted in the limiting process (e.g., increases in amplitude might be suppressed more than decreases) so the reduced AM sidebands are likely to be accompanied by other sidebands at offsets that are harmonics of the input modulation frequency. These are shown in Fig. 8.6.

As before, the FM sidebands on the fundamental are not changed:

$$m_{\mathrm{FM}o} = m_{\mathrm{FM}i}. \tag{8.7}$$

8.4 MIXERS, THROUGH THE LO PORT

Here we consider how a contaminant on the LO is transferred to the IF (Egan, 2000, pp. 78–84).

8.4.1 AM Suppression

Figure 8.7 shows a mixer conversion loss characteristic. It is labeled for conversion gain $G_c(P_L)$, a function of LO power, P_L. Since the IF output power is

$$P_o(P_L) = G_c(P_L) + P_i, \tag{8.8}$$

when the IF input power P_i is fixed, $P_o(P_L)$ equals $G_c(P_L)$ plus a constant. [For example, the axis values for $P_o(P_L)$ and $G_c(P_L)$ are numerically the same if $P_i = 0$ dBm.] Therefore, the slope of $P_o(P_L)$ and of $G_c(P_L)$ at a given value of P_i are the same, and the AM transfer characteristic, from the LO to the IF, is given by Eq. (8.6), written for this case as

$$\frac{m_{\mathrm{AM,IF}}}{m_{\mathrm{AM,LO}}} = \frac{dG(P_L)}{dP_L}. \tag{8.9}$$

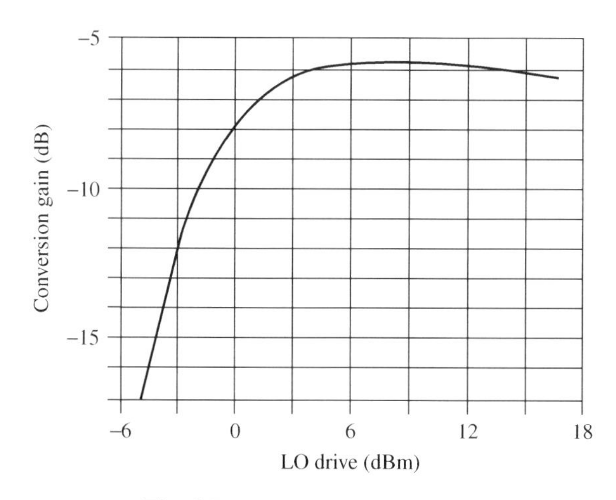

Fig. 8.7 Conversion loss curve.

Since the mixer will usually be operated for low conversion loss, this will usually be a small number ($\hat{S} \ll 1$), and AM on the LO is suppressed when it appears in the IF.

8.4.2 FM Transfer

According to Eq. (7.1), any frequency change in the LO will cause a change of equal magnitude in the IF. Thus both the peak frequency deviation Δf and its modulation frequency f_m will transfer from the LO to the IF, and the FM sidebands in the IF will have the same relative level as the FM sidebands in the LO (i.e., $m/2$). In the IF, FM from the LO is indistinguishable from FM from the RF (Fig. 8.8). (FM will be greater on LO harmonics than on its fundamental, but we are primarily interested in the effect on the desired 1×1 products in the IF.)

8.4.3 Single-Sideband Transfer

If the conversion gain curve is horizontal at the operating point, the transfer from LO to IF will be like transmission through a hard limiter. A sideband of relative value m on the LO will produce two FM sidebands of amplitude $m/2$ each in the IF (as in Fig. 8.4). If the conversion loss curve has a positive slope, the amplitude of the sideband in the IF that is at the frequency to which the single sideband (SSB) on the LO would have converted will increase, and the amplitude of the other will decrease, due to the addition and subtraction, respectively, of the FM and attenuated AM sidebands. (If the slope were 1, there would be no other sideband, just a single frequency conversion of the SSB.) A negative

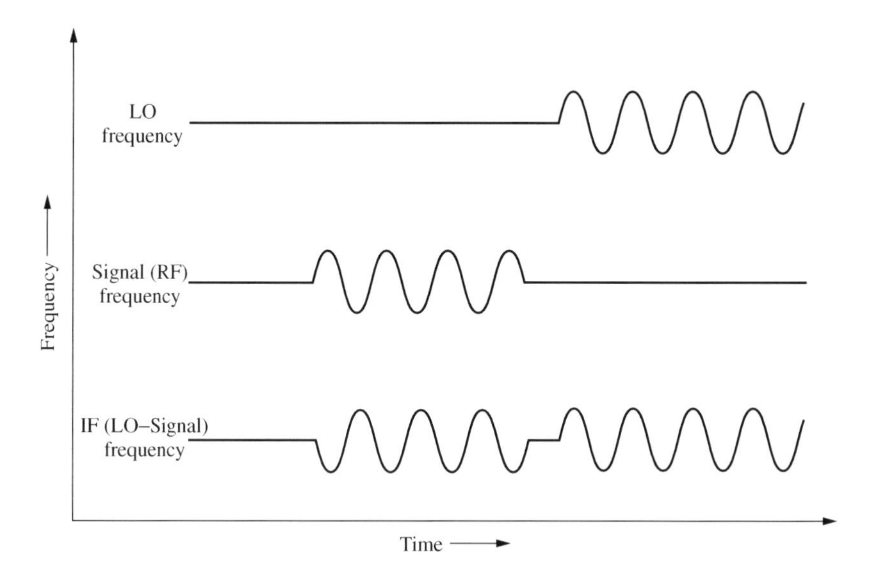

Fig. 8.8 Transfer of frequency modulation through a mixer. (From Egan, 2000.)

slope on the conversion loss curve would cause the larger sideband to be on the other side.

Example 8.1 Transfer from the LO The desired LO is accompanied by a contaminating sine wave that is 30 dB weaker and higher in frequency by 1 MHz. The mixer transfer loss curve increases 1 dB for each 4-dB increase in the input power at the operating point. If the IF is the sum of the RF and LO frequencies, show the IF spectrum.

See Fig. 8.9. The equivalent AM and FM sidebands on the LO are at −36 dBc. The FM sidebands are transferred to the IF along with the AM sidebands, the latter multiplied by the slope $\frac{1}{4}$. The AM sidebands in the IF therefore have amplitude one fourth of the FM sidebands. On the high side, the sum is $\frac{5}{4}$ (+1.94 dB) higher than −36 dBc and, on the low side, the sum is $\frac{3}{4}$ (−2.5 dB) relative to −36 dBc. Thus the high sideband has amplitude −36 dBc + 1.94 dB = −34.06 dBc, and the low sideband has amplitude −36 dBc − 2.5 dB = −38.5 dBc. The AM and

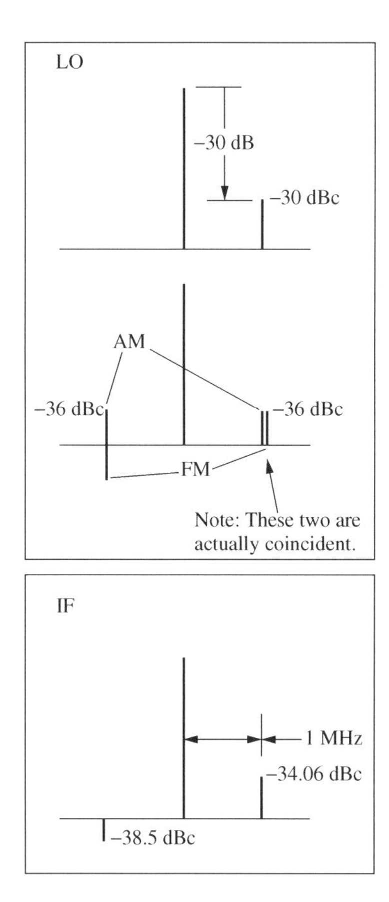

Fig. 8.9 Example 8.1.

FM sidebands reinforce on the side where the original single sideband existed in the LO and oppose on the side where there was no sideband in the LO.

8.4.4 Mixing Between LO Components

The desired LO and its undesired sideband will mix in the nonlinear elements of the mixer, producing sum and difference components and other, usually weaker, components. When a mixer is balanced to cancel the LO in the IF, these products will also tend to cancel.

If we considered the contaminating sideband on the LO as decomposed into AM and FM, we find that the two FM sidebands cancel on translation to their difference frequency f_m while the AM sidebands at f_m reinforce.[2] It does not seem surprising that an LO that is frequency modulated does not produce mixing products. Looking at it this way, there is only one signal; its frequency is just changing. On the other hand, with AM, the IF signal at f_m is basically the result of a detection process. We get the same results whether we decompose the single sideband or not. The difference is important when we do have AM or FM rather than a decomposition involving both. Even if the contaminant is pure FM at some point, however, it can produce AM by passing through a frequency-sensitive (not flat) circuit.

In an unbalanced mixer (e.g., a single diode), the results of two signals from the LO mixing are the same as when one is considered the RF; the resulting IF is weaker than the small signal by the conversion loss. When the LO is balanced, however, we get an additional reduction due to balance, perhaps 20 dB. We will call that additional reduction the balance of the LO port.

Mixing also occurs between multiple RF signals (see Section 7.3), but their amplitudes are small enough that the result tends to be small compared to the results of mixing with the LO. Some mixing products will also be reduced by any RF balance (e.g., in a doubly balanced mixer).

8.4.5 Troublesome Frequency Ranges in the LO

We can highlight some frequency ranges in the LO that have a potential for producing contaminating signals in the IF. Refer to Figs. 8.10 and 8.11.

8.4.5.1 Range 1 The problem in Range 1 in Fig. 8.10 is transfer of FM sidebands (actual or equivalent) to the IF band. This is a "single-frequency spur," requiring one RF signal to exist. At the edge of this band, the LO-induced sidebands are separated from the RF signal by the IF or RF bandwidth, whichever is greater. A contaminant in Range 1 will not always produce a contaminant in the IF. It depends on the signals that are in the RF, but this is the area of danger. See the examples below for a better understanding of this.

If the RF bandwidth is greater than the IF bandwidth, a signal on one edge of the RF filter might acquire a sideband as far away as the other edge. This

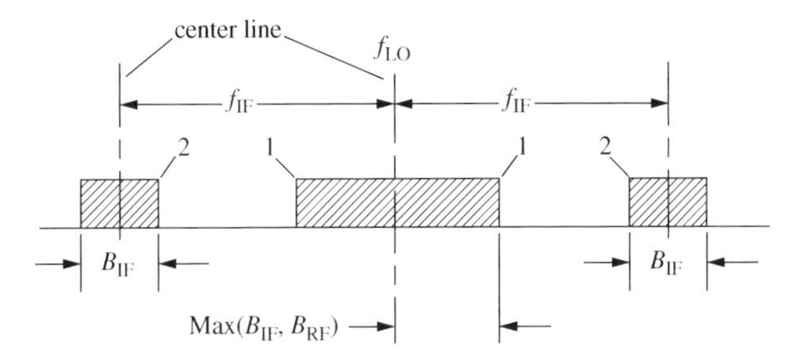

Fig. 8.10 Ranges in LO that can produce troublesome sidebands in the IF. (From Egan, 2000.)

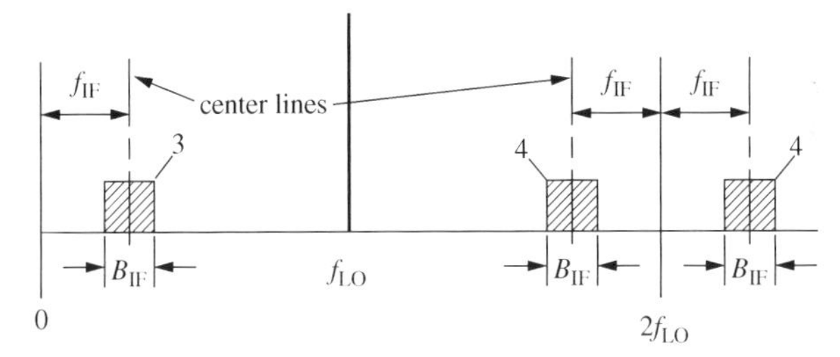

Fig. 8.11 Ranges where spurs on the LO can produce leakage to the IF. (From Egan, 2000.)

could be converted to the IF since, presumably, any part of the RF band can be converted into the IF band. This is illustrated in Fig. 8.12*a*, where the x axis represents the RF and the RF equivalent to the IF, that is, the RF frequency that would be translated, by the conversion process, to the IF shown. (We could as easily let x represent the IF frequency in the IF filter and the RF frequency that will translate to the IF.) For the case illustrated, the contaminant would enter the IF band when the latter is translated to the low end of the range shown (i.e., when the IF band slides to the left arrow point).

However, if the IF band is wider (this might occur in the mixing of two synthesized frequencies to give a wider frequency range), as shown at Fig. 8.12*b*, the sideband can be as far from the signal as the IF bandwidth because, when the edge of the RF band where the signal is located is translated to one edge of the IF band, the sideband can be at the other edge of the IF band. It is necessary for the RF signal to pass through the RF filter, but it is not necessary for the equivalent sideband to be within the RF band because it is created in the mixer after the RF filter. For the case illustrated, the contaminant enters the IF band when the tuning is as shown. When the LO changes such that the IF band moves right, converting the RF band to a different portion of the IF band, the contaminant shown will no longer be in band.

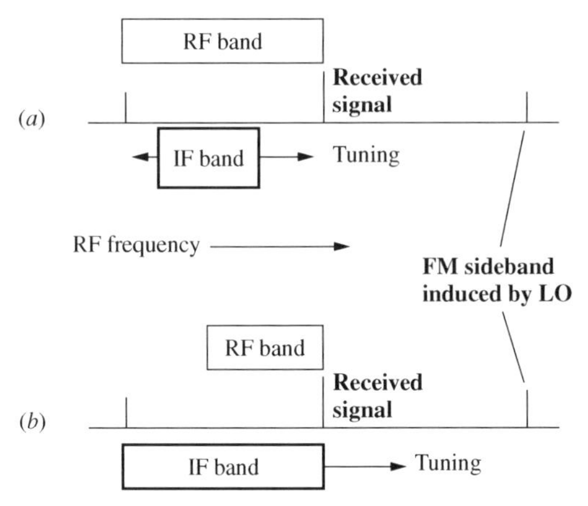

Fig. 8.12 Maximum sideband offset: determined by RF bandwidth at (*a*) and by IF bandwidth at (*b*). The IF is shown vs. equivalent RF.

If the sidebands are just *equivalent* FM sidebands (obtained from a contaminating single sideband), the sideband whose frequency could be mathematically obtained by conversion, using the contaminant as the LO, will not move with the LO. A change of Δf in the LO frequency moves the signal in the IF by Δf, leaving one of the sidebands unmoved (while the other moves by $2\Delta f$). In Fig. 8.12*b*, if the left sideband is such a contaminant, it will maintain its position in the IF as the LO moves. If, however, the right sideband is fixed, the left sideband will move when the LO tunes and it might move away from the IF band.

We will now look at an example of true FM sidebands plus four examples of equivalent FM sidebands covering the four variations of RF SSB frequencies and LO frequencies at their extremes. These will demonstrate the validity of this range as a source of spurs.

Example 8.2 FM Contaminant Transferred from LO to IF The RF band and signal are shown in Fig. 8.13*a*. Figure 8.13*b* shows the LO at the frequency that converts the upper edge of the RF band to the lower edge of the IF band. Figure 8.13*c* shows the other end of the LO range, which would convert the low end of the RF to the high end of the IF. The tuning range of the LO equals the difference between the RF and IF bandwidths.

These contaminating sidebands are true FM sidebands, and they are at the maximum problematic offset, according to Fig. 8.10, Range 1. In this case, that offset is the RF bandwidth, 15 MHz. At Fig. 8.13*b*, we see that the RF signal is converted to one edge of the IF while, at the other extreme of the LO, shown at (*c*), the contaminating signal arrives at the IF band edge, showing that the assumed sideband separation is the true limit if the modulation frequency f_m is fixed. The relative sideband amplitude in the LO, −66 dBc, is transferred to the IF.

Example 8.3 SSB Contaminant on Verge of Transfer from LO to IF Figure 8.14*c* is like Fig. 8.13*c* but the upper FM sideband has been dropped

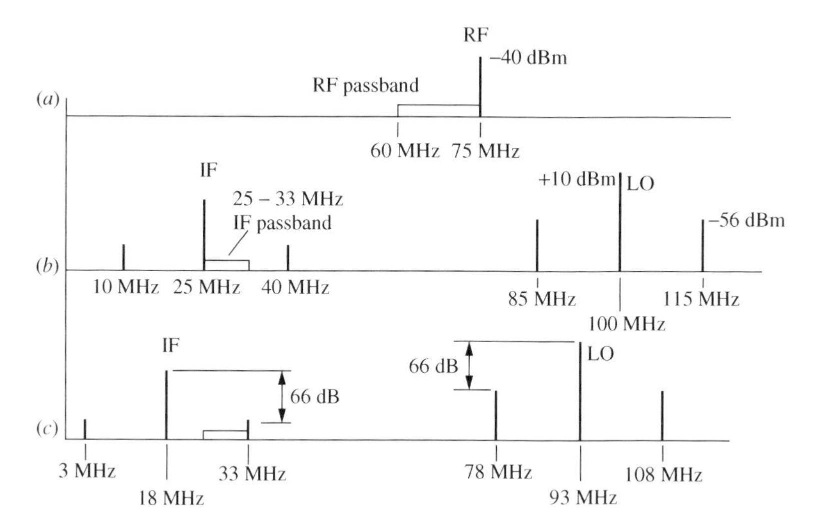

Fig. 8.13 FM contaminant transferred from LO to IF.

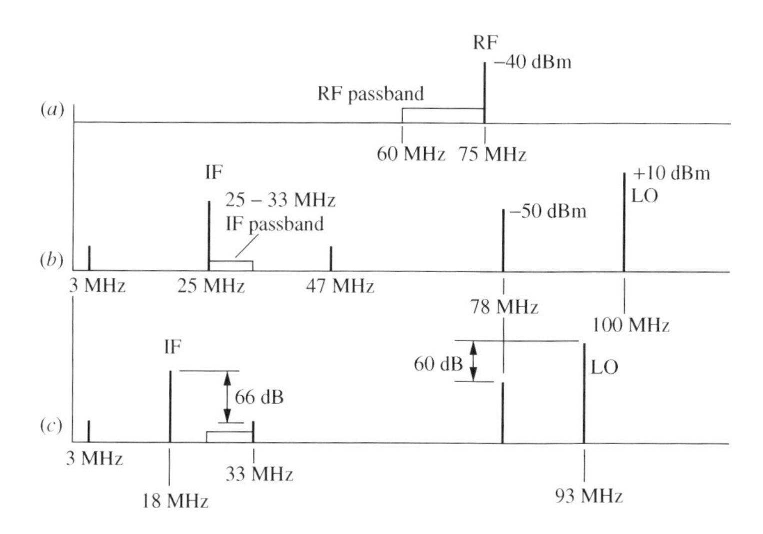

Fig. 8.14 SSB contaminant transferred from LO to IF, RF at high end.

and the remaining sideband is at −60 dBc relative to the LO. The equivalent FM sidebands are therefore the same for the two figures, with a 6-dB level reduction occurring in going from single sideband to equivalent FM sidebands. The difference is that the SSB contaminant is fixed in frequency so, when the LO moves, the modulation frequency f_m of the equivalent FM, which equals the separation between the LO and the contaminant, changes. As a result, when the LO goes higher in frequency from (*c*) to (*b*), the sideband moves away from the IF band twice as fast as it did with fixed f_m in Fig. 8.13. Nevertheless, as in the previous case, the minimum separation equal to the RF bandwidth barely allows the contaminant into the IF band.

Note that one of the sidebands in the IF does not move. This is at the IF frequency that would be obtained by using the SSB on the LO as an LO for converting the RF. The IF signal is offset from this stationary sideband by the separation between the LO and its single sideband, and the other sideband is offset by twice that frequency difference.

Example 8.4 SSB Contaminant Not Transferred In Fig. 8.15*b*, the other sideband from Fig. 8.13*b* has been retained as a single sideband. Now the stationary sideband is at 40 MHz in the IF and no sideband gets close to the IF band, but this is because of the particular RF frequency in this example.

Example 8.5 SSB Contaminant on Verge of Transfer with Signal at Other End of RF Band The only independent condition that changes between Figs. 8.15 and 8.16 is the end of the RF band where the signal appears. The LO and its contaminating sideband are the same. The fixed sideband is now at 55 MHz in the IF. In Fig. 8.16*c*, the signal is converted to the high end of the IF band whereas it goes out of the IF band as the LO is tuned toward its high end, shown at 8.16*b*. At Fig. 8.16*b*, however, where the signal is well out of the IF band, the contaminant has just reached it. This happens just at the offset limit given by the RF bandwidth.

Example 8.6 SSB Contaminant Not Transferred with Signal at Other End of the RF Band The difference between Figs. 8.17 and 8.16 is that the contaminant has been moved to 15 MHz below the LO in 8.17*c*. The LO picture is

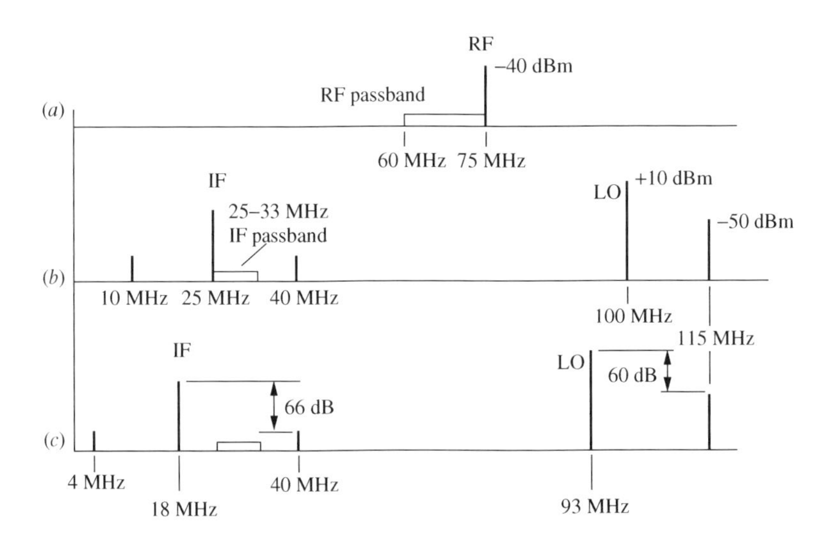

Fig. 8.15 SSB contaminant not transferred to IF, RF at high end.

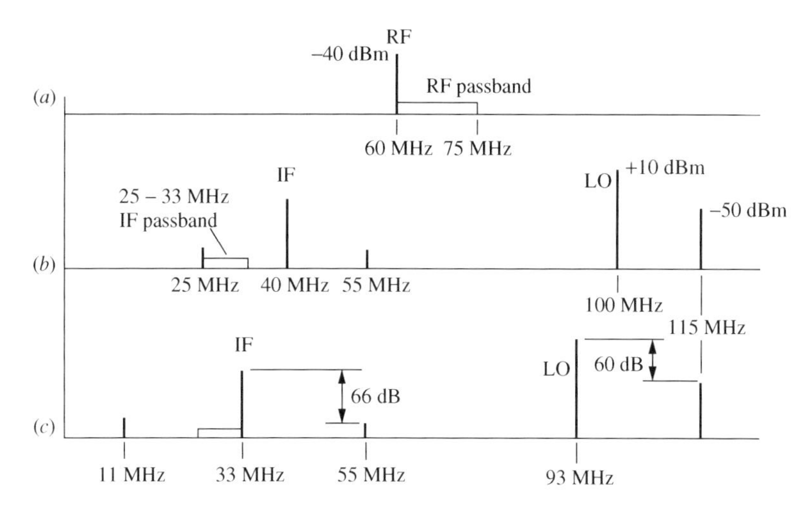

Fig. 8.16 SSB contaminant transferred from LO to IF, RF at low end.

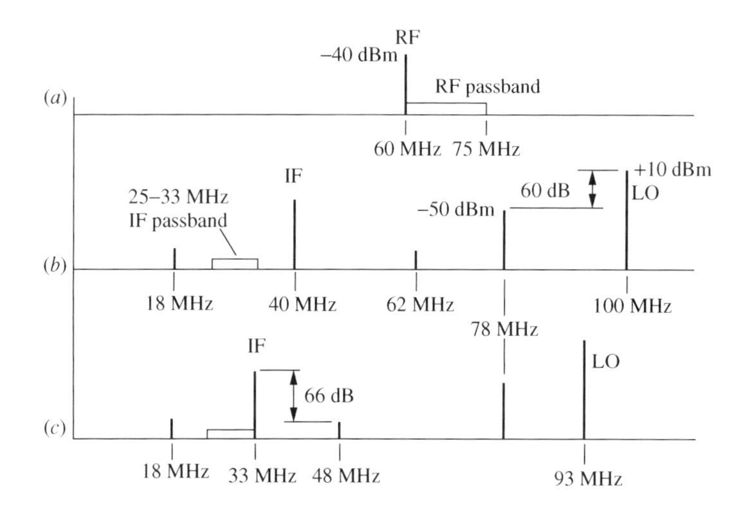

Fig. 8.17 SSB contaminant not transferred to IF, RF at low end.

the same as in Fig. 8.14. This causes the contaminant appearing at 18 MHz in the IF to be stationary as the LO moves away from the contaminant, going from Fig. 8.17*c* to 8.17*b*. Results are similar to Fig. 8.15 for the other end of the RF band. The contaminant does not enter the IF band.

Considering these four examples, if the SSB contaminant remains more than the RF bandwidth from the LO, we will not get a contaminant in the IF; if it is closer than that, we can get a contaminant in the IF. Therefore, Range 1 is a valid danger area for both SSB and FM sidebands when the RF bandwidth exceeds the IF bandwidth. A similar set of examples, for the case where the IF bandwidth exceeds the RF bandwidth, confirms the IF bandwidth as the maximum contaminant offset for that case. (AM suppression was assumed for these examples.)

8.4.5.2 Range 2 A contaminant in Range 2 (Fig. 8.10) can mix with the LO to produce a signal in the IF passband that might leak through. This is an internal spur because it exists independently of the presence of a signal at the RF. The amplitude of the contaminant will be reduced by the conversion loss and balance of the LO port.

The balance for the contaminant probably has to be obtained from the LO-to-IF isolation I_{LI} (a positive number of dB), although the frequency difference of these two signals might affect the accuracy of the approximation. A pair of AM sidebands appearing in both parts of Region 2 will increase the contaminant in the IF by 6 dB over a single sideband of the same amplitude but FM sidebands in this region (possibly produced by limiting) will not produce an IF contaminant.

Example 8.7 LO Contaminant Converted into IF The LO and its contaminant are shown in Fig. 8.18 along with the IF passband and the contaminant induced into the IF by this process. The 89-MHz LO and 115-MHz contaminant mix to give 26 MHz. After a 7-dB conversion loss, the resulting 26-MHz signal has level (−50 dBm − 7 dB) = −57 dBm. After rejection by 23 dB of balance, the level is (−57 dBm − 23 dB) = −80 dBm in the IF.

8.4.5.3 Range 3 This is also an internal spur, requiring no RF signal. Frequencies in Range 3 (Fig. 8.11) are in the IF passband so only balance protects the IF from signals at these frequencies (although limiting in the LO circuitry could reduce an incoming single sideband as much as 6 dB).

Example 8.8 LO Contaminant Leaking into IF The same contaminant in the IF as in Example 8.7 could be produced by a contaminant at the LO port at frequency 26 MHz and amplitude (−80 dBm + 23 dB =) −57 dBm, where 23 dB is the LO port balance.

Fig. 8.18 Contaminant transferred from LO to IF.

8.4.5.4 Range 4 A contaminant in Range 4 (Fig. 8.11) has equivalent AM sidebands on the LO that may be attenuated by limiting action, leaving a partially uncanceled equivalent FM sideband in Range 3. It will be at least 6 dB weaker than the original. From there it can leak to the IF, reduced by balance, to produce an internal spur. The process depends on limiting in the LO circuitry (perhaps within the mixer). This is not the same as the limiting shown in Fig. 8.7, which applies to the transfer of amplitude modulation to a signal. If there is no limiting on the LO signal itself, the process of this region will not occur.

Example 8.9 LO Contaminant Equivalent Sideband Leaking into IF *Given:* Hard LO limiting, 23-dB LO-to-IF balance, and two contaminants as shown in Fig. 8.19.

The 87-MHz contaminant is decomposed into sidebands at $\pm(87 - 50 =)$ 37 MHz offset from the 50-MHz LO. The lower of these sidebands is at $(50 - 37 =)$ 13 MHz in the IF passband. The level of the contaminant is -50 dBm so the equivalent FM sidebands are at -56 dBm. This is attenuated 23 dB by the LO-to-IF balance, causing it to arrive at $(-56\text{ dBm} - 23\text{ dB}) = -79$ dBm in the IF.

The 118-MHz contaminant is decomposed into sidebands at $\pm(118 - 50 =)$ 68 MHz on the 50-MHz LO. The lower sideband is at $(50 - 68 =) - 18$ MHz. The absolute value is 18 MHz and a sinusoid at this frequency is produced in the IF. Since the LO contaminant is 10 dB larger than the first one considered, the level in the IF at 18 MHz is $(-79\text{ dBm} + 10\text{ dB} =) -69$ dBm.

The negative frequency only affects the phase of the signal. If we used negative frequencies, as in a proper Fourier analysis, we would see that there is a +18-MHz component produced by the negative frequencies corresponding to the positive frequencies shown in Fig. 8.19. Together with the −18-MHz component produced by the positive frequencies, these two impulses at ±18 MHz represent an 18-MHz sinusoid.

8.4.6 Summary of Ranges

Table 8.1 summarizes the characteristics of the four ranges for SSB contaminants. The LO-part balance has here been equated to LO-to-IF isolation.

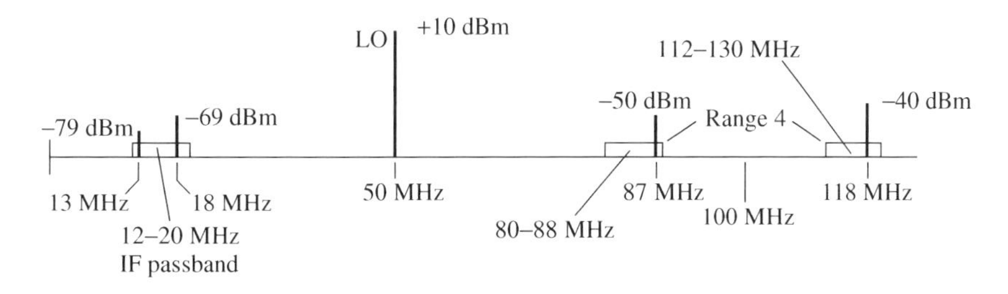

Fig. 8.19 Equivalent FM contaminant transferred from LO to IF.

TABLE 8.1 Characteristics of Troublesome Ranges in LO with Attenuation from LO to IF Shown for SSB Contaminant[a]

		Attenuations for a SSB			
Range	Process	Conversion Loss	SSB-to-FM Loss (dB)	LO-to-IF Isolation	Spur Type Produced
1	FM on signal		≈6		Single frequency
2	Mixing with LO	x	≥0	x	Internal
3	Leakage, LO-to-IF		0–6	x	Internal
4	SSB->FM on LO		≥6	x	Internal

[a]Not for modulation sidebands in pairs. Frequency-independent LO-IF isolation is assumed.

While this is an important guideline for alerting the designer to danger areas, the level of contamination experienced in practice is dependent on such things as the degree of LO limiting (possibly some of it buried within the mixer), internal mixer coupling, and response variations with frequency. It may be appropriate to determine these levels more precisely by experimentation at some stage in the design. The gathering and publication of data on IF contamination in these ranges for various mixers would be of significant value for many RF designers. In the mean time, the theory presented here gives us important information for initial system design.

One can conceive of other combinations of these processes that might produce internal spurs from contaminants on the LO, especially those in the vicinity of higher LO harmonics. However, removal of LO contaminants by means of filtering becomes relatively easy when they are well separated from the LO in frequency.

8.4.7 Effect on Noise Figure

Noise on the LO is transferred to the IF by the same processes discussed above for discrete signals. Ranges 2, 3, and 4 are portals for internal spurs, spurs that exist in the absence of an RF signal. When the contaminant is additive noise, these processes transfer noise from the LO to the IF, and thus increase the effective mixer noise without the necessity of a signal being present. This produces an increase in the mixer's noise figure. The transfer of FM noise to signals in Range 1 produces noise that varies with the strength of the signals and cannot be characterized as an increase in noise figure. It is the subject of the next chapter.

8.4.7.1 Computing the Increase If a noise power density of $k_n N_T$, where $N_T = \overline{k} T_0$ is thermal noise density and k_n is a multiplying factor, exists in Range 3 (Fig. 8.11), it will appear at the mixer IF after reduction by the LO-to-IF balance. Noise density in Range 4 will appear in the IF after being attenuated by the same amount plus at least 6 dB. Transfer of the noise in Range 2 is reduced by the conversion loss and balance. Table 8.1 applies to noise also.

If the reduction in noise power in the IF due to one of these transfer processes is k_{rj}, the increase in noise density in the IF, due to that process, will be $(k_{nj} - 1)N_T/k_{rj}$. Without noise from the LO, the noise at the mixer output is $N_T f_m g_m$, where f_m is the mixer's noise factor and g_m is its gain, the reciprocal of it conversion loss. Often $f_m = 1/g_m$, leaving just thermal noise in the IF. The ratio of noise factor with the LO noise to the noise factor without it, for N noise processes (there can be two process for each of ranges 2 and 4), is

$$\frac{f_{m.L}}{f_m} = \frac{f_m g_m N_T + (k_{n1} - 1)N_T/k_{r1} + (k_{n2} - 1)N_T/k_{r2} \cdots + (k_{nN} - 1)N_T/k_{rN}}{f_m g_m N_T}$$

$$= 1 + \frac{1}{f_m g_m} \sum_{j=1}^{N} \frac{k_{nj} - 1}{k_{rj}}. \tag{8.10}$$

Example 8.10 Mixer Noise Factor Increase Due to LO Noise The noise floor at the output of the LO oscillator is 20 dB above thermal noise. This may be due, for example, to 16 dB available gain and 4 dB noise figure acting on thermal noise at the input to the oscillator's active device. The oscillator level is 10 dBm, but 23 dBm is required for a particular high-level mixer, so 13 dB of gain is needed after the oscillator. This is shown in Fig. 8.20. Neglecting the noise figure of the amplifier (e.g., because the input noise is so high), the noise floor at the mixer's LO port is higher than N_T by (20 dB + 13 dB =) 33 dB ($k_n = 10^{(33/10)} = 2000$). If the balance from the LO port is 30 dB ($k_r = 10^{(30/100)} = 1000$), and if the mixer noise factor equals its conversion loss

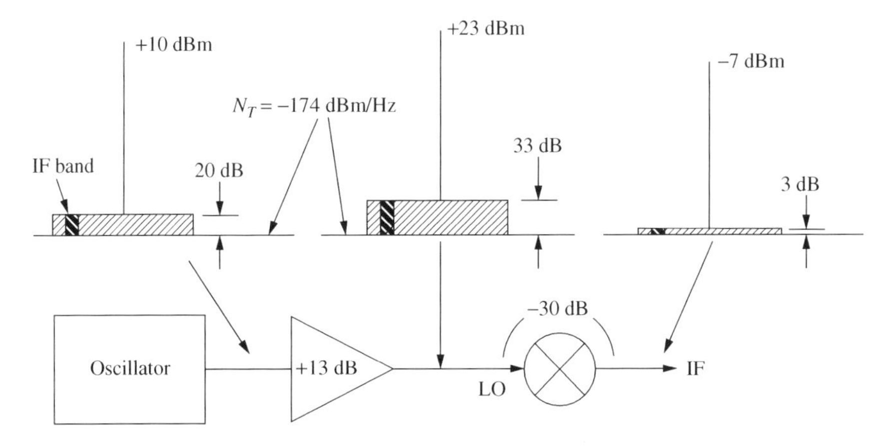

Fig. 8.20 Example of noise figure increase from broadband noise in LO.

($f_m g_m = 1$), Eq. (8.10) gives the increase in mixer's noise factor from Range 3 (LO to IF leakage) as

$$\frac{f_{m,L}}{f_m} = 1 + \frac{k_n - 1}{k_r}\frac{1}{f_m g_m} = 1 + \frac{1999}{1000} = 3 \Rightarrow 4.8 \text{ dB}. \tag{8.11}$$

Noise would also be added to the IF through the processes of Ranges 2 and 4. There are two frequencies in each of these ranges that will convert into the IF so there are four such processes in addition to the one from Range 3. For purposes of illustration, we will assume minimum loss from SSB-to-FM conversion, even though that assumption is not compatible for the various ranges. The reduction factor k_{rj} for Range 4 is then 6 dB greater than for Range 3 while k_{rj} for Range 2 is greater by the conversion loss, say 8 dB (a factor of 6.3 for power). Taking these five LO noise sources into account, Eq. (8.11) becomes

$$\frac{f_{m,L}}{f_m} = 1 + \sum_{j}^{N} \frac{k_{nj} - 1}{k_{rj}} \approx 1 + 1999\left[\frac{1}{1000}\left(1 + \frac{2}{4} + \frac{2}{6.3}\right)\right] = 4.63 \Rightarrow 6.7 \text{ dB}. \tag{8.12}$$

If the oscillator power had been only 0 dBm, requiring more LO amplifier gain, the increase in mixer noise figure would have been 15.7 dB. This shows the importance of starting with a high-power (as well as low-noise) oscillator. The difference between the value in Eq. (8.12) and that in Eq. (8.11) shows the importance of filtering out some of these ranges.

8.4.7.2 *Filtering the Noise* It may be possible to remove noise sources by filtering before the LO port. These ranges can be filtered most effectively if there is a relatively large frequency separation between them and the LO, so we can simultaneously attenuate frequencies in the range of concern and pass the LO. In performing a high-ratio downconversion (i.e., where the IF is much lower than the RF), the LO frequency will be close to the RF. Then the separation between Range 2 and the LO will be relatively (on a percentage basis) small so a high-Q filter will be required to reduce noise in Range 2. If, on the other hand, there is a high-ratio upconversion, the LO will be close to the IF, making Range 3 and the lower-frequency part of Range 4 difficult to filter.

8.4.7.3 *Oscillator Noise Sidebands* The situation can be further aggravated by an increase of noise near the LO center frequency. Oscillator power spectrums are not single-frequency lines (Egan, 2000, pp. 94–101, 106–118). They have finite widths due to noise modulation. Most of the oscillator's sideband noise power is FM; there is a fundamental process that causes this. However, there may also be AM noise caused, for example, by power supply noise that modulates the oscillator's amplitude. This may be difficult to observe because it is often masked by the FM noise when spectral power is observed, as on a spectrum analyzer. To separate the two noise types, the modulations, AM or FM, must first be separately detected.

Range 2 is not sensitive to FM noise on the oscillator. FM sidebands do not mix with the main LO power, but AM sidebands can. This is another reason why Range 2 may cause a problem in high-ratio downconversions, where it is close to the LO frequency.

AM sidebands may be attenuated by limiting in the LO circuitry, but FM sidebands will appear in the IF at the same level, relative to the LO leakage, that they had in the LO (assuming frequency-independent balance). Therefore, FM sidebands are reduced by the LO-to-IF isolation. Those that extend to Range 3 will enter the IF passband. They will not look like FM sidebands there; they will just be added noise. This is another reason that Range 3 may cause a problem in high-ratio upconversions, especially since FM sidebands on oscillators can be large.

Example 8.11 Noise with High-Ratio Upconversion A band near 1 MHz is to be converted to 200 MHz. The LO frequency is 201 MHz. The 10-dBm oscillator has a loaded Q of 25, an FM noise floor of -160 dBc/Hz, and noise sidebands falling -6 dB/octave in the range of interest. The mixer has LO-to-IF isolation of 26 dB.

FM noise typically falls at -6 dB or -9 dB/octave of frequency offset from the oscillator center frequency until it reaches the FM noise floor at the center frequency divided by $2Q$. Therefore, the FM noise will climb toward spectral center at offsets less than $(201\ \text{MHz}/50 =)\ 4$ MHz in this case. The LO power appearing in the IF is $(10\ \text{dBm} - 26\ \text{dB} =) - 16$ dBm, and the noise floor from the oscillator there is at $(-16\ \text{dBm} - 160\ \text{dBc/Hz} =) - 176$ dBm/Hz (2 dB below the available thermal noise). See Fig. 8.21. The IF is 1 MHz from the LO (and thus two octaves below the 4-MHz noise corner), so, with a slope of -6 dB/octave, the noise will be 12 dB greater than the noise floor, or -164 dBm/Hz, at the IF. This is 10 dB above the available thermal noise. Therefore, k_n/k_r in Eq. (8.10) is 10 and, if the mixer's noise figure equals its conversion loss without this noise, Eq. (8.10) shows a noise factor increase of $(1 + 10 =)$ 11, or 10.4 dB.

We can, alternatively, compute k_n and k_r separately. We obtain k_r as $(10^{26\ \text{dB}/10\ \text{dB}} =)400$. The oscillator noise floor is $(10\ \text{dBm} - 160\ \text{dBc/Hz} =) - 150$ dBm/Hz. The 12-dB increase from 4-MHz to 1-MHz offset gives -138 dBm/Hz at ± 1 MHz. This is $(-138\ \text{dBm/Hz} + 174\ \text{dBm/Hz} =)$ 36 dB above thermal noise. Therefore, k_n is $(10^{36\ \text{dB}/10\ \text{dB}} =)$ 4000 and $k_n/k_r = 4000/400 = 10$, as before.

To keep the noise density 6 dB below thermal noise (to give only a 1-dB increase in the total), a 201-MHz bandpass filter with 16 dB of attenuation at a 1% bandwidth (1 MHz from center) would be required. This might have significant loss at the center frequency.

Range 4 is important for SSB contaminants because they can convert to FM in Range 3, but it need not be considered for FM sidebands. They would already

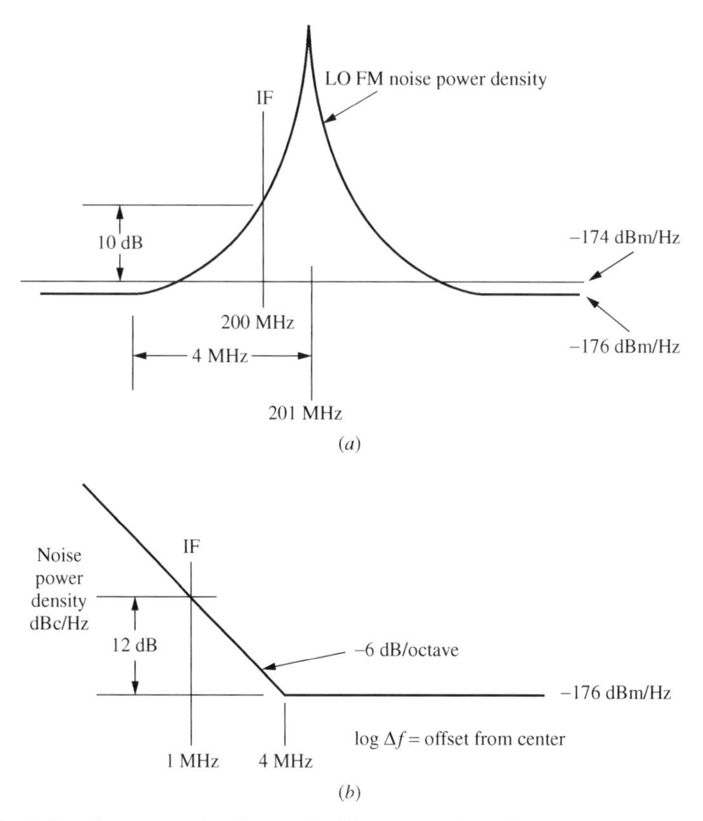

Fig. 8.21 LO noise power density at the IF port: (*a*) is a linear plot against frequency and (*b*) is a log plot of the tangential approximation.

be in Range 3, and we have found (Section 8.4.4) that FM sidebands do not mix with the LO.

8.5 FREQUENCY DIVIDERS

Frequency dividers (Egan, 2000, pp. 87–94) reduce the frequency of a signal by a constant factor. The most common type is a digital divider, consisting basically of bistable flip-flops and other logic circuits. The divider changes state when the effective edge (increasing or decreasing voltage) of the input signal passes threshold. There is essentially no transfer of AM through the divider; it acts as an ideal limiter. (AM-to-PM conversion can occur if the input is biased so switching occurs at a level other than the average input level.)

8.5.1 Sideband Reduction

The divider does pass, but modify, FM. The modulation frequency f_m is the same at input and the output; so is any change in the time of occurrence of a zero crossing. In other words, cause and effect happen at the same frequency and with only a time delay between them. What *is* different between output and input is the frequency; the output frequency being lower by the divide ratio, N. The frequency is divided by N uniformly so frequency deviation Δf is also divided by N. Thus the FM modulation index,

$$m_{\text{FM}} = \frac{\Delta f}{f_m}, \tag{8.13}$$

is smaller by a factor N at the output,

$$m_{\text{FM}o} = m_{\text{FM}i}/N. \tag{8.14}$$

As a result, the FM sidebands are smaller at the output by this factor.

Another way to look at this is that m is peak phase deviation, and it is related to the frequency f and the peak deviation of the zero crossing Δt by

$$m = f \times \Delta t, \tag{8.15}$$

so m is smaller by N at the output where the frequency is smaller by N.

A single-sideband contaminant, at the input to a divider, that has amplitude r relative to the signal there, has equivalent FM sidebands with relative amplitude of $r/2$ and will produce sidebands of relative amplitude $r/(2N)$ at the divider output. The output, and possibly the input, will not be a sinusoid. This relative sideband level applies to the harmonics at the output as well as the fundamental, but the effective divide ratio to a harmonic is smaller by the harmonic number [e.g., the relative amplitude at the third harmonic of the output would be $r/(2N/3)$].

Example 8.12 Frequency Divider Spectrums at Input and Output Figure 8.22 shows the spectrums at the input (*a*) and the fundamental output (*b*) of a $\div 5$ frequency divider with FM on the input. The input FM sidebands are at -46 dBc. This might represent decomposition of a -40-dB single sideband. The modulation frequency is $f_m = 1$ kHz, so the spectral lines are offset 1 kHz from the input carrier. The output carrier is at one fifth the input frequency. The FM sidebands there are still offset 1 kHz since the modulation frequency does not change in frequency division. Since the deviation has been reduced by 5, however, the modulation indexes are reduced by 5 also, so the relative sidebands are lower by

$$20 \text{ dB} \log_{10}(5) = 14 \text{ dB}. \tag{8.16}$$

8.5.2 Sampling

While reduction of FM sidebands by N is the primary effect observed at low modulation frequencies, a more complicated analysis must be applied at frequencies that are not low compared to the divider output frequency. We can receive

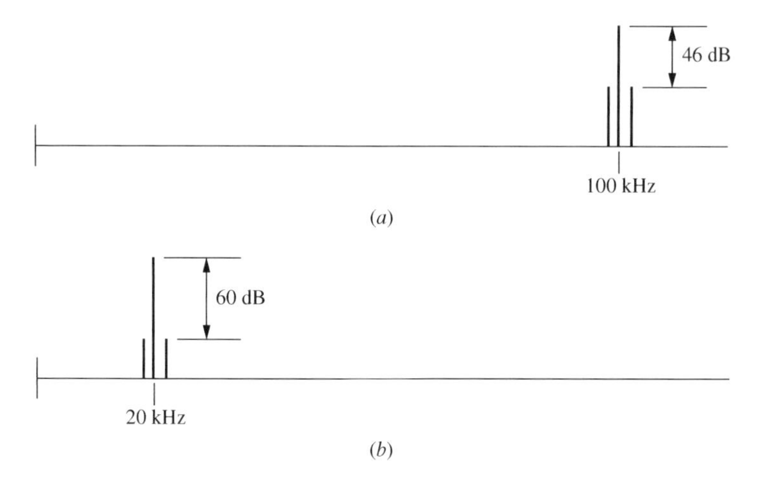

Fig. 8.22 Spectrums at input and output of a ÷5.

information about the phase or frequency of the effective edge at the output only when it occurs, once each cycle of the output frequency. The phase and frequency of the output are effectively being sampled at that rate. This has important consequences. For example, if a divider has an output frequency of 10 kHz and the input is frequency modulated at a rate of 30.1 kHz, the output frequency will have FM at a modulation frequency f_m of (30.1 kHz – 3 × 10 kHz) = 100 Hz. The modulation index at the output will be $m_{\text{out}} = m_{\text{in}}/N$, where m_{in} is the modulation index at the input, and all the modulation sidebands on the fundamental will be smaller by N than they are at the input. Refer to Egan (2000) or Egan (1981) for more information on this sampling process.

8.5.3 Internal Noise

Like most RF components, frequency dividers have internal noise, but the level is often low enough to be ignored. See Egan (2000, 123–127) for some divider noise levels.

8.6 FREQUENCY MULTIPLIERS

Frequency multipliers combine some of the features of frequency dividers and of soft limiters. They tend to be operated near saturation, so AM transfer is reduced. We can use the method described for soft limiters to compute by how much. While frequency dividers decrease frequency by a factor N, resulting in attenuation of FM sidebands by that same factor, the multiplier increases frequency by the multiplication factor M and increases the amplitude of FM sidebands by the same factor. We can combine the output AM and FM as we did in Example 8.1. Since AM is often attenuated and FM is multiplied by M, the effect of AM at the output can often be ignored. However, input AM can be converted to PM (delay through the multiplier can decrease at higher signal levels) and the resulting FM will be multiplied. As the frequency is multiplied to higher values, the FM deviation may increase to the point where the spectrum can no longer be represented by the simplified approximation of the Bessel functions (e.g., $J_1 \approx m/2$).

Example 8.13 Frequency Multiplier Spectrum at Input and Output Refer to Fig. 8.22 again but this time use the lower spectrum (*b*) as the input to a $\times 5$ multiplier. The output sidebands are increased in relative level by the value given by Eq. (8.16). If the FM were an equivalent due to a single sideband at one of the sideband frequencies, the single sideband would have amplitude -54 dBc. If there were AM at the same modulation frequency, possibly as a result of decomposition of a single sideband, the slope of the power transfer curve would tell us how much of the original AM would be transferred to the output, and we would superimpose these sidebands on the FM sidebands shown. We would have to estimate their relative phase, hopefully from physical considerations. However, if they were attenuated, they would be more than 14 dB smaller than the FM sidebands. The significance of the AM sidebands should tend to decrease as the multiplication factor M increases.

When additive noise is processed by a multiplier, the output noise variance can increase due to mixing between noise components. The effect is sometimes represented by division of the output phase variance by a "squaring loss," $S_L \leq 1$ (Egan, 1998, p. 389; Lindsey and Simon, 1973, pp. 57–75). See Egan (2000, pp. 126–128), for some frequency multiplier internal noise levels, which can be quite low.

8.7 SUMMARY

- A small signal can be considered a single sideband of the large signal that accompanies it.
- A small single sideband can be decomposed into AM and FM (or PM) sidebands on the large-signal carrier.
- The AM and FM sidebands have half the amplitude of the SSB.
- The AM and FM sidebands add to form the SSB and cancel on the other side of the large signal.
- Hard limiting eliminates the AM, leaving FM sidebands on both sides of the carrier.
- Soft limiting attenuates the AM. The AM gain is obtained from the slope of a gain curve that has axes in dB.
- The LO appears at the IF port of the mixer, reduced by the LO-to-IF isolation.

- Mixers tend to be operated with the LO in saturation. This reduces AM transfer to the IF.
- FM is transferred from the LO to the IF. Contaminants near the LO in frequency (Range 1) are transferred to the IF as FM or equivalent FM.
- FM transfer creates single-signal spurs.
- Contaminants separated from the LO frequency by the IF frequency (Range 2) can mix with the LO to create contaminants at the IF frequency. These appear in the IF, reduced by balance and conversion loss.
- Contaminants in the LO at the IF frequency (Range 3) pass to the IF reduced by balance.
- Contaminants offset from twice the LO frequency by the IF frequency (Range 4) can form equivalent FM sidebands that are at the IF frequency if the LO is limited. These appear in the IF attenuated by the LO-to-IF isolation and at least 6-dB loss in converting from SSB to FM.
- Contaminants in Ranges 2 through 4 create internal spurs.
- Noise is transferred to the IF in a manner similar to discrete contaminants and increases the mixer's noise figure.
- Noise can be filtered before entering the LO port if the problem ranges are not too close to the LO.
- Components of FM noise sidebands on the LO that are in the IF frequency band can add noise to the IF. This noise will be reduced by the LO-to-IF isolation.
- Frequency dividers limit AM. FM is reduced from input to output by the divide ratio.
- Sampling effects can produce new frequencies at the output of a frequency divider.
- Frequency multipliers limit AM to a degree. The slope of the gain curve in dB can be used to determine the degree of transfer of AM.
- Frequency multipliers increase FM deviation by the multiplication factor.

ENDNOTES

[1]Egan (1981, p. 800); Egan (2000, pp. 72–78); Egan (1998, p. 353); Goldman (1948).

[2]Information about sum-frequency products can be discerned through similar analysis, but that frequency region, near the second harmonic of the LO, is not of interest in most practical frequency translators.

CHAPTER 9

PHASE NOISE

We have considered, in Chapter 3 and beyond, noise that is added to RF signals in a system and that is characterized by noise figure. Because this noise is additive, the detriment to a signal that is imbedded in it can be lessened by an increase in signal power. However, there is another kind of noise that is basically multiplicative and whose harmful effects are therefore not reduced by an increase in signal power. The process by which this noise affects a signal is modulation, rather than addition, and the type with the more serious effect is generally phase, rather than amplitude, modulation. Phase noise is the subject of this chapter.

9.1 DESCRIBING PHASE NOISE

An undesired phase modulation of a signal is phase noise (Egan, 1998, Chapter 11). Phase modulation (PM) implies frequency modulation (FM). If we have phase modulation with a peak deviation of m and a modulation frequency of f_m, we have frequency modulation with a peak deviation of Δf, and the same modulation frequency and the two deviations are related by Eq. (8.1). Their values differ by a factor f_m, but we cannot have one without the other. This modulation creates a pair of sidebands on the modulated carrier. If they are not too large (see Section 8.1), each sideband has amplitude, relative to the carrier, of $m/2$. The relative power in each sideband is the square of the relative sideband amplitude:

$$\frac{p_{\text{sideband}}}{p_{\text{carrier}}} = \left(\frac{m}{2}\right)^2 = \frac{m^2/2}{2} = \frac{\tilde{m}^2}{2}, \tag{9.1}$$

where $\tilde{m}$ is the rms phase deviation in radians. This says that the relative power of each sideband equals half the mean-square phase deviation.

Random noise is expressed as a density. The available thermal noise density is $N_T = \overline{k}T_0$. Multiplying it by the width B of a rectangular band gives noise power:

$$p = \overline{k}T_0 B. \tag{9.2}$$

If the band is not rectangular, then B is the noise bandwidth, which has a value defined by Eq. (9.2). If the noise density $N_0(f)$ is not flat, multiplying it by a differential bandwidth will give the power in that bandwidth:

$$d\,p(f) = N_0(f)\,d\,f. \tag{9.3}$$

We can then integrate $d\,p(f)$ over a range of frequencies to get the noise in that range, or bandwidth:

$$p|_{f_1}^{f_2} = \int_{f_1}^{f_2} d\,p = \int_{f_1}^{f_2} N_0(f)\,d\,f. \tag{9.4}$$

The density of mean-square phase deviation is called phase-power spectral density (PPSD). Its symbol is S_φ. The mean-square phase deviation for modulation frequencies from f_1 to f_2 is

$$\sigma_\varphi^2|_{f_1}^{f_2} = \tilde{m}^2|_{f_1}^{f_2} = \int_{f_1}^{f_2} S_\varphi(f_m)\,d\,f_m. \tag{9.5}$$

We will use the symbol L for sideband relative (to carrier power) power density. It is called single-sideband density because it only relates to the power in one of the two sidebands, rather than the sum of powers on both sides of the carrier. Its units are reciprocal hertz (Hz^{-1}), since it expresses the ratio of a power density to the carrier power. When we are speaking only of phase noise, we will use the symbol L_φ. The relative sideband power over a frequency range is obtained by integrating $L(f)$ over that range. The term L_φ is related to PPSD, S_φ, in rad^2/Hz by

$$L_\varphi(\Delta f) = S_\varphi(f_m)/2, \tag{9.6}$$

where Δf, the frequency offset from spectral center, equals the modulation frequency f_m.

We can verify this by multiplying both sides of this equation by df to give

$$\frac{d\,p_{\text{sideband}}}{p_{\text{carrier}}} = d\tilde{m}^2/2, \tag{9.7}$$

which is the same as Eq. (9.1) for a differential bandwidth.

In decibels, we would write

$$L_\varphi(\Delta f)|_{\text{dBc/Hz}} = S_\varphi(f_m)_{\text{dBr/Hz}} - 3\text{ dB}. \tag{9.8}$$

Here dBc/Hz is an abbreviation for decibels relative to the carrier per hertz bandwidth, and dBr/Hz is an abbreviation for decibels relative to a square radian per hertz bandwidth.

We have seen how a small single sideband (SSB) can be decomposed into AM and FM sidebands. By the same method we can show that additive noise, the kind that is added to a signal, such as thermal noise, can be decomposed into effective AM and FM sidebands on that signal (Egan, 1998, Chapter 13). Due to the random nature of the phase of noise sidebands, half of the noise power becomes AM sideband power and half becomes FM sideband power. Thus, for small additive random noise, the SSB relative power spectral density (PSD) L is related to the relative PSD due to PM, L_φ, and the relative PSD due to AM, L_A, by

$$L = 2L_\varphi = 2L_A. \tag{9.9}$$

These are all ratios to the carrier power. Comparing Eqs. (9.6) and (9.9), we see that S_φ equals L in the case of small additive noise.

9.2 ADVERSE EFFECTS OF PHASE NOISE

Here are a few examples of the adverse effects of phase noise.

9.2.1 Data Errors

Figure 9.1 shows the constellation of data symbols (points) for a 16QAM code and the decision boundaries between them. The amplitude of the received signal is coherently detected against two quadrature carriers that are in synchronism with the unmodulated signal (the carrier). Two data bits (four values) are obtained in each normal direction. In the absence of noise, each received symbol matches one of the constellation points; its coordinates are the outputs from the two detectors. With additive noise, the received signal is described by a two-dimensional Gaussian probability density about each point. Figure 9.2 shows this distribution from the top at (*a*) and the Gaussian distribution along a cut through that at (*b*). The weaker is the signal, relative to the noise, the wider will be the density function. That is, the circle representing a given probability density grows in diameter when the signal becomes weaker. Probability of error is determined by integrating the probability density that falls outside the decision boundaries.

Phase noise produces uncertainty in only the phase, as shown in the upper right of Fig. 9.1. The probability density is maximum at the data point, and error is again determined by integrating the probability along the part of the arc that is outside the decision boundaries. (The arc continues with decreasing probability beyond what is shown.) The distribution along the arc would look like Fig. 9.2*b*.

The combination of the two types of noise stretches out the circularly symmetric distribution due to additive noise along the arc, as shown in the upper left of Fig. 9.1. The distribution around all of the points is affected by both additive and phase noise, but the effect of phase noise increases farther from the origin

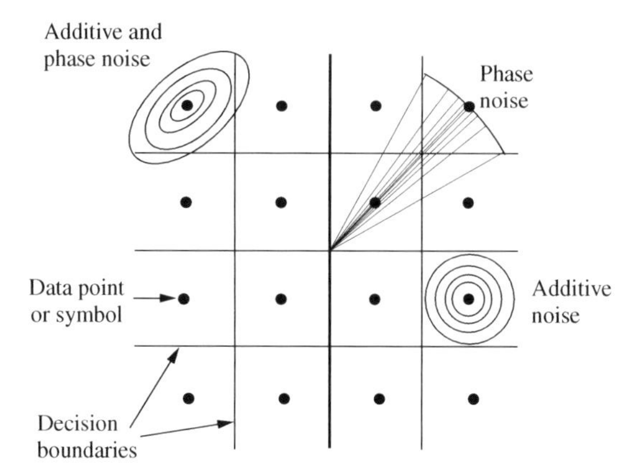

Fig. 9.1 Constellation with noise-induced probability distributions.

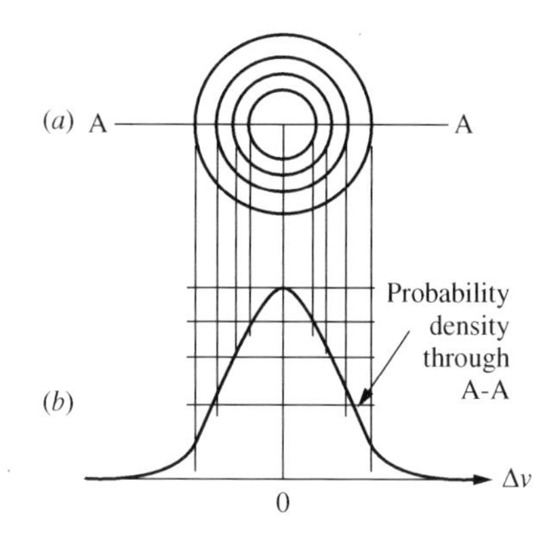

Fig. 9.2 Probability distribution.

since the spaces between decision boundaries subtend smaller angles there. The probabilities of error about all of the constellation points are added (after weighting by the probability of a point being transmitted) to give the overall probability of symbol (data point) error.

The phase noise reduces the amount of additive noise that is allowed before the probability of error becomes excessive. Distortion can also contribute to the error (Johnson, 2002).

9.2.2 Jitter

Figure 9.3 illustrates a jittery sine wave, synchronized at the start of each sweep of an oscilloscope. The evident change in frequency and phase from sweep to sweep can cause timing inaccuracies and even lead to clocking incorrect data (acquiring a data bit twice or not at all) when the data does not have matching jitter.

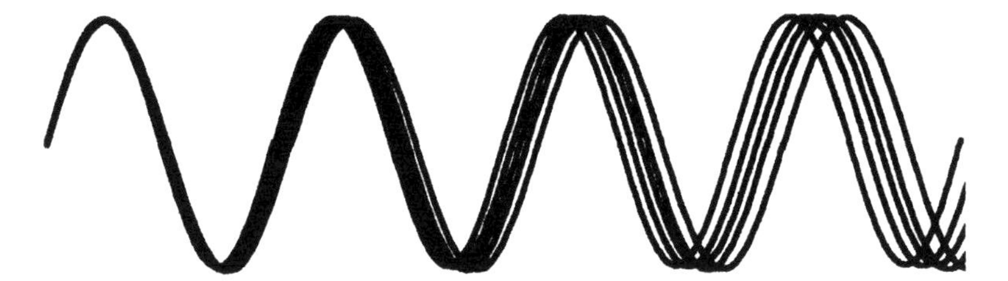

Fig. 9.3 Jitter.

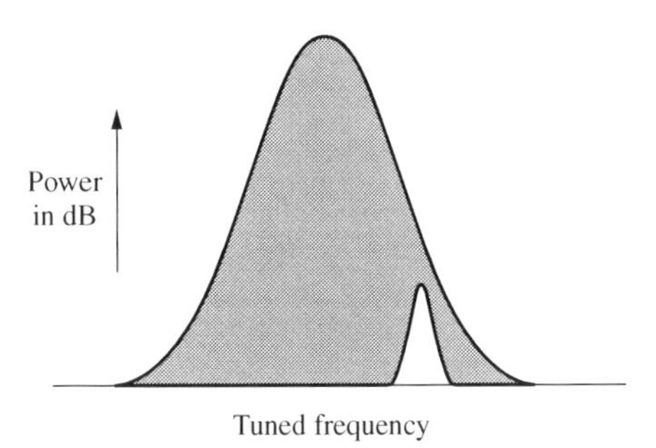

Fig. 9.4 Phase noise on interferer covers desired signal.

9.2.3 Receiver Desensitization

Figure 9.4 shows the spectrum of a large signal and a nearby small signal. This presentation depends on the bandwidth used to observe the signal; assume it is the receiver's final IF bandwidth so the power levels represent what a detector at the system output would sense. We see here that the phase noise on the large signal covers the small signal, burying it in noise (Egan, 1998, pp. 317, 318; Egan, 2000, pp. 109, 110). This is one form of desensitization (reduction in sensitivity), one that depends on the separation between signals. The phase noise could be on the transmitted signal or might have been added to the signals in our processing. When it is added during a frequency conversion, as has been described in Section 8.4.5.1, the process is sometimes called "reciprocal mixing."

Example 9.1 Desensitization A receiver specification calls for less than 1-dB desensitization within 200 kHz of a 7-dBm signal. The receiver noise figure is 10 dB. What restriction does this place on the LO spectrum?

Added noise that is about 4 dB higher than thermal noise (Fig. 9.5) will produce a 1-dB reduction in sensitivity by increasing the noise figure from 10 to 11 dB:

$$10 \text{ dB} \log_{10}[10^{10 \text{ dB}/10 \text{ dB}} + 10^{4 \text{ dB}/10 \text{ dB}}] = 10 \log_{10}[10 + 2.5] = 10.97 \text{ dB}. \tag{9.10}$$

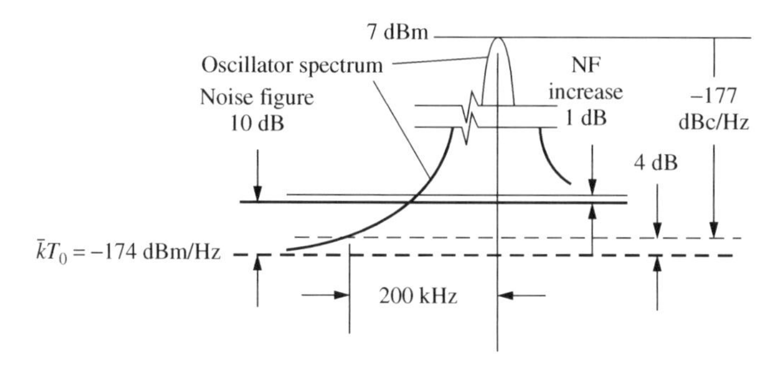

Fig. 9.5 Example 9.1.

Available thermal noise power density is −174 dBm/Hz so the phase noise sidebands on the strong signal cannot be higher than (−174 dBm/Hz + 4 dB =) −170 dBm/Hz at a separation of 200 kHz from spectral center. This is [+7 dBm − (−170 dBm/Hz)] = 177 dBc/Hz below the power of the strong signal. Therefore, our LOs must have phase noise, at a 200-kHz offset, lower than

$$L_{\varphi}(200 \text{ kHz}) = -177 \text{ dBc/Hz}$$

so less than the allowed level will be transferred to a clean +7-dBm interferer.

9.3 SOURCES OF PHASE NOISE

Essentially all signals originate in oscillators and all oscillators produce signals with phase noise. It is an inherent process. Phase noise can also be added to the signal after it is created, especially as a component of additive noise, or by vibration of filters or other parts that influence phase (including the oscillators) (Robins, 1984).

9.3.1 Oscillator Phase Noise Spectrums

Figure 9.6 illustrates typical oscillator phase noise density (Egan, 2000, pp. 106–118; Leeson, 1966). These are straight-line tangential approximations on a log plot. At large modulation frequency f_m, corresponding to large sideband separation Δf from spectral center, the noise is just amplified additive noise. The active device is acting essentially as an amplifier, raising the noise power density at its input by its gain and noise figure. As noted above, $S_{\varphi} = L$ here (they both equal twice L_{φ}). At an offset from spectral center of $f_m = \Delta f = f_{\text{osc}}/(2Q)$, where f_{osc} is the frequency of the center of the oscillator's spectrum and Q is the loaded quality factor of its resonator, S_{φ} begins to rise with decreasing f_m. This is due to feedback in the oscillator circuit, the thing that makes the oscillator

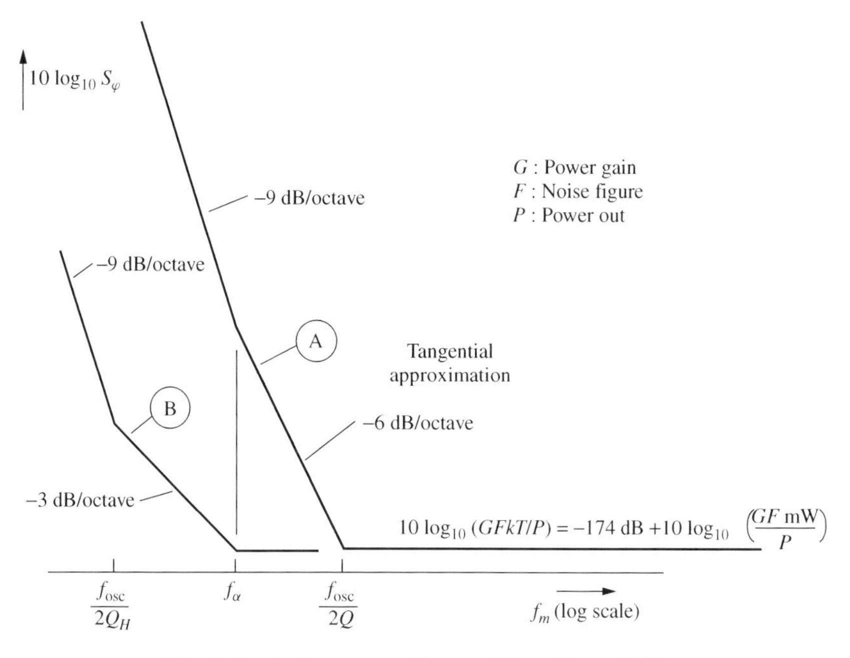

Fig. 9.6 S_φ for two oscillators. (From Egan, 2000.)

oscillate. This mechanism causes the phase noise to acquire an additional slope of −6 dB/octave at $f_m < f_{\text{osc}}/(2Q)$.

Close to f_{osc} there is an increase in noise due to modulation of the active device by flicker ($1/f$) noise, a noise type that is ubiquitous at low frequencies. This causes an additional −3 dB/octave slope. For many voltage-controlled oscillators (VCOs), this slope change occurs closer to spectral center than the corner at $f_{\text{osc}}/(2Q)$. However, if Q is high enough (e.g., in crystal oscillators), it may occur at a frequency higher than $f_{\text{osc}}/(2Q)$. Therefore, we show two curves in Fig. 9.6. They both show the effect of flicker noise at the lowest modulation frequencies, giving a slope of −9 dB/octave there. As f_m increases, when the effective flicker noise drops below the additive noise, the slope of curve A changes from −9 dB/octave to −6 dB/octave and, at higher f_m, falls below the flat additive noise. Curve B, however, representing a high-Q oscillator, reaches $f_{\text{osc}}/(2Q_H)$ in the region affected by flicker noise, so its slope changes from −9 to −3 dB/octave and then to zero slope where the effective flicker noise drops below the flat additive phase noise.

A slope of −12 dB/octave (called random-walk FM) has sometimes been observed at very low modulation frequencies and may be related to the physical environment, for example, vibration, temperature, and the like (Howe, 1976, p. 13).

This provides a good theoretical model that seems to match measured results well, although some deviations can be observed. It provides a guide for straight-line representations of the, normally noiselike, noise curves.

Discrete modulation at power-line frequencies and their harmonics and lumps caused by vibration (Egan, 2000, p. 118) are other features sometimes seen in plots of S_φ.

9.3.2 Integration Limits

How much phase deviation does a given PPSD plot represent? If we integrate the whole spectrum to find mean-square phase deviation, the answer will be infinite (Egan, 1998, pp. 302–304). The question is not practical. We can ask how much phase deviation exists with modulation frequencies above 1 Hz. We can ask what will be the rms phase deviation measured over a given time or the rms phase change during a given time. Answering these questions precisely requires more complex mathematics, but the answers, generally obtained by multiplying the PPSD by some function that decreases at low frequencies, are finite.

Without the qualifications, if we consider what the question means, an infinite answer seems justified. To answer the unqualified question by measurement, we would measure the phase and then come back later to see how much it had changed. As we continued this process over a long time, we would observe a continually increasing change, approaching an infinite change as our measurement time approached infinity.

This means that we must ask the question to which we really need the answer. When that is properly formulated, which can be difficult in some cases, the answer will be finite.

9.3.3 Relationship Between Oscillator S_φ and L_φ

Equation (9.6) indicates that the SSB relative PSD L_φ will have the same shape as the PPSD S_φ, but Eq. (9.6) is only valid for a small modulation index, that is, for $\sigma_\varphi \ll 1$ rad (Egan, 2000, pp. 104–106). For oscillator spectrums, where the integrated phase is infinite, this has been interpreted to mean that the spectrums of S_φ and L_φ match beyond a modulation frequency $f_m = f_{\min}$ for which

$$\int_{f_{\min}}^{\infty} S_\varphi \, d f_m \ll 1 \text{ rad}^2. \tag{9.11}$$

At lower modulation frequencies the failure of Eq. (9.6) to hold allows the oscillator power to remain finite while the integrated S_φ approaches infinity. Because people have often ignored this limitation, the traditional symbol for L_φ, which was $\mathscr{L}$, now officially equals half of S_φ, rather than being the SSB relative power density due to phase noise (Hellwig et al., 1988).

9.4 PROCESSING PHASE NOISE IN A CASCADE

Phase noise is unaffected by linear processes (amplifiers). However, increases in noise figure along the cascade will increase the effective phase noise, based on Eqs. (9.6) and (9.9). Filters can reduce phase noise at sufficiently high f_m; this is also based on Eq. (9.6), which shows the equivalence of S_φ and the sidebands that the filters reduce. Mixers can add phase noise from their LO ports (see Section 8.4.5.1). Various processing components can add jitter, and thus phase noise. Power supply noise seems more likely to create AM, but there can be AM-to-PM conversion in some components (e.g., improperly biased frequency dividers). Frequency multipliers multiply phase noise (see Section 8.6) by the multiplication ratio M, increasing it by 20 dB $\log_{10}M$. Frequency dividers divide phase noise (see Section 8.5) by the divide ratio N, changing it by -20 dB $\log_{10}N$.

9.4.1 Filtering by Phase-Locked Loops

Phase-locked loops (PLLs) filter phase noise directly (Egan, 1998, Chapter 12). The phase of the output, from a VCO, is made to follow the phase of the loop input (reference) by a control system that has a defined response to modulation. The response of a PLL to PM on an input reference signal is low-pass, so it attenuates high-f_m noise on the reference. However, the output of a PLL is from an oscillator that is constrained to follow the input only at low frequencies so the loop acts like a high-pass filter to noise introduced by its VCO. This is illustrated in Fig. 9.7. The bandwidths of these two processes are approximately equal, so the filters shown in the figure act almost like a diplexer.

Figure 9.8 shows how a phase-locked frequency synthesizer processes phase noise. The phase noise shown at (Fig. 9.8*a*) is the VCO noise and the multiplied reference (input) noise. The latter is multiplied by the divide ratio N because the output is divided by N, and the loop forces the divided signal to follow the input, thus causing the output to change phase N times more than the input.

The effect of the loop is shown at Fig. 9.8*b*. Here the VCO noise has been suppressed by the loop at modulation frequencies well below the loop bandwidth (f_m at unity open-loop gain). The input reference noise is similarly suppressed well above the loop bandwidth. (Actually, without the loop, there would be no reference noise at the output. The "suppression" is relative to what an infinite bandwidth would produce.) Near the loop bandwidth, the details of the response depend on other loop parameters.

For more details, including the effects of noise in other loop components, see Egan, 2002, pp. 106–133.

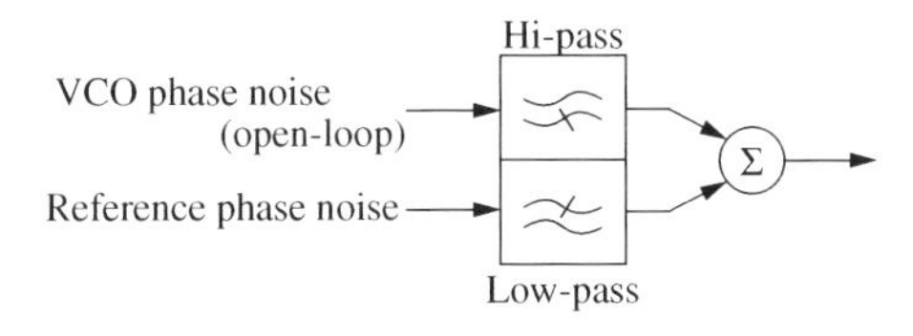

Fig. 9.7 Equivalent filtering action of PLL.

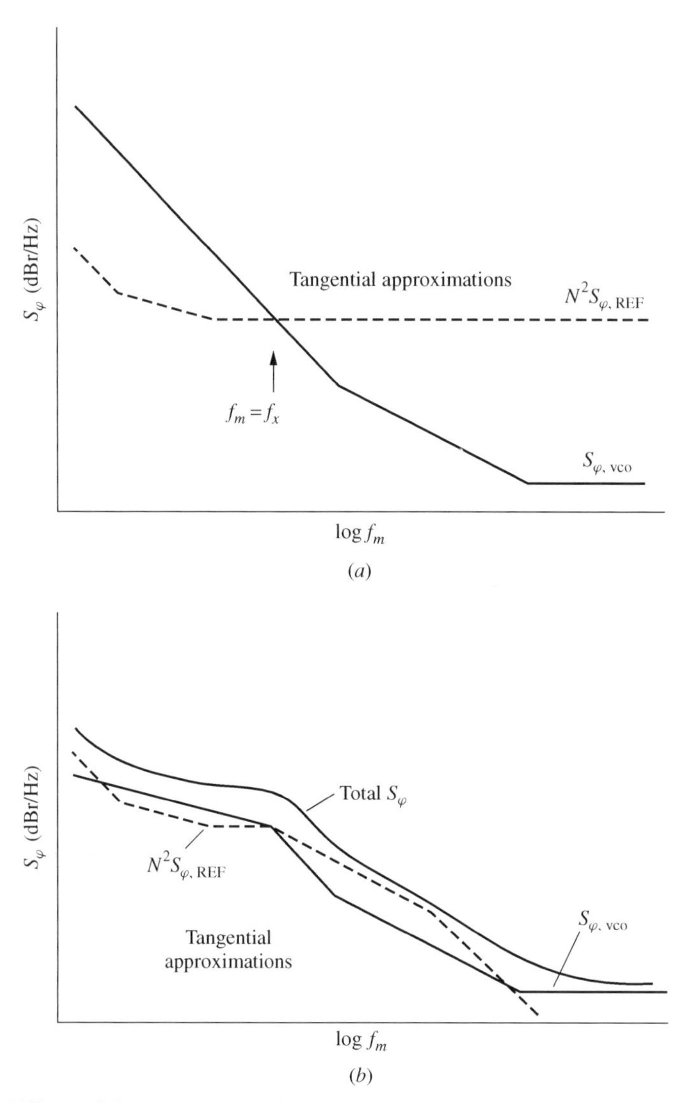

Fig. 9.8 Effect of loop on synthesizer broadband output noise: (*a*) open loop and (*b*) closed loop. (From Egan, 2000.)

9.4.2 Filtering by Ordinary Filters

To find how ordinary filters affect phase noise, we represent the input $S_{\varphi 1}$ by the equivalent $L_{\varphi 1}$ and apply the filter's frequency response to $L_{\varphi 1}$ (Egan, 2000, pp. 349–351). Then we can translate the resulting $L_{\varphi 2}$ back to $S_{\varphi 2}$, using Eq. (9.6) both ways.

If the filter is symmetrical about the signal at f_c, the response $F(f_c + f_m)$ of the bandpass filter can be represented by the response $F(f_m)$ to the modulation by an equivalent low-pass filter, where

$$F(f_m) = F(f_c + f_m)/F(f_c). \tag{9.12}$$

That is, the response of the equivalent low-pass filter at a frequency f_m is the same as the relative (to center frequency) response of the bandpass filter at f_m from its center frequency.

At the other extreme of symmetry, if the filter essentially eliminates power at $\Delta f = k_1$ but not at $\Delta f = -k_1$, $S_\varphi(f_m = k_1)$ will be only a quarter of the value it would have had with two equal sidebands. (Only half the power exists and half of that is AM.) In this case, we must decompose the single remaining sideband into AM and FM before using Eq. (9.6) to convert $L_{\varphi 2}$ back to $S_{\varphi 2}$.

In general, if the signal, after passing through the bandpass filter, no longer represents pure FM, the sidebands must be separated into L_φ and L_a before L_φ is reconverted to S_φ. This can be done by treating each sideband as a single sideband and adding the FM sidebands after decomposition. The process is represented in Fig. 9.9.

9.4.3 Implication of Noise Figure

Noise figure tells us how far the (available) noise floor is above $\overline{k}T_0$. If we have a signal of power P_s at some point in the cascade, the relative sideband density due to thermal noise will be

$$L = F - 174 \text{ dBm/Hz} - P_s, \tag{9.13}$$

where P_s is in dBm, F is noise figure in dB, and L is relative sideband density in dBc/Hz. Since this is random additive noise, the equivalent PPSD is $S_\varphi = 2L_\varphi = L$.

9.4.4 Transfer from Local Oscillators

Some of the most significant phase noise sources are LOs used in frequency conversion. We have seen that FM noise on an LO transfers directly to the signal as it appears in the IF (Section 8.4.2). We then have the phase noise of one oscillator modulating the spectrum of another oscillator, which itself has phase noise. The result may be similar to a convolution of the power spectrums. However, considered from the standpoint of PPSD S_φ, since the frequency deviations at a

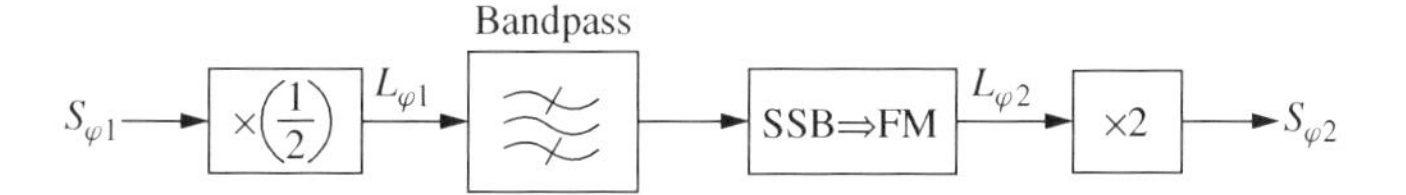

Fig. 9.9 Effect of bandpass filter on PPSD.

given f_m add, if their phases are random the PPSDs of the two oscillators add at any value of f_m:

$$S_\varphi(f_m) = S_{\varphi 1}(f_m) + S_{\varphi 2}(f_m). \tag{9.14}$$

Thus the final value phase noise spectrum will be the sum of the PPSDs of the incoming signal plus all of the LOs used in frequency conversion (plus possibly other phase noise discussed in this chapter).

Eq. (9.14) represents the addition of two uncorrelated noise sources. In some cases a single noise source may arrive at a destination by more than one path so that the resulting deviations in phase are correlated. In such cases the deviations, not their variances, must be added (allowing the possibility of cancellation). A transfer function must be written for phase modulation from source to destination. S_φ is then determined by multiplying S_φ at its source by the squared magnitude of that transfer function. The resulting S_φ can then be added to other randomly related components according to Eq. (9.14).

For example, two frequency synthesizers, each using the same frequency reference, may generate a first and second LO. The transfer function for phase deviations from the reference to some point after the second mixer will include the effects of the paths through the two synthesizers, including delays, frequency (phase) multiplications, and possibly the filtering effects of PLLs. Low-frequency phase deviations processed through these two paths may add or subtract, depending on the frequency conversion plan. Since phase undergoes the same processing as frequency, their transfer function will match the frequency conversion plan at zero modulation frequency, but not generally at high modulation frequencies.

Other examples of coherent addition of noise can be seen in Section 9.5.3 below and in Egan, 1998, pages 329–331.

9.4.5 Transfer from Data Clocks

Phase noise residing on the clock of an analog-to-digital converter (ADC) or of a digital-to-analog converter (DAC) is transferred to the converted data or waveform.

A phase change $\delta_{\varphi c}$ on the clock is equivalent to a time change

$$\delta_{Tc} = \delta_{\varphi c}/f_c, \tag{9.15}$$

where f_c is the clock frequency. (Cycles divided by cycles/second or radians divided by radians/second produce seconds). In an ADC, a clock delay of δ_{Tc} causes the value of the sampled signal to be the same as if the signal had been early by δ_{Tc} (Fig. 9.10). In a DAC it causes the reconstructed data to be late by δ_{Tc} (Fig. 9.11). Therefore, the apparent time change on the signal is

$$\delta_{Ts} = \pm\delta_{Tc}, \tag{9.16}$$

and the equivalent signal phase change is

$$\delta_{\varphi s} = \delta_{Ts} f_s, \tag{9.17}$$

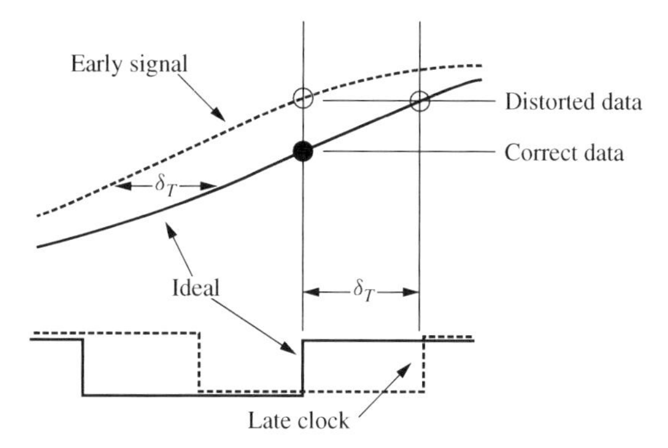

Fig. 9.10 ADC with clock jitter.

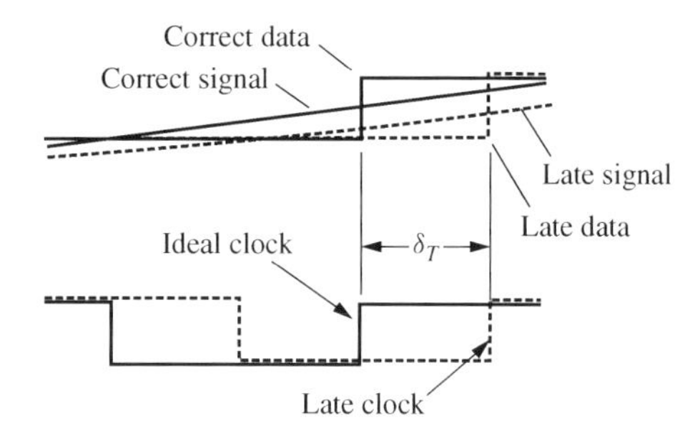

Fig. 9.11 DAC with clock jitter. Here $\delta_T = |\delta_{Tc}| = |\delta_{Ts}|$.

where f_s is the frequency of the signal. Putting these last three relationships together, the phase deviation on the converted signal is

$$\delta_{\varphi s} = \pm \delta_{\varphi c} \frac{f_s}{f_c}. \tag{9.18}$$

Thus the phase deviation is transferred to the signal, as in the previous section, but is reduced by the ratio of the clock frequency to the signal frequency. This applies to the frequency f_s of each component of a complex signal. Equation (9.18) also implies that the phase power spectral density of the clock is transferred to the signal:

$$S_{\varphi s}(f_s) = S_{\varphi c} \left(\frac{f_s}{f_c} \right)^2. \tag{9.19}$$

Again, the transferred phase change is attenuated, and the degree of attenuation changes with the frequency of the signal components. If there is phase modulation on the signal being converted, $S_{\varphi s}$ adds to it, as in the previous section.

We can determine the resulting phase variance by integrating S_φ, as described in the next section, or, if we can approximate the frequency of the signal as a single value, we can write

$$\sigma_{\varphi s} = \sigma_{\varphi c}\frac{f_s}{f_c}. \tag{9.20}$$

9.4.6 Integration of Phase Noise

We must integrate PPSD over frequency to find the corresponding phase variance, σ_φ^2. We can often approximate PPSD by a set of straight lines on a log plot (S_φ in dBr/Hz vs. log f_m), as in Fig. 9.6. We can integrate under these curves using (Egan, 1998, p. 300)

$$\sigma_\varphi^2|_{f_{m1}}^{f_{m2}} = \frac{f_{m1}S_\varphi(f_{m1})}{b+1}\left[\left(\frac{f_{m2}}{f_{m1}}\right)^{b+1} - 1\right] = \frac{f_{m2}S_\varphi(f_{m2})}{b+1}\left[1 - \left(\frac{f_{m1}}{f_{m2}}\right)^{b+1}\right], \tag{9.21}$$

where b is the slope of S_φ on the log plot,

$$S_\varphi = Kf_m^b. \tag{9.22}$$

A slope designated by b corresponds to $3b$ dB/octave = $10b$ dB/decade.

This equation is indeterminate for $b = -1$, for which special case we use instead

$$\sigma_\varphi^2|_{f_{m1}}^{f_{m2}} = f_{m1}S_\varphi(f_{m1})\ln\left(\frac{f_{m2}}{f_{m1}}\right) = f_{m2}S_\varphi(f_{m2})\ln\left(\frac{f_{m2}}{f_{m1}}\right). \tag{9.23}$$

9.5 DETERMINING THE EFFECT ON DATA

Figure 9.1 illustrates how phase noise can increase the probability of a bad decision on the received data symbol. Figure 9.12 illustrates the effect of phase noise on a data-error probability curve. Without phase noise, the error probability becomes continuously lower as the ratio of signal (carrier) power-to-noise in the receiver bandwidth increases. With phase noise, a limit is reached where the errors induced by phase noise dominate and no further improvement is seen with signal power increases (Reuter, 2000).

9.5.1 Error Probability

Error probability under given noise conditions can be computed by obtaining the error probability for each symbol (point) in a constellation, multiplying by the probability of transmission of that symbol (e.g., 1/16 for simple 16QAM) and adding all of the resulting probabilities of error. The error probability, in the absence of phase noise, for a particular symbol is computed by integrating the bivariate Gaussian probability density function over the region outside the decision boundary for the symbol. A table of functions of the normal probability distribution would be used. Computation of the error probability with both phase

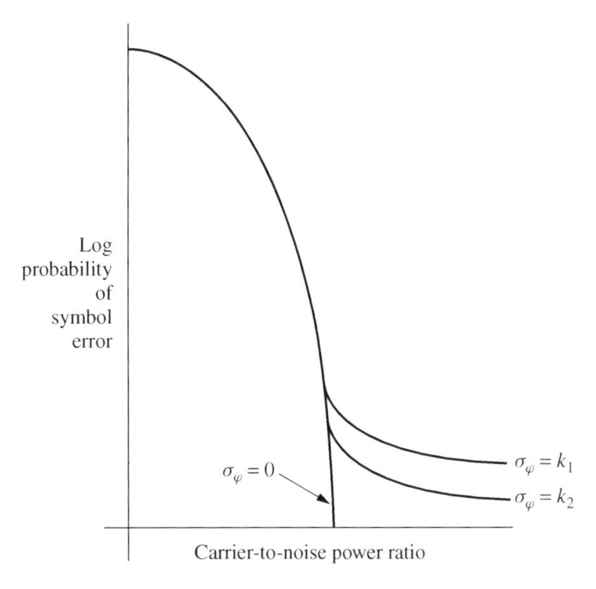

Fig. 9.12 Effect of phase noise on error probability.

noise and additive noise, in order to make an accurate version of a plot like Fig. 9.12, is more complex due to the approximately elliptical distribution about the point with the major axis perpendicular to the radius. Fortunately, many of the cells in most constellations have similarities that lead to the same error probability for a subset, thus reducing the number of independent computations. For example, the error probability for each data point in the 8PSK constellation in Fig. 9.13 is the same.

While an approximate error probability can be computed for a combination of additive and phase noise, it is much easier to get a worst-case value for phase noise alone, something that serves as the upper limit of error due to phase noise. We would pick the point that is farthest from center, since the angle that causes an error is smallest there, and assume a Gaussian distribution of phase. Using the normal probability curves, we would compute the probability of the phase being outside the decision boundaries for a given phase variance.

9.5.2 Computing Phase Variance, Limits of Integration

The width of the distributions that are used to determine error probability depend on the phase variance, σ_φ^2, which is obtained by integrating S_φ over frequency; but how do we establish limits of integration? At the high end, the modulation frequency f_m should get high enough for the effect to be averaged out over a symbol period, so there should be some high limit on f_m based on processing. In the absence of more sophisticated analysis, we might decide to terminate the integration at $f_m = 1/T_s$, the reciprocal of the symbol period; but what do we use for a low-end limit? We know that σ_φ^2 will be infinite if we just integrate from $f_m = 0$.

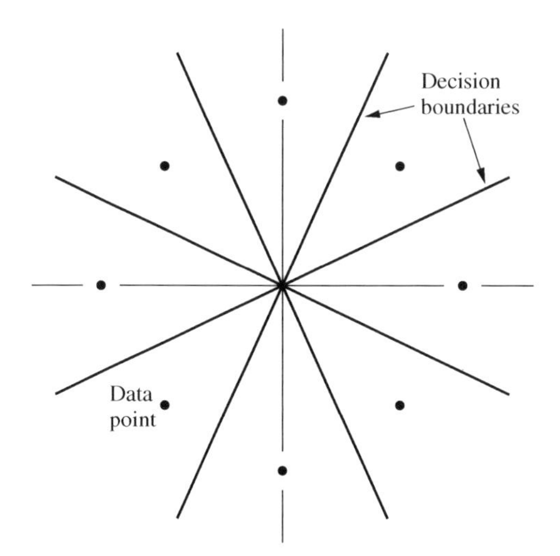

Fig. 9.13 8PSK constellation.

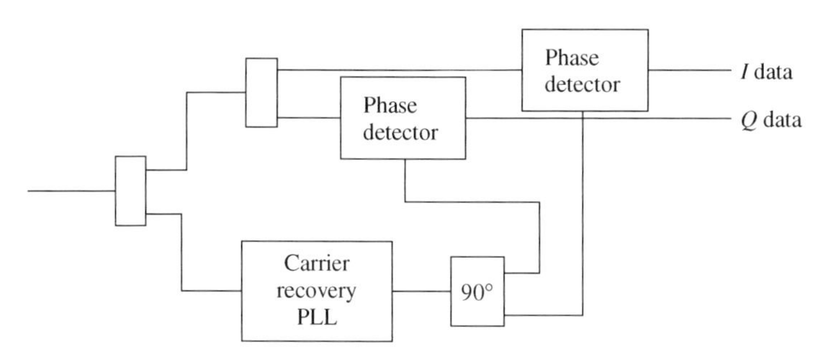

Fig. 9.14 QAM detection.

As we have seen, in a practical problem, an infinite answer implies the wrong question. The system could not decode phase without a reference. The phase is normally detected against a recovered carrier signal. The latter is obtained by locking to the signal with a phase-locked loop that uses a phase detector that is able to ignore the data transitions (Fig. 9.14) or with some other circuit that has similar properties (Egan, 1998, pp. 247–249) — we will assume a loop is used. If this carrier-recovery loop causes the phase of the recovered carrier to follow the phase of the signal's carrier, a phase variation in the data will not be seen when the data is compared to the recovered carrier.

9.5.3 Effect of the Carrier-Recovery Loop on Phase Noise

The phase noise at the output of the carrier-recovery loop equals the signal phase noise at its input φ_{in} multiplied by the loop's forward transfer function, $H(f_m)$.

The phase detected in the data phase detectors (excluding the data modulation) is the difference between the input and output of the carrier-recovery loop:

$$\Delta\varphi(f_m) = \varphi_{\text{in}}(f_m)[1 - H(f_m)]. \tag{9.24}$$

The factor $[1 - H(f_m)]$ is the error response of a phase-locked loop (Egan, 1998, Chapter 7), in this case of the carrier recovery loop. Since this is true of the phase at any frequency, it must also be true of differential bands of phase noise, leading to a similar expression for PPSD:

$$S_{\Delta\varphi}(f_m) = S_{\varphi,\text{in}}|1 - H(f_m)|^2. \tag{9.25}$$

Note that, if we write the error response in dB, we do not have to consider whether or not it is squared, since decibels are defined differently and appropriately for both phase [20 dB $\log_{10}(\varphi_1/\varphi_2)$] and phase squared [10 dB $\log_{10}(\varphi_1^2/\varphi_2^2)$]. Of course, Eq. (9.25) in dB would become a summation, rather than a product.

The magnitude of the response, $1 - H(f_m)$, is shown in Fig. 9.15. The response is

$$1 - H(f_m) = \frac{1}{1 + G_{\text{OL}}(f_m)}, \tag{9.26}$$

where G_{OL} is the loop's open-loop transfer function. At high frequencies, $G_{\text{OL}}(f_m)$ becomes small and the error response approaches unity (there is no functioning loop there, so the entire excitation becomes the error). At low frequencies,

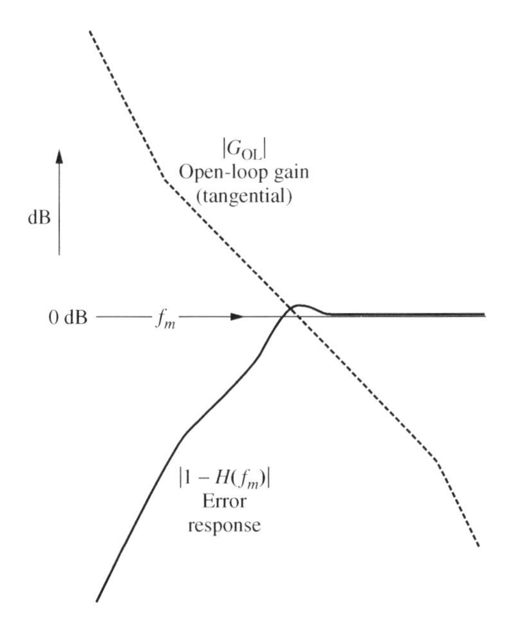

Fig. 9.15 Error response of a loop.

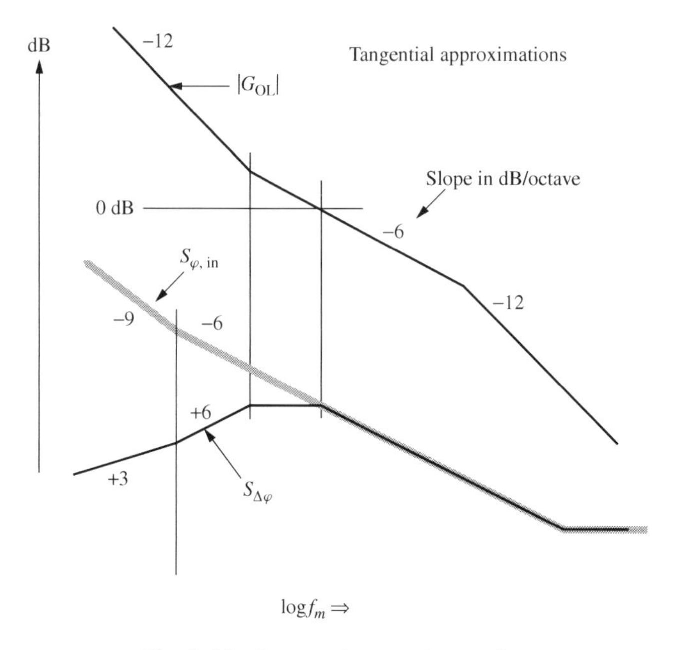

Fig. 9.16 Loop acting on phase noise.

$G_{OL}(f_m)$ is large giving

$$1 - H(f_m) \approx 1/G_{OL}(f_m). \tag{9.27}$$

The open-loop gain $|G_{OL}|$ will often have a slope of -12 dB/octave at low frequencies (Fig. 9.16). Therefore, its reciprocal and, as a result, the magnitude of the error response, will have a slope of $+12$ dB/octave. When this response multiplies the input PPSD $S_{\varphi,\text{in}}$, which typically has a -9 dB/octave slope at low frequencies, the resulting slope becomes positive; $S_{\Delta\varphi}(f_m)$ decreases as f_m goes toward zero, and the integral of the resulting S_φ becomes finite.

The higher is the gain of the carrier-recovery loop, the more the low-frequency noise will be suppressed and the smaller will be $S_{\Delta\varphi}(f_m)$, which determines the phase noise ultimately seen on the data. At low f_m, the loop is tracking the phase noise that is on the data so this noise appears on both inputs to the data-measuring phase detector, producing no phase difference and, therefore, no output noise.

At very low frequencies the error will again increase with decreasing frequency due to an inevitable reduction of the gain slope and, possibly, due to -40 dB/decade noise. Noise at very low frequencies may represent a very long term process corresponding to finite operational lifetime.

9.5.4 Effect of the Loop on Additive Noise

Widening the loop has the additional advantage of attenuating its VCO noise, which is also multiplied by the loop's error response. However, the wider bandwidth also has the disadvantage of increasing $S_{\varphi a}$, the equivalent phase noise due to additive noise, at the carrier-recovery output. This is the product of the equivalent phase noise at the input and $H(f_m)$. Since $H(f_m)$ is a low-pass function, increasing bandwidth lets more noise through, shifting the $\sigma_\varphi = 0$ curve in Fig. 9.12 to the right.

If tracking the input phase noise S_φ is good, why is it not good to track the phase component of the input additive noise? It appears on the data also. However, in the process of ignoring phase steps due to the data, $S_{\varphi a}$ on the recovered carrier becomes uncorrelated from $S_{\varphi a}$ on the data. Thus the $S_{\varphi a}$ that is tracked by the carrier-recovery loop adds to $S_{\varphi a}$ of the data, increasing the total effective phase noise.

To understand how one kind of noise can cancel while the other adds, consider that a phase deviation of $\Delta\varphi$ shifts both the carrier phase and the data phase by $\Delta\varphi$. Even after the data causes a transition of φ_d, its phase will be increased by $\Delta\varphi$. If the loop is wide enough, the data and the recovered carrier will have the same $\Delta\varphi$. However, the *equivalent* phase noise that is derived from additive noise depends on the relationship between the noise sidebands and the signal and, when the signal phase steps by φ_d, that relationship changes. The equivalent phase noise on the data undergoes steps, but the equivalent phase noise on the recovered carrier does not. Thus the *equivalent* phase changes do not cancel.

9.5.5 Contribution of Phase Noise to Data Errors

Example 9.2 Figure 9.17 shows the SSB phase noise density of a received carrier that is modulated by $16QAM$. This signal will be frequency converted

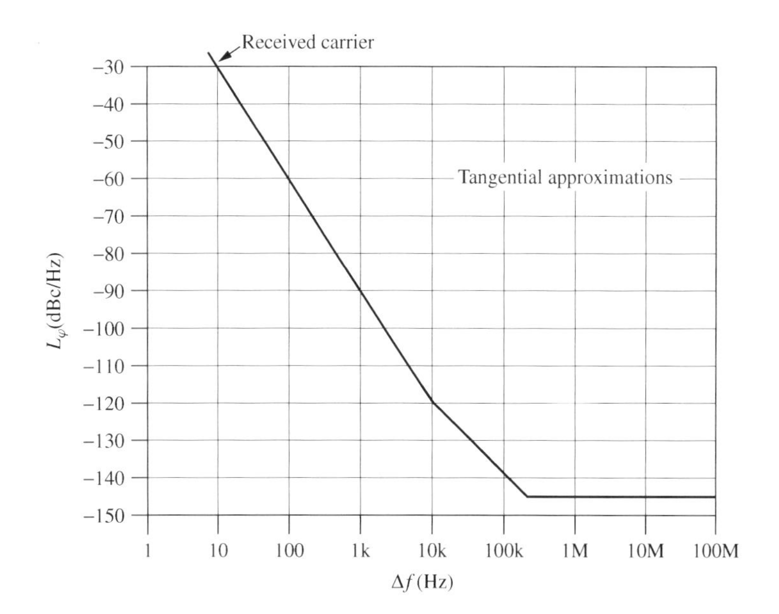

Fig. 9.17 SSB PN density of received carrier.

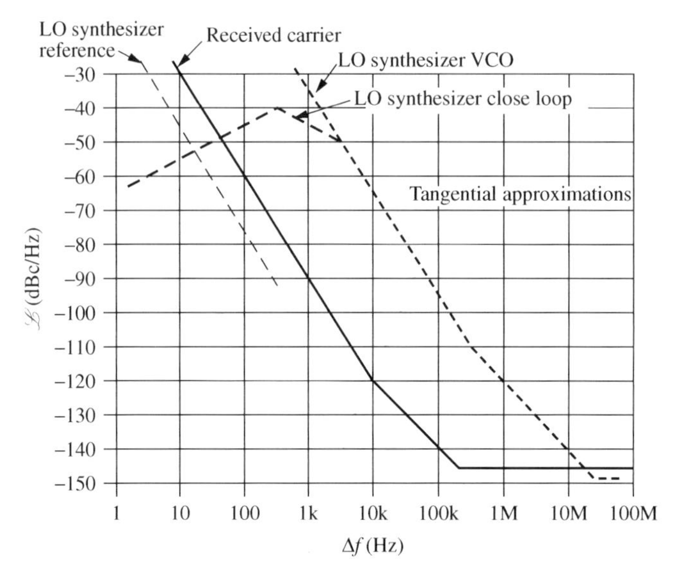

Fig. 9.18 SSB PN density of LO synthesizer. Received carrier PN from Fig. 9.17 is also shown as is PN for synthesizer's reference.

using an LO with the spectrum shown in Fig. 9.18. This shows the open-loop VCO phase noise and the noise level after the synthesizer loop attenuates the VCO noise at frequency offsets that are lower than the loop bandwidth. We will use tangential (straight lines on a log plot) approximations throughout. At about 20 Hz, the synthesizer's reference noise pokes out above the suppressed VCO noise, but we will ignore it because it is about 15 dB lower than the received-signal noise. Notice that we have changed the ordinate variable from L_φ to $\mathscr{L}$. This is done because the open-loop VCO phase noise begins to have values at low frequencies that are too large for the small-modulation-index approximation so L_φ would deviate from the straight-line shape that is characteristic of S_φ (and, therefore, of $\mathscr{L}$) there. We are really interested in S_φ anyway and are using single-sideband density as a way of representing it.

The error response of the carrier-recovery loop is shown in Fig. 9.19. Also shown is the product of this response with the total noise due to signal and synthesizer. (Here we have used a tangential approximation even for the total, as opposed to the more accurate process illustrated in Fig. 9.8.) Figure 9.20 shows the integration of this total noise. It employs the spreadsheet IntPhNs, where the vertices of the overall response are entered in the first two columns. Equation (9.21) or (9.23) is used, even though the input is for $\mathscr{L}$ rather than S_φ; the spreadsheet does the translation. The total rms phase noise (standard deviation) is $\sigma_\varphi = 6.3°$. We notice from the last line, fourth column, in Fig. 9.20, that very little of the total noise is coming from the highest values of f_m, so we are not very concerned about exactly where to discontinue the integration (e.g., near the data rate).

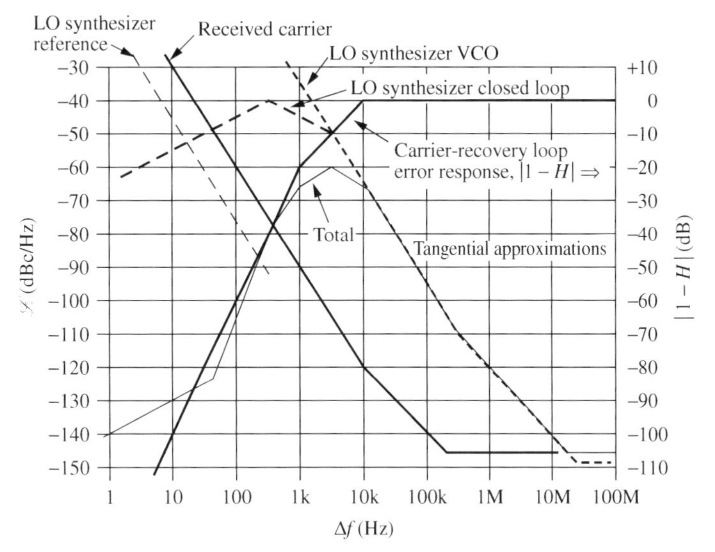

Fig. 9.19 Carrier-recovery loop and total phase noise. Labels on right axis apply only to $|1 - H|$ curve.

INTEGRATED PHASE DENSITY

Enter SSB density representing phase noise to obtain phase deviation.

ENTER DATA		Do not enter data below. Copy last line OK.				
Mod. Freq. Hz	SSB density dBc/Hz	slope	Integrated segment rad^2	sum rad^2	RMS Phase rad	degrees
1.00E+0	−139.7					
4.17E+1	−123.5	1	1.862E−11	1.862E−11	4.3E−06	0.00025
3.24E+2	−79	4.998	1.36E−06	1.36E−06	0.00117	0.06682
1.00E+3	−64.3	3.003	0.0001836	0.0001849	0.0136	0.77917
2.70E+3	−60	0.997	0.0023322	0.0025171	0.05017	2.87456
1.00E+4	−65.7	−1	0.0070593	0.0095764	0.09786	5.60691
3.00E+5	−110	−3	0.0026898	0.0122662	0.11075	6.34567
1.70E+7	−145	−2	5.915E−06	0.0122721	0.11078	6.3472
1.00E+8	−145	0	5.249E−07	0.0122726	0.11078	6.34733

Fig. 9.20 Integration of phase noise. Columns 3 and 4 pertain to columns 1 and 2 between previous line and current line.

To evaluate the effect of this much phase noise, we look at the farthest constellation symbol in Fig. 9.1, shown if Fig. 9.21. We calculate the angle at the decision boundary from the geometry as

$$\varphi_b = \sin^{-1}\left(\frac{1}{1.5\sqrt{2}}\right) = 28.1^\circ, \tag{9.28}$$

where the units are the distance between decision boundaries. (The computation would be much simpler for Fig. 9.13.) The phase error that will produce a

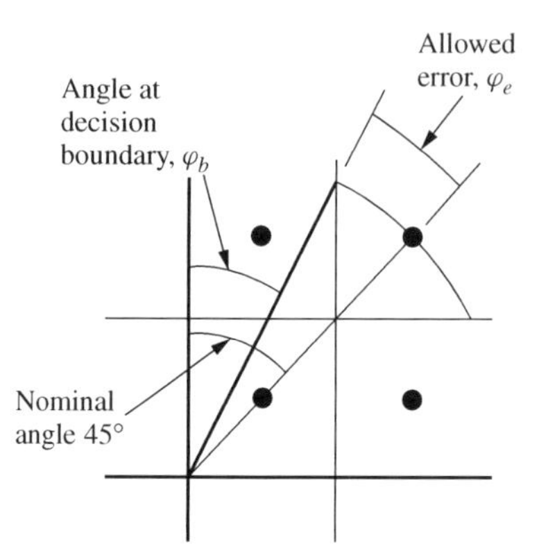

Fig. 9.21 Phase error at decision boundary.

decision error is

$$\varphi_e = 45^\circ - \varphi_b = 16.9^\circ. \tag{9.29}$$

Thus an error occurs for normalized phase deviations greater than

$$\varphi_e/\sigma_\varphi = 16.9^\circ/6.3^\circ = 2.68. \tag{9.30}$$

From tables for the normal distribution $Z(x)$ (Gaussian distribution for $\sigma = 1$) we find that

$$\int_{-\infty}^{2.68} Z(x)\,dx = 0.9963\ldots, \tag{9.31}$$

which tells us that the integral from $-\infty$ to -2.68, or from 2.68 to ∞, is (1 − 0.9963 =) 0.0037. This is the probability of a phase error beyond the decision boundary on one side. The probability for both sides is twice this, or 0.0074. This is much too high for most systems. Specifications of 10^{-6} or 10^{-8} are more likely. We must decrease σ_φ.

Two methods of decreasing σ_φ are suggested by Figs. 9.19 and 9.20. We see from the latter (column 4) that most of the noise is at frequencies between 1 kHz and 300 kHz. We see from the former that the noise in this region can be decreased by increasing the bandwidth of the synthesizer loop (now 2.7 kHz) or by increasing the bandwidth of the carrier-recovery loop (now 10 kHz). The former may be constrained by the required step size of the synthesizer. The latter may be constrained by the data rate or additive noise; increasing the carrier-recovery loop bandwidth will reduce attenuation of perturbations that occur at the symbol rate and will tend to cause a higher error rate due to additive noise. We will increase the carrier-recovery loop bandwidth by a factor of 10 to see what effect it will have.

The new curves are shown in Fig. 9.22. The integration of the phase noise, shown in Fig. 9.23, now produces a standard deviation of 0.77°. Equation (9.30)

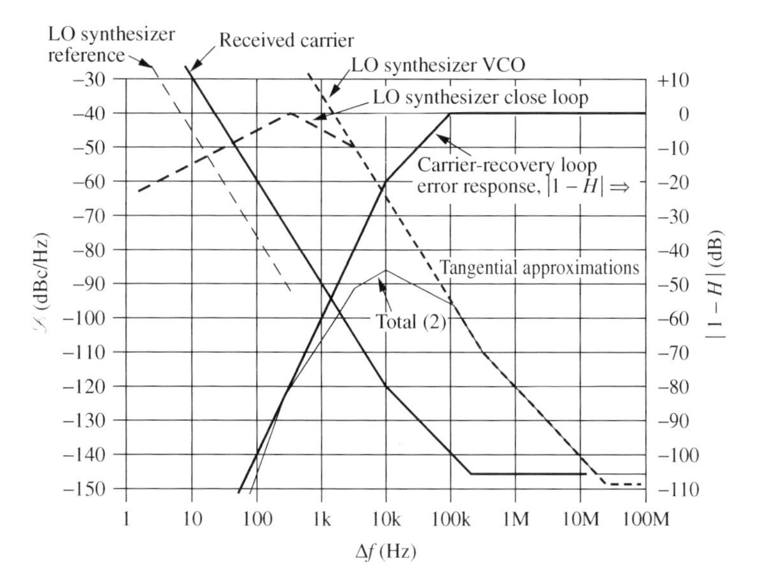

Fig. 9.22 PN SSB density, second design.

INTEGRATED PHASE DENSITY

Enter SSB density representing phase noise to obtain phase deviation.

ENTER DATA		Do not enter data below. Copy last line OK.				
Mod. Freq. Hz	SSB density dBc/Hz	slope	Integrated segment rad^2	sum rad^2	RMS Phase rad	degrees
1.00E+2	−147.5					
2.56E+2	−123	6.001	3.66E−11	3.66E−11	6E−06	0.00035
3.00E+3	−90.93	3	1.211E−06	1.211E−06	0.0011	0.06304
1.00E+4	−85.7	1	2.449E−05	2.57E−05	0.00507	0.29047
1.00E+5	−95.7	−1	0.0001239	0.0001497	0.01223	0.70091
3.00E+5	−110	−3	2.395E−05	0.0001736	0.01318	0.75492
1.70E+7	−145	−2	5.915E−06	0.0001795	0.0134	0.76767
1.00E+8	−145	0	5.249E−07	0.00018	0.01342	0.76879

Fig. 9.23 Phase integration, second design.

now gives a ratio of

$$\varphi_e/\sigma_\varphi = 16.9^\circ/0.77^\circ \approx 22. \tag{9.32}$$

The normal distribution table gives

$$1 - \int_{-\infty}^{22} Z(x)\,dx = 10^{-106.8}, \tag{9.33}$$

which is half of the error probability and is infinitesimal, so we need not be concerned about errors due to phase noise with this wider carrier-recovery loop.

Assuming that the carrier-recovery loop bandwidth is still low compared to the data bandwidth, since the carrier-recovery loop low-passes the phase noise component of the additive noise and does not pass the AM component, we expect the amount of additive noise on the recovered carrier to be small compared to the amount on the data.

9.5.6 Effects of the Low-Frequency Phase Noise

Example 9.2 (continued) Let us conclude the example by considering an effect that we have ignored, an increasing level of phase noise at low frequencies. While we will come to the conclusion that the additional noise is ignorable, we should understand the reason. (We will refer to the loop represented by Fig. 9.19 rather than our final loop in the example because it shows some features of interest in this discussion better than does Fig. 9.22.)

We see (Fig. 9.19) that the error response of the carrier-recovery loop has a slope of 20 dB/decade just below the loop bandwidth (between 1 kHz and 10 kHz). This is due to a −20 dB/decade slope in the open-loop gain G_{OL} in Eq. (9.27), which, in turn, is due to an integration process that is inherent in the loop. That integration represents a necessary conversion of the loop variable from frequency to phase, and it applies at all modulation frequencies (f_m or Δf). Below 1 kHz we see an additional +20 dB/decade slope in the error response. This is characteristic of a type 2 PLL and is produced by an integrator circuit in the loop filter. If it were not for this circuit, the total phase noise would not drop as fast at low frequencies and, in fact, it would have a negative slope, −10 dB/decade, below the corner at 41.7 Hz. This would present a theoretical problem because the phase noise obtained by integrating to zero would be infinite.

Unfortunately, the integrator circuit does not eliminate the theoretical problem; it just moves it to a lower frequency. This is because the gain of a true integration process is inversely proportional to frequency but the actual circuit will not be able to provide the increasing gain below some frequency. Below that frequency—let us use 1 Hz for this example—the slope of the carrier-recovery loop's error response will revert to +20 dB/decade and, in this example, the total phase noise will take on a slope of −10 dB/decade. This results in a theoretically infinite phase variance because we have not restricted the observation period; the variance grows to unacceptable values only at extremely low frequencies representing changes over an extremely long time.

Let us use Eq. (9.23) to determine how low in frequency we can set our lower limit of integration to produce a negligible phase error variance of 0.1° squared ($= 3 \times 10^{-6}$ rad^2) from this region. We let f_{m1} represent that lower limit while $f_{m2} = 1$ Hz. From Fig. 9.19, $S_\varphi(1\text{ Hz}) = 2L_\varphi(1\text{ Hz}) = 2 \times 10^{-14}$ rad^2/ Hz, so

$$3 \times 10^{-6}\text{rad}^2 = \sigma_\varphi^2 = 1\text{ Hz} \times 2 \times 10^{-14}\frac{\text{rad}^2}{\text{Hz}} \ln\left(\frac{1\text{ Hz}}{f_{m1}}\right), \tag{9.34}$$

$$f_{m1} = 1\text{ Hz} \times \exp[-1.5 \times 10^8] = 10^{-6.52 \times 10^7}\text{ Hz}. \tag{9.35}$$

This lower limit of integration, $10^{-6.5\times 10^7}$ Hz, has a period of $10^{6.5\times 10^7}$ (1 with 6.5×10^7 zeros after it) seconds, which is also approximately $10^{6.5\times 10^7}$ years.[1] The phase noise that we are not including, below f_{m1}, has even longer periods, leading us to believe that we will not see this phase deviation in our (or Earth's) lifetime.

The proper question is: What will be the variance measured during a finite time, say 100 years? The answer to that question (Egan, 2000, p. 504) is obtained by multiplying S_φ by $[1 - \text{sinc}^2(x)]$ where $\text{sinc}(x) \triangleq \sin(\pi x)/(\pi x)$, $x = f_m T$ rad/cycle, and $T = 100$ years $\approx 3 \times 10^9$ seconds. This factor becomes approximately unity at frequencies significantly above (e.g., 10 times) 1 cycle$/T \approx 3 \times 10^{-10}$ Hz, so it has no effect on the variance that we computed in the previous section. However, it is approximately equal to $(\pi x)^2/3$ at frequencies well below 3×10^{-10} Hz. Since x is proportional to f_m, this means that the response slope will increase by 40 dB/decade in that region, effectively eliminating contributions to σ_φ^2 below about $f_m = 0.55$ cycle$/T = 2 \times 10^{-11}$ Hz. (This is where the low-frequency 40-dB/decade slope meets the high-frequency flat response equal to unity, the corner of a rough tangential approximation.) Since this is many times $10^{-6.5\times 10^7}$ Hz, we know that the extra phase deviation will be much smaller than 0.1° (actually 4×10^{-5} degrees) when observed over a period of 100 years.

9.6 OTHER MEASURES OF PHASE NOISE

Here we discuss two other common measures of phase noise, jitter and Allan variance.

9.6.1 Jitter

Phase noise implies jitter in a signal's zero crossings (Egan, 2000, pp. 504–506). The standard deviation σ_T of such zero crossings is related to the standard deviation σ_φ of phase at frequency f by

$$\sigma_T = \sigma_\varphi / f, \tag{9.36}$$

where σ_φ uses cycle units for phase to cancel the same units in f, giving σ_T time units.

We again must consider how we should obtain these values from S_φ. If the jitter of interest is a comparison between zero crossings seen on an oscilloscope at some delay T_m after a synchronized zero crossing, as in Fig. 9.24, then the phase noise of interest is that which changes the phase over T_m. (T_m may be just one period of the wave.) This has been given as (Egan, 2000, p. 506; Drakhlis, 2001)

$$\sigma_T^2 \approx \frac{4}{\omega_{\text{avg.}}^2} \int_{f_{m.\min}}^{f_{m.\max}} S_\varphi(f_m) \sin^2\left(\frac{\pi f_m T_m \text{ rad}}{\text{cycle}}\right) d f_m, \tag{9.37}$$

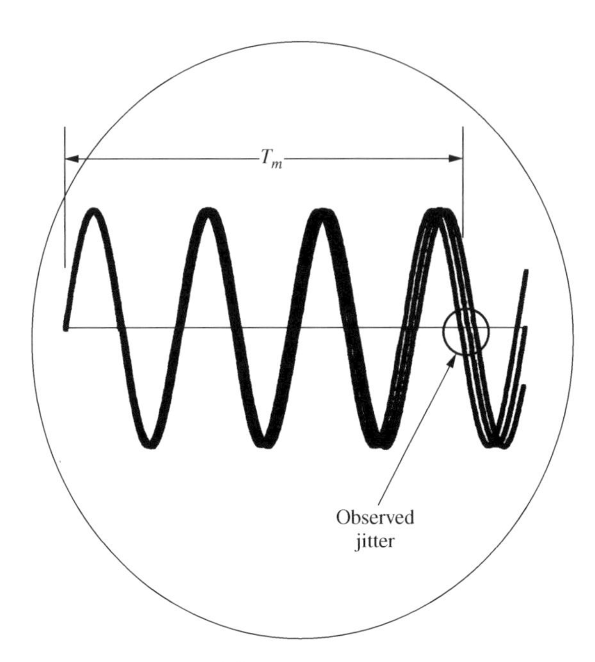

Fig. 9.24 Observed jitter.

where S_φ is in rad^2/Hz, ω_{avg} is the average radian frequency, f_m is noise modulation frequency, and T_m is the time from synchronization (where the phases are equal, on the left of Fig. 9.24) to observation. One can multiply a plot of S_φ by the $\sin^2$ and integrate. At higher frequencies, where S_φ changes little over a frequency of $\Delta f_m = 1/T_m$, the average value of the $\sin^2$, 0.5, can be used for that function. Note that Eq. (9.37) indicates there is no effect from modulation frequencies that are multiples of $1/T_m$. This is as it should be since they cause no change in phase over a period T_m.

The upper limit $f_{m,\max}$ may depend on the response of the using equipment to high-frequency variations. It may not be critical if the level of S_φ is low there. The lower limit is important, since S_φ often climbs at low f_m, in which case Eq. (9.37) might produce an infinite answer if $f_{m,\min}$ were zero. However, the question asked in that case is: What is the jitter variance over an infinite time? There is nothing to limit the period of observation, as there is in practice. Long-term drift is included in that question. In the absence of a better procedure, we might set $f_{m,\min}$ to the reciprocal of the observation time or some fraction of that reciprocal, in order to ignore modulations that change little over an observation period.

An equation that gives finite results without requiring finite duration of observation is available for jitter between the output and input of a PLL that is caused by the loop's VCO noise (Egan, 2000, p. 505).

Drakhlis (2001) provides a detailed study of the computation of jitter from S_φ and comparisons to measurements using a digital storage oscilloscope (DSO) as well as discussions of measurement errors.

9.6.2 Allan Variance

Allan variance $\sigma_y^2(T)$ is a measure of short-term frequency stability (not long-term drift) (Egan, 2000, pp. 499–503; Allan, 1966). It is measured by making a large number M of adjacent frequency counts of duration T and averaging them according to

$$\sigma_y^2(T) \approx \frac{1}{2(M-1)f^2} \sum_{i=1}^{M-1} \Delta_i^2(T), \tag{9.38}$$

where the equation is approximate due to the finite number of measurements. Here f is the average frequency and $\Delta_i(T)$ is the change in frequency between adjacent averaging (counting) periods. This variance can be computed from S_φ by the formula

$$\sigma_y^2(T) = 2\left(\frac{2}{T\omega_{\text{avg}}}\right)^2 \left[\int_0^{f_{\max}} S_\varphi(f_m)\sin^4\left(\frac{\pi f_m T \,\text{rad}}{\text{cycle}}\right) d f_m + \sum_{k=1}^{N} \tilde{\varphi}_k^2 \sin^4\left(\frac{\pi f_k T \,\text{rad}}{\text{cycle}}\right)\right], \tag{9.39}$$

where S_φ is in rad^2/Hz, ω_{avg} is the average radian frequency, and f_m is noise modulation frequency. In addition to the density S_φ, each of N discrete components is represented in Eq. (9.39) by its modulation frequency f_k and its mean-square phase deviation $\tilde{\varphi}_k^2$. Note that phase deviations occurring at frequencies that are multiples of

$$f_m, f_k = \text{cycle}/T \tag{9.40}$$

do not contribute to the variance. They would not affect the difference in average frequency between adjacent counts since they would affect one count the same as the next. Note, also, that the $\sin^4(\pi f_m T$ rad/cycle) is proportional to f_m^4 for $f_m T \ll 1$ cycle, so the integral will be finite with S_φ as steep as f_m^{-4} at low frequencies. The integration can be done graphically at low frequencies but, when f_m gets high enough so S_φ changes little over a cycle of the sine, $\sin^4$ can be approximated by its average value, $\frac{3}{8}$, to simplify the integration. An algorithm for computer computation is available (Egan, 1988).

9.7 SUMMARY

- Phase noise is unwanted phase modulation.
- Phase modulation (PM) implies frequency modulation (FM) and visa versa.
- Noiselike PM is described by its phase-power spectral density (PPSD) $S_\varphi(f_m)$.

- $S_\varphi(f_m)$ can be integrated over modulation frequency f_m to give phase variance σ_φ^2.
- If limits of integration are not correctly chosen, computed σ_φ may be infinite.
- Single-sideband spectral density $L_\varphi(\Delta f = f_m)$, corresponding to $S_\varphi(f_m)$, is half of the value of $S_\varphi(f_m)$ if the total phase variance is small compared to a square radian.
- Additive noise (e.g., thermal noise) can be considered half AM and half PM on a carrier.
- Phase noise can cause errors in interpretation of data. Widening the carrier-recovery loop, which is used as a reference in phase detection, can reduce the effective phase noise.
- Jitter, which is essentially phase noise, can also cause data errors.
- Phase noise on a receiver's LO can desensitize the receiver by broadening the spectrum of strong received signals, causing them to cover over small signals.
- Oscillators are the source of most phase noise.
- Phase noise can be reduced by filtering, particularly by phase-locked loops.
- The variances of uncorrelated phase noise introduced at various stages in a cascade are additive.
- Allen variance is a measure of phase noise. Its value can be computed from a plot of $S_\varphi(f_m)$.

ENDNOTE

[1]Note that there are 6.5×10^9 zeros in these numbers; if we divide by 1,000,000, the effect won't be seen in this notation since $6.5 \times 10^9 \approx 6.5 \times 10^9 - 6$.

APPENDIX A

OP AMP NOISE FACTOR CALCULATIONS

This appendix details the effects of certain changes in the representation of the circuit discussed in the Example 3.8 (Section 3.12) and shown in Fig. 3.18. Results are discussed in Section 3.12.1.

A.1 INVARIANCE WHEN INPUT RESISTOR IS REDISTRIBUTED

The cascade is not changed if we consider part of the input resistor to op amp 2 or 3 to be part of the output resistance of the previous stage. This is just a matter of redrawing the boundaries between stages. Therefore, the cascade noise figure should be unaffected. To verify this, Fig. A.1 is another spreadsheet representing Fig. 3.18 except that 1 kΩ of the input resistors of op amps 1 and 2 are moved to the previous stages. The last three stages, so considered, are shown in Fig. A.2. The output resistances $R_{22}(k\text{-})$ seen by the last two stages now increase by 1 kΩ, and the noise figure of each stage is changed because of the new values of $R_{22}(k\text{-})$ and of gain and of the resistor noises that change with the resistor configuration. The change in noise factor for Op Amp 1 is fairly easily computed, being due only to the addition of the noise of a 1-kΩ output resistor. Section A.3 gives a model for the other two op amps that is used to compute their noise factors before and after the change. Cells G48–H50 in Fig. A.1 contain the ratios of the cascade noise factors in Fig. A.1 to those in Fig. 3.19 and the overall noise factor can be seen to be the same, even though the parameters of the various amplifiers change considerably. (Compare cells F13–F15 on the two spreadsheets.)

	A	B	C	D	E	F	G	H
12	Filter	−7.0 dB	0.3 dB	R_0	R_{22k-}	$1/g$	c_k	a_k
13	Op Amp 1	*12.0 dB*		2000 Ω	2000 Ω	6.5303 dB	1	4
14	Op Amp 2	*−0.1 dB*		2000 Ω	1020 Ω	11.1394 dB	0.49505	2
15	Op Amp 3	*19.9 dB*		2000 Ω	1020 Ω	8.6889 dB	0.49505	20
16			DERIVED (*italics* above are derived also)					
17			Gain			NF using mean NFs (see Note*)		
18		mean	max	min	±	mean G	max G	min G
28	Op Amp 1	12.04 dB	12.04 dB	12.04 dB	0.00 dB	6.53 dB	6.53 dB	6.53 dB
29	Op Amp 2	−0.09 dB	−0.09 dB	−0.09 dB	0.00 dB	8.52 dB	8.52 dB	8.52 dB
30	Op Amp 3	19.91 dB	19.91 dB	19.91 dB	0.00 dB	6.30 dB	6.30 dB	6.30 dB
31			CUMULATIVE					
32			Gain			NF using mean NFs		
33	at output of	mean	max	min	±	mean G	max G	min G
42	Filter	7.43 dB	14.77 dB	0.09 dB	7.34 dB	3.09 dB	2.47 dB	5.1914 dB
43	Op Amp 1	19.47 dB	26.81 dB	12.13 dB	7.34 dB	4.27 dB	2.74 dB	8.2798 dB
44	Op Amp 2	19.39 dB	26.72 dB	12.05 dB	7.34 dB	4.38 dB	2.77 dB	8.5150 dB
45	Op Amp 3	39.30 dB	46.64 dB	31.96 dB	7.34 dB	4.44 dB	2.79 dB	8.6377 dB
46							NF ratio to previous	
47							at min G	
48	Op Amp 1						1.004567	0.01979 dB
49	Op Amp 2						1.004413	0.01912 dB
50	Op Amp 3						1.000000	0.00000 dB
51			*Note: Cable NF depends on SWR, which is assumed to be fixed.					

Fig. A.1 Alternate spreadsheet for Fig. 3.18. Here each of the last two amplifiers is partitioned in the middle of its input resistor. Missing lines are identical to those with the same number in Fig. 3.19.

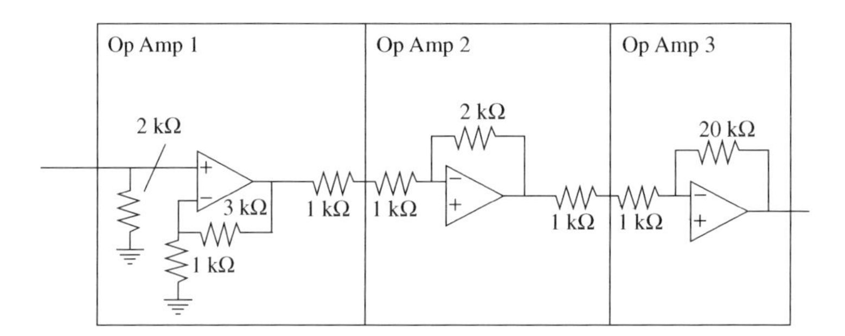

Fig. A.2 Last stages with resistors reassigned.

A.2 EFFECT OF CHANGE IN SOURCE RESISTANCES

We have used 20 Ω as the output resistance of the op amps. What is the effect of an inaccurate value for this output resistance?

From Eq. (3.73), we find that the contribution of each stage to total noise factor is proportional to $R_{22(k-)}$. However, Eq. (3.66) shows that f for each stage is inversely proportional to $R_{22(k-)}$ so the dependencies cancel except for the -1 in Eq. (3.73). This represents subtraction of the noise attributed to the source, so it will not appear both as part of the noise of the preceding stage and of the source for the stage in question. Thus an error in the value of $R_{22}(k\text{-})$ causes some error in overall noise factor due to the noise difference between the correct and erroneous values at that point in the circuit. In addition, $R_{22}(k\text{-})$ has some influence on the preceding gain through c_k, but these values will often be close to unity for circuits with output impedance that are small compared to the driven impedances.

Figure A.3 is like Fig. 3.19 except the output resistances $R_{22(k\text{-})}$ of the last three op amps are changed from 20 to 40 Ω, and noise factors have been recomputed for those values. We can see, from cell H58, that this 2:1 change in assumed output impedance causes only 0.008 dB change in overall noise figure in this particular cascade, even though the noise figure of the last two amplifiers have changed nearly 3 dB.

Figure A.4 shows that a change of 10% in source impedance for the first op amp (cell E13) results in only a 0.03-dB change in overall noise figure for this example.

	A	B	C	D	E	F	G	H
12	Filter	-7.0 dB	0.3 dB	*R0*	*R22k-*	*1/g*	Ck	Ak
13	Op Amp 1	*12.0 dB*		2000 Ω	2000 Ω	6.5006 dB	1	4
14	Op Amp 2	*-0.2 dB*		2000 Ω	40 Ω	24.9289 dB	0.98039	1
15	Op Amp 3	*19.8 dB*		2000 Ω	40 Ω	22.8160 dB	0.98039	10
16	Op Amp 1				*1/g*	0.0006 dB		
17	Op Amp 2				changes:	-2.9386 dB		
18	Op Amp 3					-2.9485 dB		
19			DERIVED (*italics* above are derived also)					
20		Gain				NF using mean NFs (see Note*)		
21		mean	max	min	±	mean G	max G	min G
30	Filter	-7.00 dB	-6.70 dB	-7.30 dB	0.30 dB	7.00 dB	6.70 dB	7.30 dB
31	Op Amp 1	12.04 dB	12.04 dB	12.04 dB	0.00 dB	6.50 dB	6.50 dB	6.5006 dB
32	Op Amp 2	-0.17 dB	-0.17 dB	-0.17 dB	0.00 dB	8.57 dB	8.57 dB	8.5744 dB
33	Op Amp 3	19.83 dB	19.83 dB	19.83 dB	0.00 dB	6.82 dB	6.82 dB	6.8170 dB
34							NF ratio to previous	
35							at min G	
36	Op Amp 1						1.000140	0.00061 dB
37	Op Amp 2						1.012932	0.05580 dB
38	Op Amp 3						1.009249	0.03998 dB
39			CUMULATIVE					
40		Gain				NF using mean NFs		
41	at output of	mean	max	min	±	mean G	max G	min G
50	Filter	7.43 dB	14.77 dB	0.09 dB	7.34 dB	3.09 dB	2.47 dB	5.1914 dB
51	Op Amp 1	19.47 dB	26.81 dB	12.13 dB	7.34 dB	4.26 dB	2.74 dB	8.2604 dB
52	Op Amp 2	19.30 dB	26.64 dB	11.96 dB	7.34 dB	4.37 dB	2.77 dB	8.4997 dB
53	Op Amp 3	39.13 dB	46.47 dB	31.79 dB	7.34 dB	4.44 dB	2.79 dB	8.6458 dB
54							NF ratio to previous	
55							at min G	
56	Op Amp 1						1.000091	0.00040 dB
57	Op Amp 2						1.000882	0.00383 dB
58	Op Amp 3						1.001877	0.00814 dB
59			*Note: Cable NF depends on SWR, which is assumed to be fixed.					

Fig. A.3 Effect of assumed output resistance. Modification of Fig. 3.19 where output resistances of Op Amps 1 through 3 change from 20 Ω to 40 Ω. Missing lines are as in Fig. 3.19.

	A	B	C	D	E	F	G	H
12	Filter	-7.0 dB	0.3 dB	*R0*	*R22k-*	*1/g*	Ck	Ak
13	Op Amp 1	*12.0 dB*		2200 Ω	2200 Ω	6.5000 dB	1	4
14	Op Amp 2	*-0.1 dB*		2200 Ω	20 Ω	27.8674 dB	0.9901	1
15	Op Amp 3	*19.9 dB*		2200 Ω	20 Ω	25.7646 dB	0.9901	10
16			DERIVED (*italics* above are derived also)					
17		Gain				NF using mean NFs (see Note*)		
18		mean	max	min	±	mean G	max G	min G
27	Filter	-7.00 dB	-6.70 dB	-7.30 dB	0.30 dB	7.00 dB	6.70 dB	7.30 dB
28	Op Amp 1	12.04 dB	12.04 dB	12.04 dB	0.00 dB	6.50 dB	6.50 dB	6.5000 dB
29	Op Amp 2	-0.09 dB	-0.09 dB	-0.09 dB	0.00 dB	8.17 dB	8.17 dB	8.1653 dB
30	Op Amp 3	19.91 dB	19.91 dB	19.91 dB	0.00 dB	6.45 dB	6.45 dB	6.4533 dB
31							NF ratio to previous	
32							at min G	
33	Op Amp 1						1.000000	0.00000 dB
34	Op Amp 2						0.921877	-0.35327 dB
35	Op Amp 3						0.928185	-0.32365 dB
36			CUMULATIVE					
37		Gain				NF using mean NFs		
38	at output of	mean	max	min	±	mean G	max G	min G
47	Filter	7.43 dB	14.77 dB	0.09 dB	7.34 dB	3.09 dB	2.47 dB	5.1914 dB
48	Op Amp 1	19.47 dB	26.81 dB	12.13 dB	7.34 dB	4.26 dB	2.74 dB	8.2600 dB
49	Op Amp 2	19.39 dB	26.72 dB	12.05 dB	7.34 dB	4.36 dB	2.77 dB	8.4749 dB
50	Op Amp 3	39.30 dB	46.64 dB	31.96 dB	7.34 dB	4.42 dB	2.78 dB	8.6046 dB
51							NF ratio to previous	
52							at min G	
53							1.000000	0.00000 dB
54							0.995195	-0.02092 dB
55							0.992429	-0.03301 dB
56			*Note: Cable NF depends on SWR, which is assumed to be fixed.					

Fig. A.4 Effect of 10% change in cascade source resistance. Modification of Fig. 3.18. Missing lines are as in Fig. 3.19.

A.3 MODEL

Refer to Fig. A.5. The mean-square equivalent input noise voltage due to the resistors is

$$e_R^2 = 4kT_0B\left[R_s + R_{\text{in}} + (R_f + R_{\text{out}})\left(\frac{R_s + R_{\text{in}}}{R_f}\right)^2\right], \tag{1}$$

where the last factor is the reciprocal of the op-amp gain (squared) and refers the output noise to the input. The term R_{out} is the output resistance of the op amp, as reduced by the gain of its feedback loop. Its value will be frequency dependent because the op-amp gain is frequency dependent.

The mean-square input voltage that produces the same current through R_f as does i_n is

$$e_i^2 = i_n^2(R_s + R_{\text{in}})^2. \tag{2}$$

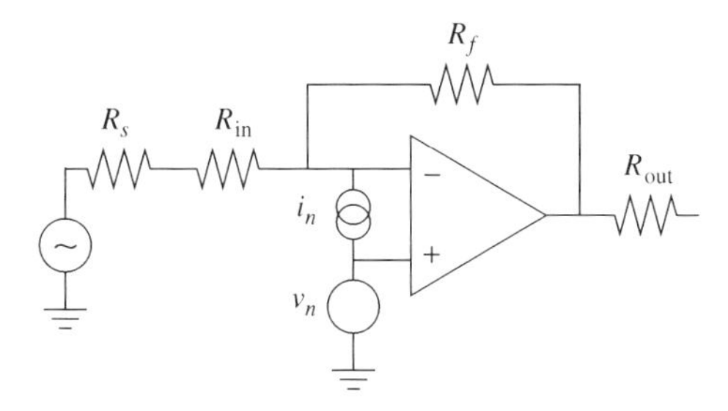

Fig. A.5 Op amp with noise sources.

The mean-square input voltage that produces the same output as does v_n is

$$e_v^2 = v_n^2 \frac{(1+a)^2}{a^2} = v_n^2 \left(1 + \frac{1}{a}\right)^2, \tag{3}$$

where a is the voltage gain of the op amp, $R_f/(R_s + R_{\text{in}})$. Here v_n multiplied by the gain from the noninverting input (i.e., $1 + a$) equals e_n multiplied by a, the gain from the inverting input.

Adding these three equivalent mean-square input voltages and dividing by the voltage equivalent to the available power from the source resistor, we obtain the

TABLE A.1 Op Amp Noise Factors for Various Parameter Values

	A	B	C	D	E	F
2	Nt	3.98107E-21 W				
3	Rs		1,020 Ω	1,020 Ω	20 Ω	20 Ω
4	Rin		1,000 Ω	1,000 Ω	2,000 Ω	2,000 Ω
5	Rf		20,000 Ω	2,000 Ω	20,000 Ω	2,000 Ω
6	a		9.9010	0.9901	9.9010	0.9901
7	In	4.00E-12 A				
8	Vn	4.00E-09 V				
9	eR term	from R's (less Rout)	2.18041176	3.98058824	111.201	203.01
10	ev term	From Vn	1.19408284	3.97971466	60.89822488	202.965447
11	ei term	From In	4.01941231	4.01941231	204.9900279	204.990028
12	f without Rout		7.39390692	11.9797152	377.0892528	610.965475
13						
14	Rout		20	20	20	20
15	f with Rout		7.39410694	11.9997172	377.0994538	611.985575
16	NF		8.68886 dB	10.7917 dB	25.76456 dB	27.8674 dB
17						
18	Rout		1020	1020	1020	1020
19	f with Rout		7.40410792	12.9998152	377.6095038	662.990575
20	NF		8.69473 dB	11.1394 dB	25.77043 dB	28.2151 dB

noise factor (Steffes, 1998):

$$f = \frac{e_R^2 + e_i^2 + e_v^2}{4kT_0BR_s} \tag{4}$$

$$= \left(1 + \frac{R_{\text{in}}}{R_s}\right)\left(1 + \frac{R_s + R_{\text{in}}}{R_f} + \frac{R_{\text{out}}}{R_f}\frac{R_s + R_{\text{in}}}{R_f}\right) + \frac{e_i^2 + e_v^2}{4kT_0BR_s} \tag{5}$$

$$= \left(1 + \frac{R_{\text{in}}}{R_s}\right)\left(1 + \frac{1}{a} + \frac{R_{\text{out}}}{R_f a}\right) + \frac{i_n^2(R_s + R_{\text{in}})^2 + v_n^2(1 + 1/a)^2}{4kT_0BR_s}. \tag{6}$$

By adding the mean-square voltage due to the noise current and the noise voltage sources, we are assuming their independence. (We can do that for an example, but the issue can be important in practice where correlation may have to be taken into account.) The noise factor for various values of these variables is shown in Table A.1.

Note, from Eq. (6) and from the table, how high gain improves the noise figure. It would seem to be better to get all the gain in one op amp, rather than two as is done in Fig. 3.18. There could be other requirements, however, such as a specific gain required at an intermediate output after Op Amp 2, or wide bandwidth (which is adversely affected by high closed-loop gain in a classic op amp), or the desire to study the effects of changes in Op Amp 3 in an example.

APPENDIX B

REPRESENTATIONS OF FREQUENCY BANDS, IF NORMALIZATION

B.1 PASSBANDS

The passband is plotted on a normalized graph, where a fixed value of LO, RF, and IF is represented by a point (Fig. B.1). The point can represent more than one set of LO, RF, and IF values, however, since it only specifies the normalized values, y and x.

If only the LO changes, its range is represented by a vertical line (Fig. B.2). If only the IF changes, x and y change equally, leading to a diagonal line (Fig. B.3).

At a given LO and IF, a range of RF values is represented by a horizontal line (Fig. B.4). As the IF takes on various values, still at the same LO frequency, the same RF range is represented by multiple horizontal lines (Fig. B.5). We can represent the RF band over this range of IF values by a parallelogram (Fig. B.6) that connects the two horizontal lines that are at the extreme IF values.

If the IF is held constant, the RF range is represented at various values of LO by horizontal lines (Fig. B.7). We can connect the horizontal lines at the extreme values of LO (Fig. B.8) to form a rectangle representing the RF and LO ranges together at the fixed IF. This passband, representing a continuum of RF and LO values, can be drawn for several values of IF (Fig. B.9). The combined RF, LO, and IF ranges can be represented by a polygon formed by connecting the rectangles at the IF extremes (Fig. B.10).

B.2 ACCEPTANCE BANDS

Between the passband, where low attenuation is required, and the rejection band, where a given attenuation is required, is a region where we require neither. We

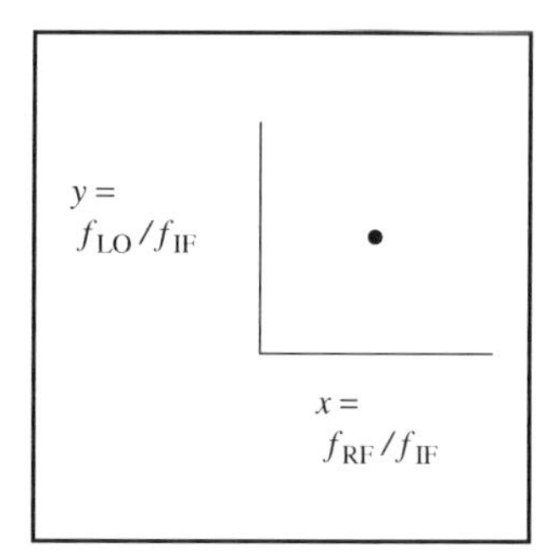

Fig. B.1 Normalized-to-IF graph.

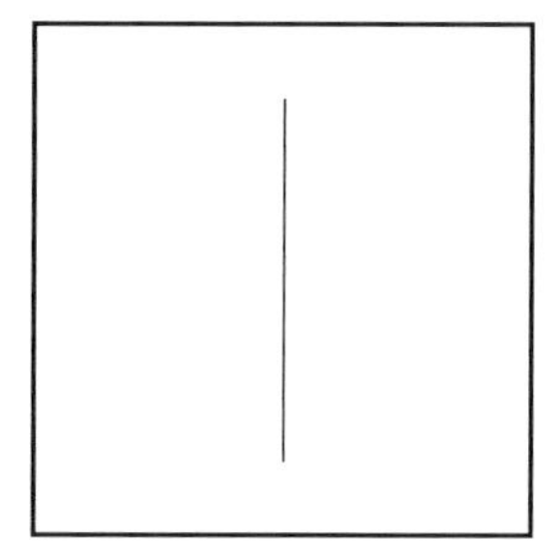

Fig. B.2 LO range.

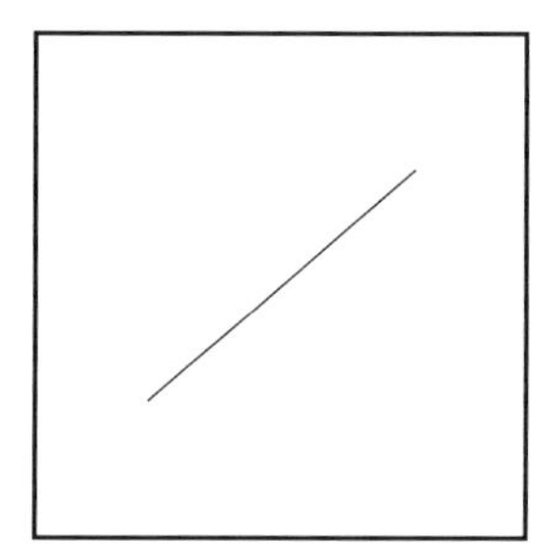

Fig. B.3 IF range.

Fig. B.4 RF range.

Fig. B.5 RF range at various IFs.

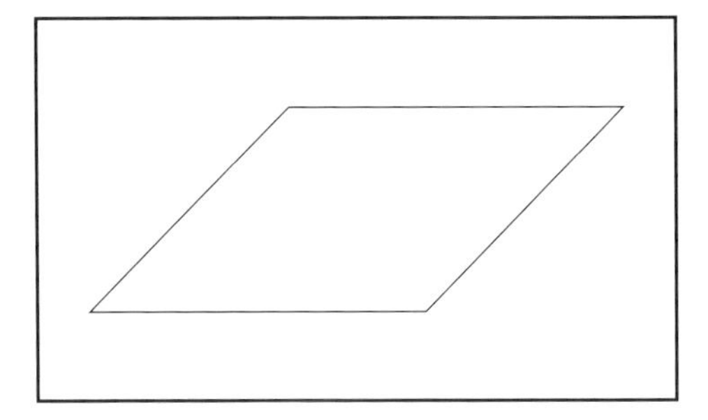

Fig. B.6 RF and IF ranges.

Fig. B.7 RF range at various LOs.

will call this the acceptance band. In Figs. 7.23 and 7.25, for example, we could designate a 14-dB rejection band for shape factors greater than 2.33. Then the acceptance band would extend from the passband edge, at a shape factor of 1, to a shape factor of 2.33.

If the required attenuation shown at a shape factor of 9 were greater than 14 dB, we could define an additional rejection band starting at 9, and the corresponding additional acceptance band would extend to there. (Also, a rejection of

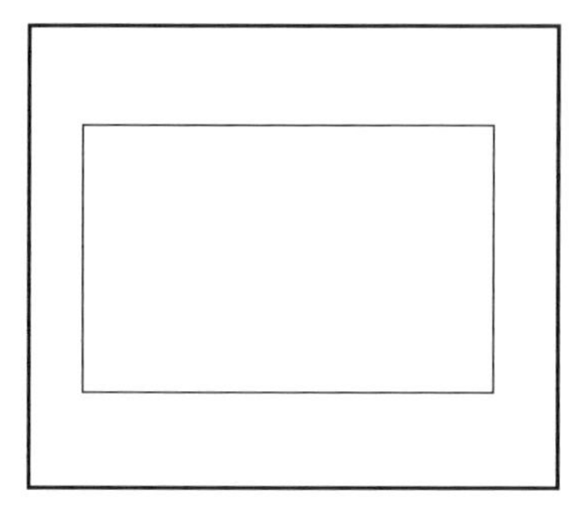

Fig. B.8 RF and LO ranges.

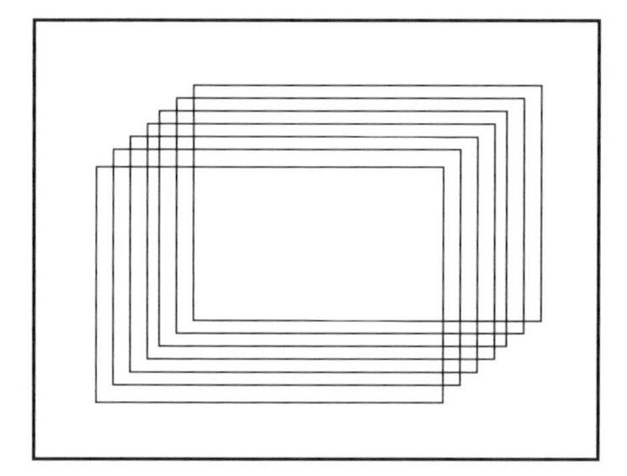

Fig. B.9 RF and LO ranges for several IFs.

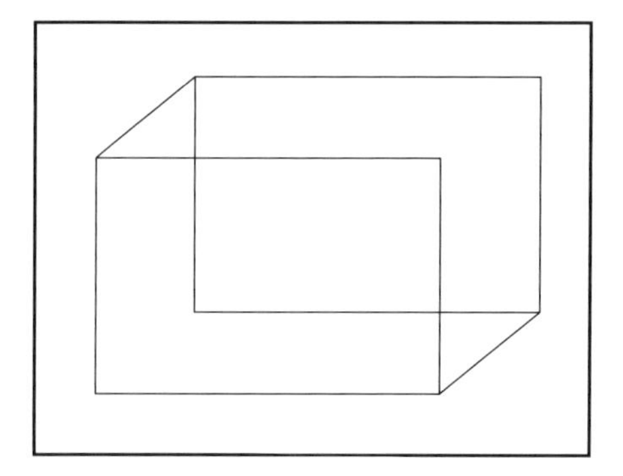

Fig. B.10 RF, LO, and IF ranges.

75 dB is required at 12.33.) The acceptance band is just the region between the passband and the point where a given rejection is required.

We can expand the polygon representing the passband to represent an acceptance band. While our intention is to reject *spurs* beyond the acceptance band, we would test the response by determining how much attenuation a *signal* receives as it moves from the passband, through the acceptance band, into the rejection band. The rejection r (R in dB) is the ratio of the attenuation in the rejection band to that in the passband.

With LO and IF fixed, a rejection (attenuation) of R will occur when the RF is above the upper edge of the RF filter acceptance band or below its lower edge. This is represented by extending the horizontal lines corresponding to the RF range (Fig. B.4) to the point where the filter attenuation reaches R. As a result, the three-dimensional passband is expanded as shown by the dashed lines in Fig. B.11. Such an acceptance band, if it is arithmetically symmetrical, can be described by an RF filter shape factor (Section 7.7).

With the LO and RF fixed, a rejection of R dB will occur when the IF is above the upper edge of the IF filter acceptance band or below its lower edge. This is represented by extending the diagonal lines corresponding to the IF range (Fig. B.3) and leads to an extension of the passband as shown in Fig. B.12. Such an acceptance band, if it is arithmetically symmetrical, can be described by an IF filter shape factor.

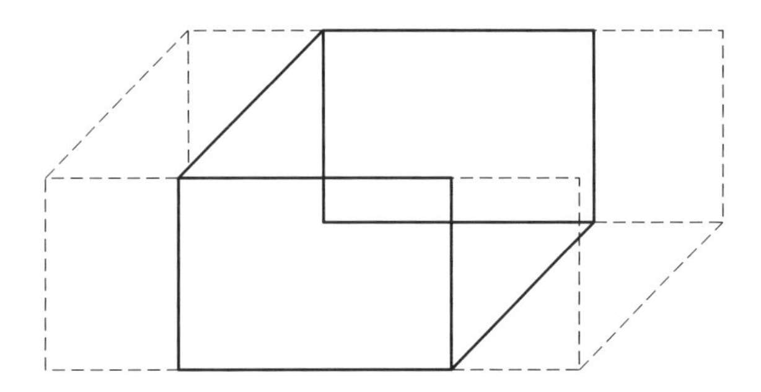

Fig. B.11 RF acceptance band.

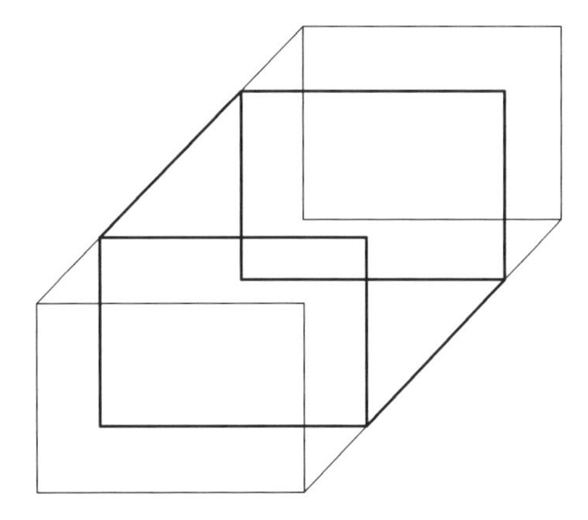

Fig. B.12 IF acceptance band.

Combining these, we obtain Fig. B.13. This represents the region beyond which mixer products are reduced by at least R. If the product being rejected is not of first order in the RF ($n > 1$), the RF-filter rejection at the edges need be only R/n whereas the IF-filter rejection would be R. These correspond to attenuation before and after the point where the spur is created.

The extended RF region in Fig. B.13 is within the IF passband; so attenuation in that region is due solely to the RF filter. The extended IF region is within the RF passband; so attenuation there is due only to the IF filter. Beyond the corners of these regions, attenuation is produced by both the RF and the IF filters. Thus the region beyond which spurs are reduced by R might actually be as shown in Figs. B.14 or B.15, depending on the details of the filters.

We can plot the attenuation due to both filters along a particular spur curve. We might plot the spur attenuation seen in the IF due to the RF filter response,

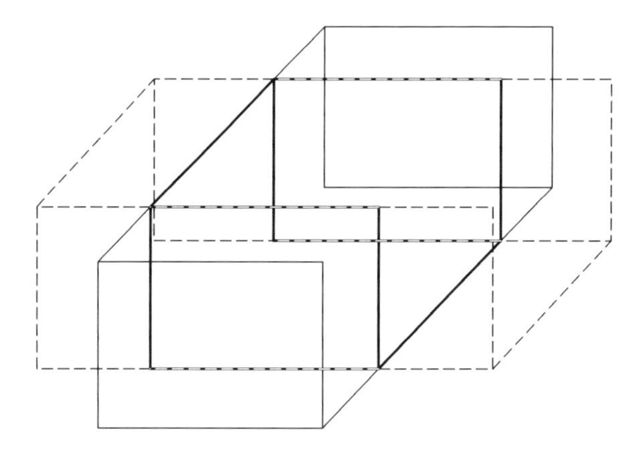

Fig. B.13 RF and IF acceptance bands.

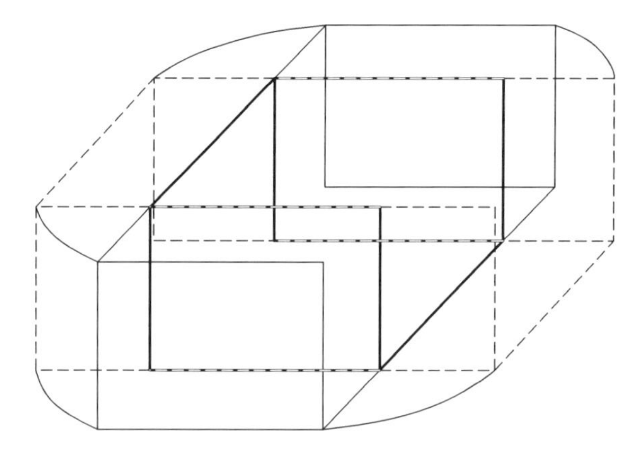

Fig. B.14 Total acceptance band.

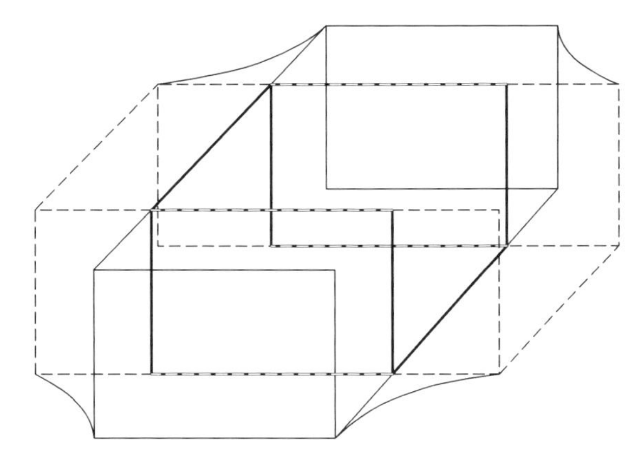

Fig. B.15 Another total band.

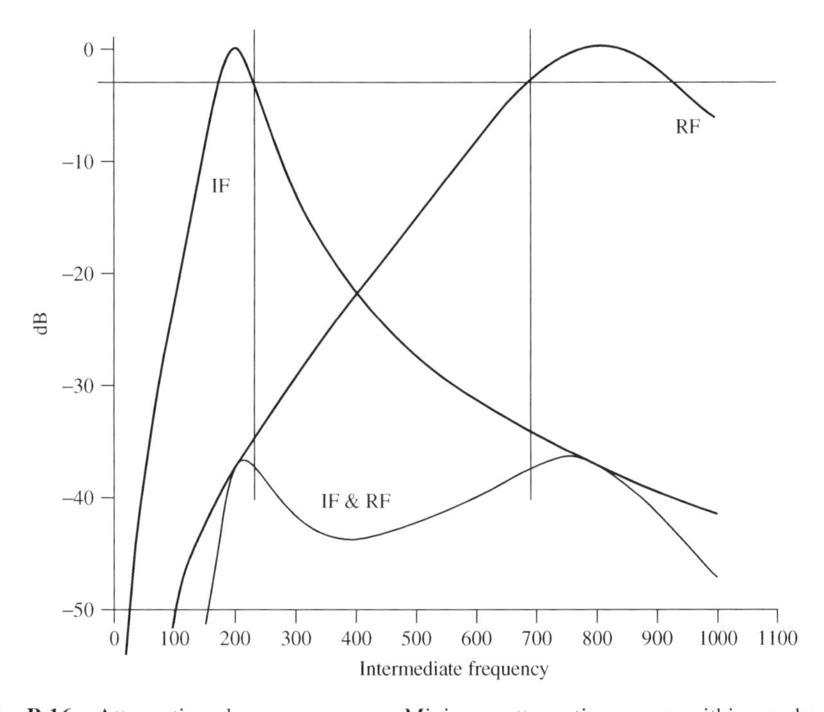

Fig. B.16 Attenuation along a spur curve. Minimum attenuation occurs within passband.

$-G_{\mathrm{IF},R}$, as a function of the IF:

$$G_{\mathrm{IF},R}(f_{\mathrm{IF}} = mf_{\mathrm{LO}} + nf_R) = nG_{\mathrm{RF}}(f_{\mathrm{RF}}). \tag{1}$$

We would add this to the IF response, $G_{\mathrm{IF}}(f_{\mathrm{IF}})$. Figures B.16 and B.17 illustrate two possible results for some simple filters: one result in which the minimum net

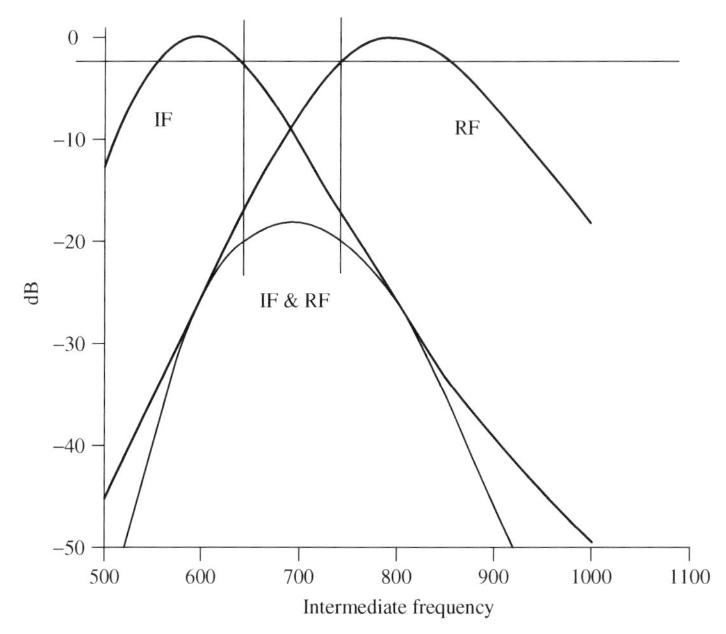

Fig. B.17 Attenuation along a spur curve. Minimum attenuation occurs between passbands.

attenuation occurs within a passband, and one in which it occurs between the IF and RF passbands. In both cases, however, the minimum attenuation is greater than both the minimum produced by the IF filter within the RF 3-dB bandwidth and the minimum produced by the RF filter within the IF 3-dB bandwidth. Thus, if the attenuation is adequate at the computed RF and IF shape factors, it is adequate in fact. Nevertheless, we can conceive of the situation in which this would not be true, especially if the passband is defined at a smaller attenuation (ripple) level and the filter attenuation increases very slowly.

Such a condition is illustrated in Fig. B.18, which shows the upper left corner of Fig. B.14. Here a spur curve passes through the line of constant rejection at a frequency that is within neither the IF nor the RF passband. This condition may not occur at all in a given application. If it does, the effect may be negligible compared to such things as the variation in mixer spurious responses.

We have introduced the concept of acceptance band to aid in our understanding of the combined effect of RF and IF filtering. We will not use it as an analysis tool per se.

B.3 FILTER ASYMMETRY

Figure B.16 illustrates filter asymmetry. Many bandpass filters are not arithmetically symmetrical. Attenuation may increase faster on the low-frequency side

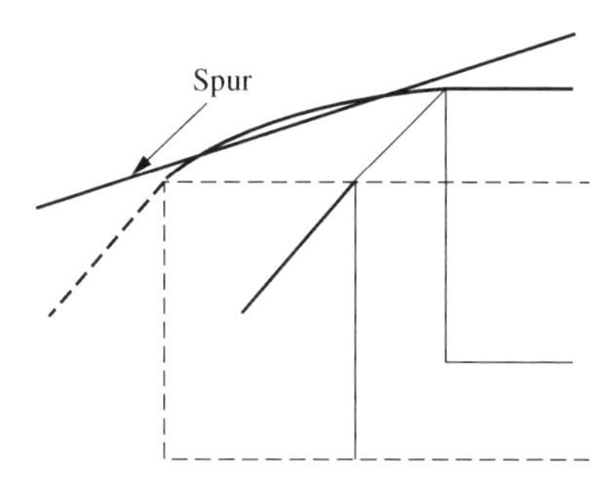

Fig. B.18 Spur minimally attenuated between pass bands.

of the filter than on the high-frequency side. Thus the filter requirements (e.g., the number of poles) depend not only on the required attenuation and the shape factor but also on whether the spur occurs on the high or low side of the filter. In Fig. 7.23, this is indicated by the sign that is attached to the shape factor.

APPENDIX C

CONVERSION ARITHMETIC

C.1 RECEIVER CALCULATOR

Figure C.1 shows a spreadsheet to aid in calculating the IF and RF filter and LO ranges, given the center frequencies of the signals at RF and at IF and their bandwidths. It also illustrates the process. Results are given for high- and low-LO architectures, that is, for architectures where the LO frequency is above the RF frequency and for those where it is below.

No units are used. We decide what units these numbers represent, and the same units (e.g., MHz or kHz) apply to all of the numbers. We are also free to define the meanings of the bandwidths (e.g., -1 dB, -3 dB, etc.). They are the ranges over which significant signals can exist.

Data is entered in the upper left box. We enter the maximum and minimum RF center frequencies and the center frequency of the signal in the IF filter. The IF bandwidth is set equal to the spectral width. If the RF filter is tuned (i.e., a tracking filter), we also enter its bandwidth at each end of the RF band (entered as 10 at both ends in this example), since the bandwidth of a tuned filter might be determined by considerations other than the spectral width.

The computed LO frequencies convert the specified RF center frequencies to the specified IF center frequency in every case. The results in the upper right are for tuned preselector (RF) filters (e.g., see Fig. 7.36), and those below are for fixed preselector filters, where the specified RF (tracking filter) bandwidths do not apply. The RF bandwidths include provisions for a finite spectral width equal to that of the IF filter.

Fixed preselectors must cover the whole signal range so the RF range given at the lower right is the range required to allow any of the given center frequencies to be converted to the center of the IF plus the additional bandwidth for signals

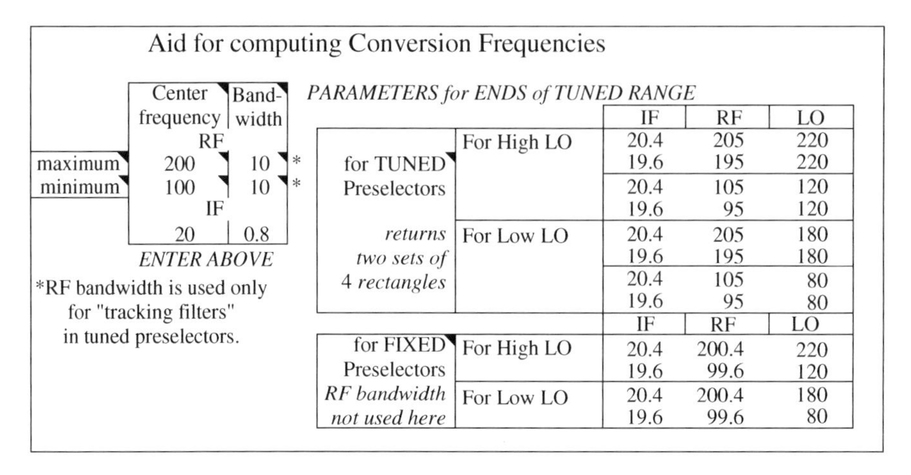

Aid for computing Conversion Frequencies

	Center frequency	Bandwidth
RF		
maximum	200	10 *
minimum	100	10 *
IF	20	0.8

ENTER ABOVE

*RF bandwidth is used only for "tracking filters" in tuned preselectors.

PARAMETERS for ENDS of TUNED RANGE

		IF	RF	LO
for TUNED Preselectors	For High LO	20.4	205	220
		19.6	195	220
		20.4	105	120
		19.6	95	120
returns two sets of 4 rectangles	For Low LO	20.4	205	180
		19.6	195	180
		20.4	105	80
		19.6	95	80

		IF	RF	LO
for FIXED Preselectors	For High LO	20.4	200.4	220
		19.6	99.6	120
RF bandwidth not used here	For Low LO	20.4	200.4	180
		19.6	99.6	80

Fig. C.1 Calculator for receiver frequency conversion.

Aid for computing Synthesis Frequencies

IF:	
max of interest	400
max synthesized	260
min synthesized	210
min of interest	160
Fixed Input =	1000
is (LO or RF)	1
(read as)	= LO

IF = LO – RF	IF	RF	LO
@ max IF	400	740	1000
of interest	400	790	1000
passband	260	740	1000
@ high IF	260	790	1000
passband	210	740	1000
@ low IF	210	790	1000
@ min IF	160	790	1000
of interest	160	740	1000

IF = RF – LO	IF	RF	LO
@ max IF	400	1260	1000
of interest	400	1260	1000
passband	260	1210	1000
@ high IF	260	1260	1000
passband	210	1210	1000
@ low IF	210	1260	1000
@ min IF	160	1210	1000
of interest	160	1260	1000

IF = LO + RF	IF	RF	LO	
@ max IF	400	–790	1000	NOT ALLOWED
of interest	400	–740	1000	NOT ALLOWED
passband	260	–790	1000	NOT ALLOWED
@ high IF	260	–740	1000	NOT ALLOWED
passband	210	–790	1000	NOT ALLOWED
@ low IF	210	–740	1000	NOT ALLOWED
@ min IF	160	–790	1000	NOT ALLOWED
of interest	160	–740	1000	NOT ALLOWED

Fig. C.2 Calculator for synthesizer frequency conversions.

that will fill the IF filter band. The entered RF bandwidths are ignored in these fixed-preselector calculations.

C.2 SYNTHESIS CALCULATOR

Figure C.2 shows a spreadsheet to aid in calculating the range of mixer inputs (RF or LO) to produce a given IF range with a given fixed second mixer input (LO or RF). The LO is fixed at 1000 in this example, and the RF inputs are to be computed. There are three sets of frequencies for the three $\pm 1 \times \pm 1$ products, two of which are valid results. Which two depends on the chosen output (IF) and fixed input frequencies.

Each of the three sets defines four straight lines, corresponding to the fixed and variable inputs at four different IFs (outputs). The middle two sets describe the synthesized range while the other two correspond to some wider range, possibly the minimum and maximum output frequencies of interest. They are not essential, but they allow us to see the IF extended above and below the passband, highlighting the region where spurs must be controlled.

The four horizontal lines (degenerate rectangles with zero height) corresponding to the upper set in Fig. C.2 are plotted in Fig. C.3. The lines connecting rectangles 1 and 4 enclose a region corresponding to the RF input at the various

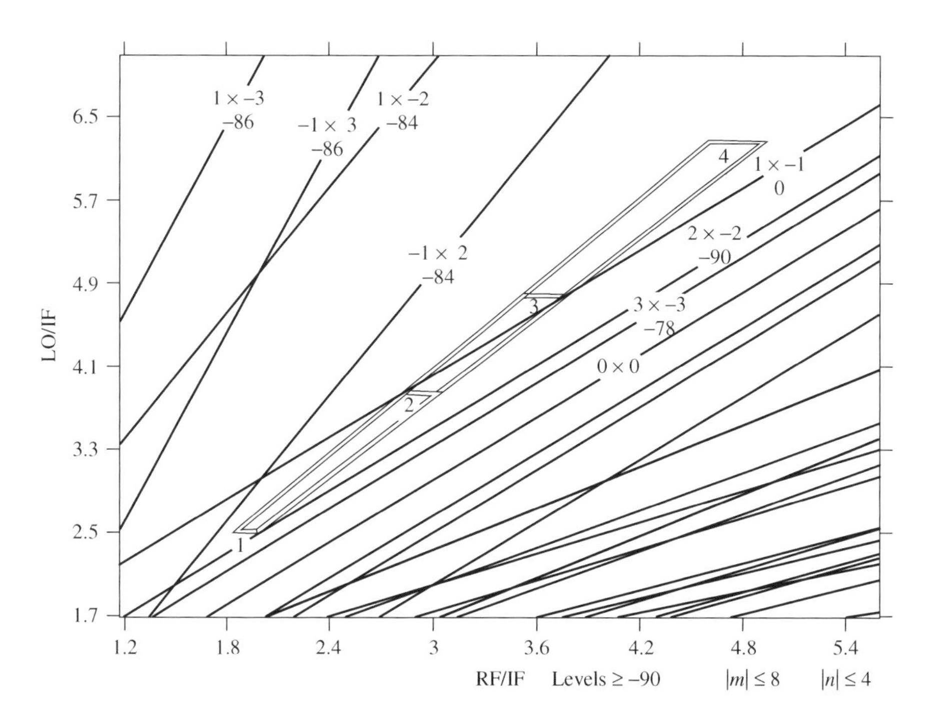

Fig. C.3 Spur graph for upper synthesizer realization of Fig. C.2.

IF values. Since the RF range is absolute (like the RF in a receiver with a "brick wall" filter where no input can occur anywhere outside of the RF passband), only the shape factor for the IF is significant. The points where spurs enter this region (or the region enclosed by further extension of these connecting lines) correspond to the IF shape factors. We can see, from Fig. C.3, that the 2×-2 spur is the closest, just touching line (rectangle) 1. The shape factor determined by the point where it enters the extended polygon will define the smallest shape factor for the IF filter.

APPENDIX E

EXAMPLE OF FREQUENCY CONVERSION

Suppose we are to convert 3.7 MHz to 1.3 MHz using a high-side LO. This is a "high-side downconversion." The LO frequency would be

$$f_{\mathrm{LO}} = f_{\mathrm{RF}} + f_{\mathrm{IF}} = 3.7\ \mathrm{MHz} + 1.3\ \mathrm{MHz} = 5\ \mathrm{MHz}. \tag{1}$$

The plot for high-side (1×-1) downconversion,

$$f_{\mathrm{IF}} = f_{\mathrm{LO}} - f_{\mathrm{RF}} \tag{2}$$

is shown in Fig. E.1 with our particular frequencies marked. Line b, representing $f_{\mathrm{RF}} = 3.7$ MHz, intersects the 1×-1 curve where $f_{\mathrm{IF}} = 1.3$ MHz (line a). This diagram indicates that, if we input 3.7 MHz to the mixer, when the LO is at 5 MHz, we get an IF output of 1.3 MHz. In a receiver, this means that we would tune the LO to 5 MHz in order to convert a 3.7-MHz RF input to our IF at 1.3 MHz. It also shows that a 1.3-MHz IF output implies a 3.7-MHz RF input; so, when we see a signal at 1.3 MHz in our IF, we interpret it as having originated at 3.7 MHz. (We are aware that the image frequency could also be implied, but that is not shown in Fig. E.1.)

In Fig. E.2, a spurious response, the -2×3 response, has also been plotted. This is the solution of

$$f_{\mathrm{IF}} = m f_{\mathrm{LO}} + n f_{\mathrm{RF}} = -2 f_{\mathrm{LO}} + 3 f_{\mathrm{RF}} = -10\ \mathrm{MHz} + 3 f_{\mathrm{RF}}. \tag{3}$$

The plot reveals that another IF output, at 1.1 MHz, is produced by this spurious response. To eliminate this undesired response, we can filter the IF to pass the desired response and reject the spurious response. The band edges of such a filter

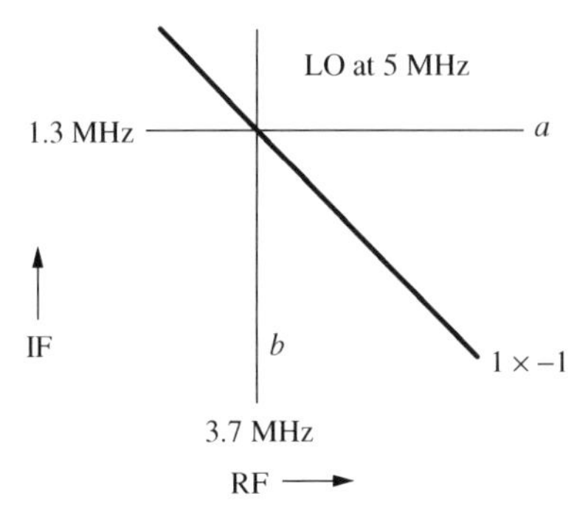

Fig. E.1 High-side downconversion from 3.7 to 1.3 MHz.

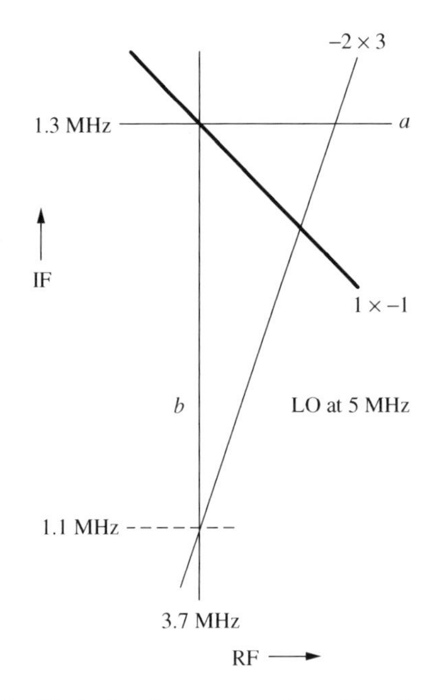

Fig. E.2 High-side downconversion with -2×3 spur.

are indicated by the double lines in Fig. E.3. This filter rejects signals that are more than 20 kHz from 1.3 MHz.

While the IF filter rejects the 1.1-MHz spurious response, it does not prevent a signal at some other RF frequency from producing a signal within the IF filter band. Figure E.4 shows such a spur in the IF caused by the -2×3 response. Here the same 1.3 MHz that we expect to obtain from 3.7 MHz RF is produced by an RF signal whose frequency is 67 kHz higher (line c). This might interfere with the desired signal or it might cause us to mistakenly believe that we are processing a 3.7-MHz RF input whereas we are actually receiving a signal at

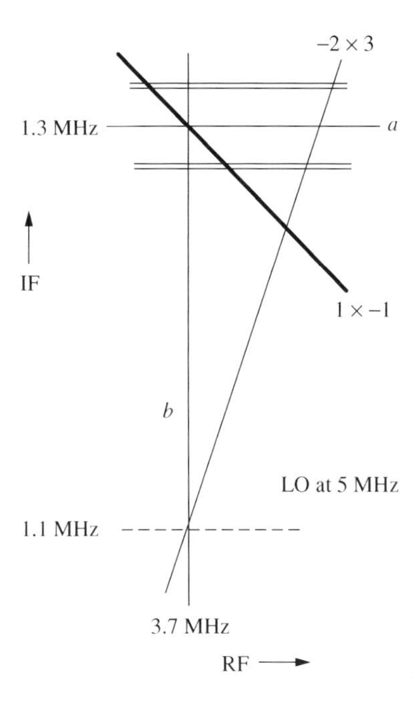

Fig. E.3 Rejecting the spurious response with an IF filter.

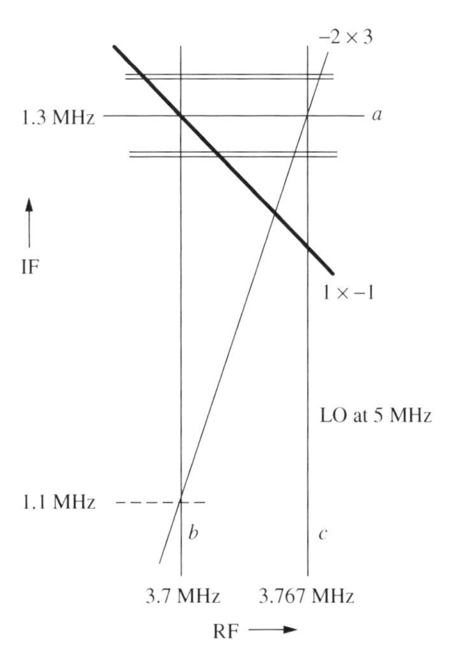

Fig. E.4 Spurious RF response.

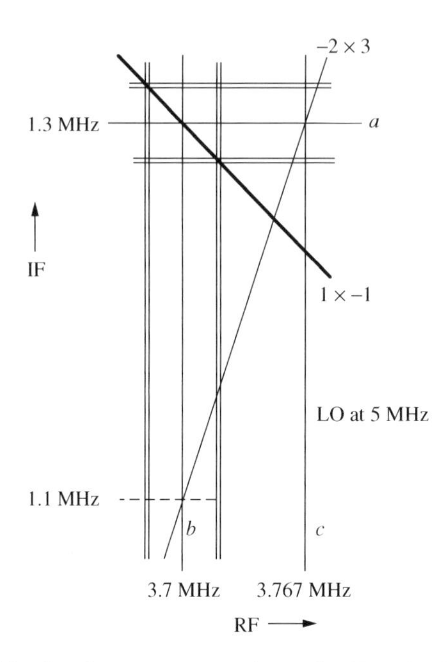

Fig. E.5 Spurious responses rejected by IF and RF filters.

another frequency. To prevent this, we use an RF filter, as shown in Fig. E.5. Only RF frequencies between the vertical double lines are passed.

Now the frequency range of both the RF and the IF are restricted, preventing (or at least attenuating) the spurious responses. We can simplify Fig. E.5 by replacing the four double lines by a rectangle outlining their crossing area, as in Fig. E.6. This rectangle is a square, representing the conversion of all 40 kHz of the RF band into the full 40-kHz-wide IF band. The 1×-1 response runs diagonally, from corner to corner.

In Fig. E.7, the RF bandwidth has been doubled by increasing the frequency of the upper edge. We note that the -2×3 spur is now at the corner of the rectangle, indicating a spurious response at the upper edge of the RF band and the lower edge of the IF band, Eq. (3) giving

$$f_{\text{IF}} = -2(5 \text{ MHz}) + 3(3.76 \text{ MHz}) = 1.28 \text{ MHz}. \tag{4}$$

Since this is at both band edges, neither filter will be effective at reducing the spur. Either we must ensure that the RF level is small enough that the spur level is sufficiently (for our application) lower than the desired IF or we will have to change the design.

Increasing the RF filter bandwidth has made a greater RF range available to us, but it is not being converted to the IF. Only the original 40-kHz-wide RF band is being converted to frequencies that are in the IF passband. To take advantage of the doubled RF range, we must use two LO frequencies. Since the expanded

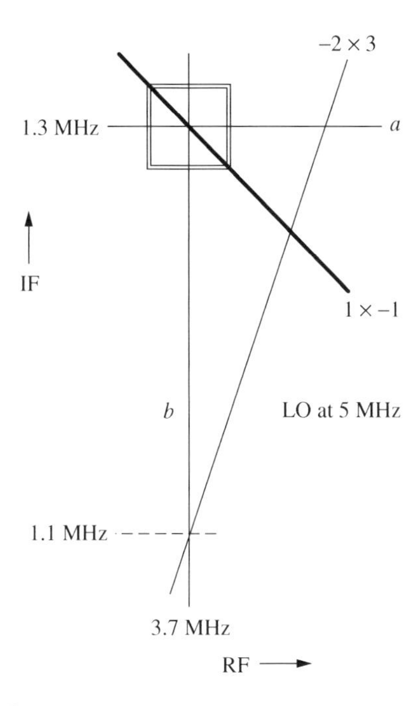

Fig. E.6 Rectangle representing the IF and RF bands.

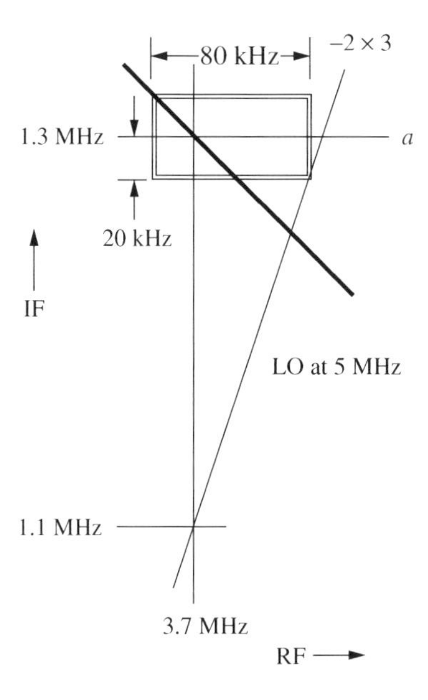

Fig. E.7 Expanded RF band.

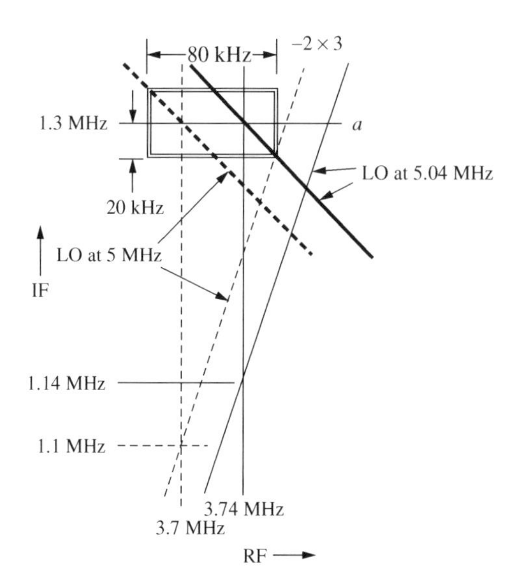

Fig. E.8 Responses with a 5-MHz LO (dashed) and with a 5.04-MHz LO.

part of the RF band is 40 kHz higher in frequency than the original part, we require an LO with a frequency 40 kHz higher to give the original IF range according to Eq. (2). The desired and spurious responses are shown in Fig. E.8. The responses with a 5-MHz LO, the same as are shown in Fig. E.7, are dashed and the responses with a 5.04-MHz LO are in solid lines.

While we have considered a conversion that uses two LO frequencies to convert the whole RF band into the IF band, we could take the two sets of curves as limits as the LO frequency is changed continuously between 5 and 5.04 MHz and imagine the space between similar response curves as representing the area occupied as a curve moves continuously from one extreme to the other.

We see that the spur in the IF will still be rejected by the IF filter (but it is at the band edge at one extreme of the LO range). We see that the lower half of the RF band (3.7 MHz $\pm$ 20 kHz) is converted to the IF band by one LO frequency and that the upper half (3.74 MHz $\pm$ 20 kHz) is converted by the other. Note that the LO range (40 kHz) plus the IF bandwidth (40 kHz) equals the RF bandwidth (80 kHz). This is common; the LO moves the center frequencies of signals in the RF band, but their finite spectral width, presumably corresponding to the IF bandwidth, requires additional RF bandwidth.

The task of plotting responses over a range of frequencies is typically more difficult than in this example because more spurious responses must be plotted. It is generally simpler to normalize the axes of the plot to correspond to a modified form of Eq. (3),

$$y = m + nx, \tag{5}$$

where $y = f_{\text{IF}}/f_{\text{LO}}$, $x = f_{\text{RF}}/f_{\text{LO}}$, and Eq. (5) is obtained by dividing Eq. (3) by f_{LO}.

The curves generated by this equation are similar to those from Eq. (3) with $f_{\text{LO}} = 1$ and the relationship between their y and x coordinates does not change with the value of the LO frequency. However, as the LO frequency changes, the rectangle does move in these normalized coordinates. Figure E.9 shows the normalized plot corresponding to Fig. E.8. Note that there is only one 1×-1 desired curve and only one -2×3 spur curve, but there are two rectangles corresponding to the two values of LO frequency. We can see again that the lower half of the RF band is converted to the IF band when the LO frequency is 5 MHz and the upper half is converted when it is 5.04 MHz. We can also see again that the spur is at the same band edges when the LO is 5 MHz but is separated from the band edge with the 5.04 MHz LO.

If the two rectangles represent extremes of a (more or less) continuous LO range, we can connect corners as in Fig. E.10. Here the outline is a polygon that encompasses the passband as the LO frequency varies.

We can see, in Fig. E.11, how two RF sub-bands (3.68 MHz–3.72 MHz and 3.72 MHz–3.76 MHz) are converted to the same IF band by the two extreme LO frequencies. Following the 1×-1 curve between the points labeled at the edges of either rectangle, we see that the lowest RF frequency is converted to the highest IF frequency. Because this is a high-side downconversion, the spectrum becomes inverted.

Alternatively, we can normalize the RF and LO frequencies to the IF, as shown in Fig. E.12. Here the rectangles represent RF and LO ranges at the extreme values of the IF, and we can visualize the IF moving from one extreme value to the other along the outlined path. If the LO has only two possible frequencies,

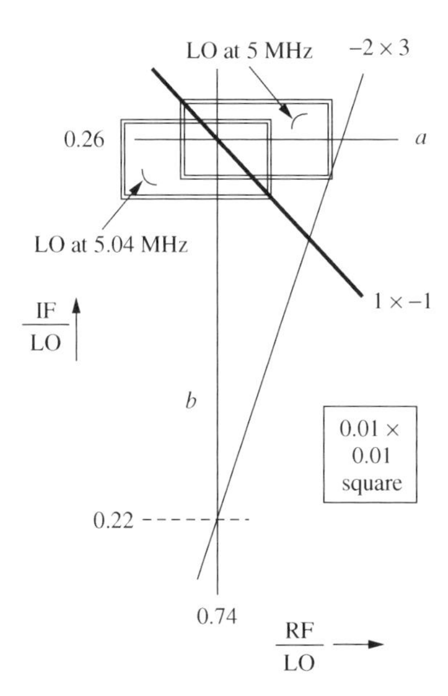

Fig. E.9 Response curves normalized to LO frequency.

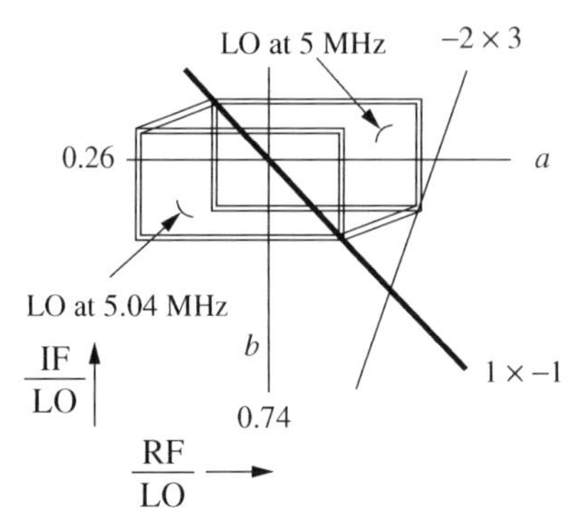

Fig. E.10 Locus of the passbands as the LO changes frequency.

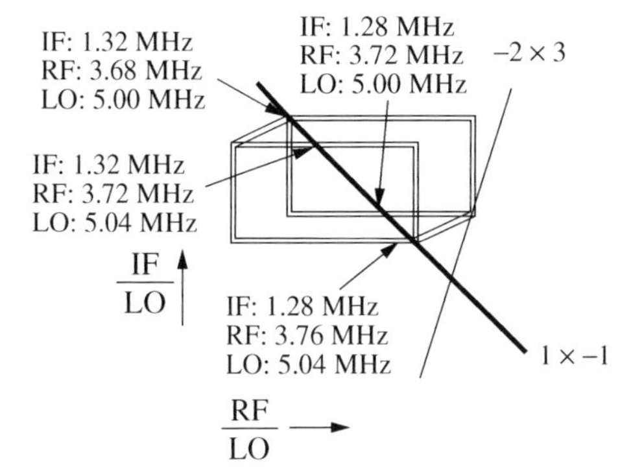

Fig. E.11 Frequencies of translation from RF to IF at one LO setting.

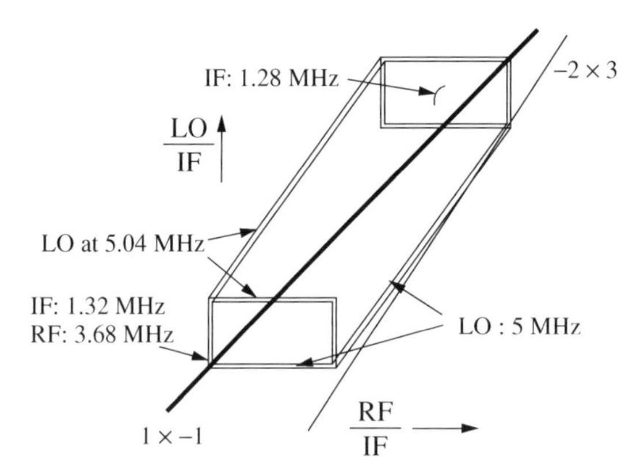

Fig. E.12 Response curves normalized to IF.

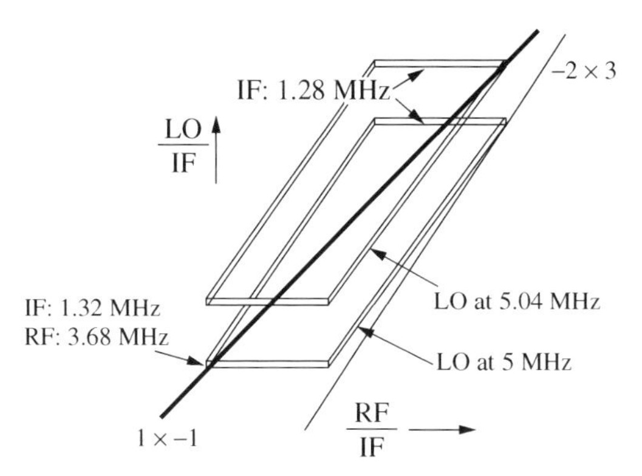

Fig. E.13 Response normalized to IF with only two LO frequencies.

rather than a continuum, the bands would be represented as in Fig. E.13. This represents the same design as does Fig. E.9.

APPENDIX F

SOME RELEVANT FORMULAS

F.1 DECIBELS

The power gain of module j is given by the ratio of power delivered to a load to power input to the module (under some specified conditions, e.g., values of source and load):

$$G_j = 10 \text{ dB} \log_{10}(p_{j+1}/p_j), \tag{1}$$

where $\log_{10}$ is logarithm to the base 10. This can also be written in terms of the effective (rms) voltages $\tilde{v}$ appearing across the (equivalent shunt) load and input resistances:

$$G_j = 10 \text{ dB} \log_{10}\left(\frac{\tilde{v}_{j+1}^2/R_{j+1}}{\tilde{v}_j^2/R_j}\right) \tag{2}$$

$$= 20 \text{ dB} \log_{10}\left(\frac{\tilde{v}_{j+1}}{\tilde{v}_j}\right) - 10 \text{ dB} \log_{10}\left(\frac{R_{j+1}}{R_j}\right). \tag{3}$$

Often the last term is ignored and voltage gain is given using

$$A_j = 20 \text{ dB} \log_{10}\left(\frac{\tilde{v}_{j+1}}{\tilde{v}_j}\right). \tag{4}$$

This practice of expressing voltage gain in dB is common and often useful but not strictly correct unless $R_{j+1} = R_j$, and it can lead to confusion since Eqs. (3) and (4) will give different values if $R_{j+1} \neq R_j$. If $R_{j+1} = R_j$, or if the voltages are normalized to R (see Section 2.2.2), then $A \equiv G$.

The advantage of using decibels is that products or ratios can be obtained by addition or subtraction of dB. If

$$g_T = g_1 g_2, \tag{5}$$

then

$$G_T \triangleq 10 \text{ dB} \log_{10}(g_T) = 10 \text{ dB} \log(g_1 g_2) \tag{6}$$

$$\equiv 10 \text{ dB} \log(g_1) + 10 \text{ dB} \log(g_2) \equiv G_1 + G_2. \tag{7}$$

Sometimes it is convenient to express even voltage and power in decibels by taking their ratios to some reference value. For example, dBm uses a 1-mW reference, dBW uses a 1-W reference, and dBV uses a 1-V reference. Thus, if

$$V_j = 20 \log_{10} \frac{\tilde{v}_j}{1 \text{ V}}, \tag{8}$$

and similarly for V_{j+1}, they would be expressed in dBV, and Eq. (4) could be expressed as

$$A_j = V_{j+1} - V_j. \tag{9}$$

F.2 REFLECTION COEFFICIENT AND SWR

The ratio of reflected wave to forward wave at the input or output port of a module is the reflection coefficient ρ_m at that port:

$$\rho_m = \frac{v_{\text{out}}}{v_{\text{in}}}, \tag{10}$$

where v_{in} is the incident wave and v_{out} is the resulting reflected wave.

The return loss (RL) is the loss of power in the reflected wave relative to the incident wave, usually expressed in dB,

$$\text{RL} = -20 \text{ dB} \log_{10} |\rho_m|, \tag{11}$$

$$|\rho_m| = 10^{\text{-RL}/(20 \text{ dB})}. \tag{12}$$

In terms of the impedance Z_m looking into that port and the standard characteristic impedance of the transmission line, $Z_0 = R_0$, we state, without proof for now, that (Pozar, 2001, p. 33)

$$\rho_m = \frac{Z_m - R_0}{Z_m + R_0} = \frac{Z_m/R_0 - 1}{Z_m/R_0 + 1}. \tag{13}$$

This coefficient ρ_m has both magnitude and phase, as Z_m has both magnitude and phase.

The standing wave caused by the reflection will have a maximum when the forward wave v_{in} and reverse wave v_{out} add in phase. This maximum will be

$$V_{max} = |v_{in}| + |v_{out}| = |v_{in}|(1 + |\rho_m|). \tag{14}$$

Similarly, the minimum will be

$$V_{min} = |v_{in}| - |v_{out}| = |v_{in}|(1 - |\rho_m|) \tag{15}$$

so the ratio of V_{max} to V_{min}, the standing-wave ratio or SWR, will be

$$\text{SWR} = \frac{V_{max}}{V_{min}} = \frac{1 + |\rho_m|}{1 - |\rho_m|}. \tag{16}$$

Solving for $|\rho_m|$ in terms of SWR, we obtain

$$|\rho_m| = \frac{1 - \text{SWR}}{1 + \text{SWR}}. \tag{17}$$

If Z_m is real, so $Z_m = R_m$, Eq. (16) can be combined with Eq. (13) to give

$$\text{SWR} = \frac{R_m + R_0 + |R_m - R_0|}{R_m + R_0 - |R_m - R_0|}, \tag{18}$$

which is

$$\text{SWR} = \text{MAX}\left\{\frac{R_m}{R_0}, \frac{R_0}{R_m}\right\}. \tag{19}$$

That is, the ratio of R_m to R_0 or its reciprocal, whichever is greater than 1.

We will now verify Eq. (13). The voltage at the module input is the sum of the forward and reverse voltages:

$$V_m = v_{in} + v_{out}. \tag{20}$$

The currents equal the voltages divided by R_0. However, the forward and reverse currents have different signs because of the differing directions of propagation. Therefore, the total current in the direction of the module is

$$I = (v_{in} - v_{out})/R_0. \tag{21}$$

The input impedance at the module is the ratio of voltage to current there:

$$Z_m = \frac{V_m}{I_m} = \frac{v_{in} + v_{out}}{(v_{in} - v_{out})/R_0} \tag{22}$$

$$= \frac{1 + \rho_m}{1 - \rho_m} R_0. \tag{23}$$

Solving this for ρ_m, we obtain Eq. (13).

F.3 COMBINING SWRs

F.3.1 Summary of Results

If two lossless reciprocal elements are connected (Fig. F.1) by a cable of undetermined length (i.e., if relative signal phases are arbitrary), their individual reflections combine such that the resulting SWR at the cascade input is (Ragan, 1948) no greater than the product of the individual SWRs:

$$\text{SWR} \leq \text{SWR}_{\text{max}} = \text{SWR}_1 \times \text{SWR}_2 \tag{24}$$

and is no less then their quotient:

$$\text{SWR} \geq \text{SWR}_{\text{min}} = \text{SWR}_1/\text{SWR}_2, \tag{25}$$

where

$$\text{SWR}_1 \geq \text{SWR}_2. \tag{26}$$

These results are probably better known than are their derivations or their range of applicability, both of which we explore here.

One can logically extend these relationships to cover N reflections. The maximum SWR is then the product of all the individual SWRs:

$$\text{SWR}_{\text{max}} = \prod_{i=1}^{N} \text{SWR}_i, \tag{27}$$

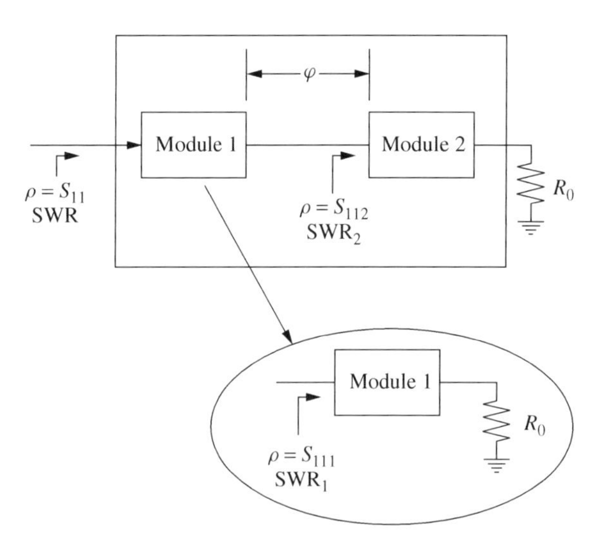

Fig. F.1 SWR of combined series elements.

and the minimum is the ratio of the largest SWR to the product of all the others, except it cannot be less than 1:

$$\mathrm{SWR_{min}} = \mathrm{MAX}\left\{ \frac{\mathrm{SWR}_1}{\prod_{i=2}^{N} \mathrm{SWR}_i}, 1 \right\}, \tag{28}$$

where

$$\mathrm{SWR}_1 \geq \mathrm{SWR}_i. \tag{29}$$

If the ratio in Eq. (28) is less than 1, then, by optimum selection of the various phases (lengths of interconnects), the reflections can be made to cancel (SWR = 1).

The lossless and reciprocal restrictions need not apply to the remotest reflection in the cascade.

F.3.2 Development

Here we follow the development by Fano and Lawson (1948) but use the S and T parameters defined in Section 2.2. Common passive networks are reciprocal so that, if a driving voltage or current at one port produces a response current or voltage, respectively, at a second port, then reversing the network will cause the same drives to produce the same responses in the opposite direction. We will use the normalized voltages of Eq. (2.16), which, by that equation, can be expressed as the square root of the product of voltage and current. Reciprocity then applies to both the currents and voltages and, as a result, to their products, giving

$$S_{12} = S_{21}. \tag{30}$$

Since the networks are lossless, the power must either be transmitted or reflected, leading to

$$|S_{11}|^2 + |S_{21}|^2 = 1 \tag{31}$$

and

$$|S_{22}|^2 + |S_{12}|^2 = 1. \tag{32}$$

Combining the last three equations, we see that

$$|S_{11}| = |S_{22}|. \tag{33}$$

Having established this necessary background, we now write the T matrix for a cascade consisting of a module 1 followed by a lossless interconnect cable and ending in a terminated module 2.

$$\mathbf{T} = \begin{pmatrix} T_{11} & T_{12} \\ T_{21} & T_{22} \end{pmatrix} = \begin{pmatrix} T_{111} & T_{121} \\ T_{211} & T_{221} \end{pmatrix} \begin{pmatrix} e^{j\varphi} & 0 \\ 0 & e^{-j\varphi} \end{pmatrix} \begin{pmatrix} T_{112} & T_{122} \\ T_{212} & T_{222} \end{pmatrix} \tag{34}$$

$$= \begin{pmatrix} T_{111} & T_{121} \\ T_{211} & T_{221} \end{pmatrix} \begin{pmatrix} T_{112}e^{j\varphi} & T_{122}e^{j\varphi} \\ T_{212}e^{-j\varphi} & T_{222}e^{-j\varphi} \end{pmatrix}. \tag{35}$$

From this we can obtain

$$T_{11} = T_{111}T_{112}e^{j\varphi} + T_{121}T_{212}e^{-j\varphi}. \tag{36}$$

Using Eq. (2.29), we can write this as

$$\frac{1}{S_{21}} = \frac{1}{S_{211}}\frac{1}{S_{212}}e^{j\varphi}(1 - S_{221}S_{112}e^{-j2\varphi}). \tag{37}$$

F.3.3 Maximum SWR

The phase φ can be chosen to maximize this expression, leading to

$$\frac{1}{|S_{21}|_{\min}} = \frac{1 + |S_{221}S_{112}|}{|S_{211}S_{212}|}. \tag{38}$$

We can use Eq. (31) to write this as

$$\frac{1}{|S_{21}|_{\min}} = \frac{1 + |S_{221}S_{112}|}{\sqrt{1 - |S_{111}|^2}\sqrt{1 - |S_{112}|^2}}. \tag{39}$$

Again using Eq. (31), as well as Eq. (33), we can write S_{11} in terms of S_{21}:

$$|S_{11}|_{\max} = \sqrt{1 - |S_{21}|^2_{\min}} = \sqrt{1 - \frac{(1 - |S_{111}|^2)(1 - |S_{112}|^2)}{(1 + |S_{111}S_{112}|)^2}} \tag{40}$$

$$= \sqrt{\frac{1 + 2|S_{111}S_{112}| + |S_{111}S_{112}|^2 - 1 + |S_{111}|^2 + |S_{112}|^2 - |S_{111}S_{112}|^2}{(1 + |S_{111}S_{112}|)^2}} \tag{41}$$

$$= \frac{|S_{111}| + |S_{112}|}{1 + |S_{111}||S_{112}|}. \tag{42}$$

Since $S_{11} = \rho_m$ for the terminated cascade, we can use Eq. (16) to give the maximum SWR for the cascade:

$$\text{SWR}_{\max} = \frac{1 + |S_{11}|_{\max}}{1 - |S_{11}|_{\max}} = \frac{1 + |S_{111}||S_{112}| + |S_{111}| + |S_{112}|}{1 + |S_{111}||S_{112}| - |S_{111}| - |S_{112}|} \tag{43}$$

$$= \frac{(1 + |S_{111}|)(1 + |S_{112}|)}{(1 - |S_{111}|)(1 - |S_{112}|)} \tag{44}$$

$$= \text{SWR}_1 \times \text{SWR}_2. \tag{45}$$

F.3.4 Minimum SWR

Now, choosing φ to minimize Eq. (37), we can find the minimum value of SWR:

$$\frac{1}{|S_{21}|_{\max}} = \frac{1 - |S_{221} S_{112}|}{|S_{211} S_{212}|} \tag{46}$$

$$= \frac{1 - |S_{221} S_{112}|}{\sqrt{1 - |S_{111}|^2}\sqrt{1 - |S_{112}|^2}} \tag{47}$$

$$|S_{11}|_{\min} = \sqrt{1 - \frac{(1 - |S_{111}|^2)(1 - |S_{112}|^2)}{(1 - |S_{111} S_{112}|)^2}} \tag{48}$$

$$= \frac{||S_{111}| - |S_{112}||}{1 - |S_{111}||S_{112}|} \tag{49}$$

$$\text{SWR}_{\min} = \frac{1 + |S_{11}|_{\min}}{1 - |S_{11}|_{\min}} = \frac{1 - |S_{111}||S_{112}| + ||S_{111}| - |S_{112}||}{1 - |S_{111}||S_{112}| - ||S_{111}| - |S_{112}||}. \tag{50}$$

Let

$$S_{11+} = \text{Maximum}(S_{111}, S_{112}) \tag{51}$$

and

$$S_{11-} = \text{Minimum}(S_{111}, S_{112}). \tag{52}$$

Then Eq. (50) becomes

$$\text{SWR}_{\min} = \frac{1 + |S_{11}|_{\min}}{1 - |S_{11}|_{\min}} = \frac{1 - |S_{11+}||S_{11-}| + |S_{11+}| - |S_{11-}|}{1 - |S_{11+}||S_{11-}| - |S_{11+}| + |S_{11-}|} \tag{53}$$

$$= \frac{(1 + |S_{11+}|)(1 - |S_{11-}|)}{(1 - |S_{11+}|)(1 + |S_{11-}|)} \tag{54}$$

$$= \frac{\text{SWR}_+}{\text{SWR}_-}, \tag{55}$$

where SWR_+ and SWR_- are the two SWRs and

$$\text{SWR}_+ \geq \text{SWR}_-. \tag{56}$$

F.3.5 Relaxing Restrictions

While we have assumed that both reflections were produced by lossless reciprocal networks, that requirement appears to be unnecessary for module 2 since the output port of module 1 reacts to the wave reflected from module 2 independently of how it was produced. In other words, if we replace a lossless reciprocal module 2 with anything else that produces the same reflection, the total reflection from the input to module 1 will not change.

F.4 IMPEDANCE TRANSFORMATIONS IN CABLES

We can write the impedance looking toward the module through a cable of length d as $Z_{\text{in}}(-d)$. This can also be expressed as

$$Z_{\text{in}}(\varphi) = \frac{V(\varphi)}{I(\varphi)} = \frac{v_{\text{in}} e^{+j\varphi} + v_{\text{out}} e^{-j\varphi}}{(v_{\text{in}} e^{+j\varphi} - v_{\text{out}} e^{-j\varphi})/R_0} \tag{57}$$

$$= \frac{1 + \rho_m e^{-j2\varphi}}{1 - \rho_m e^{-j2\varphi}} R_0, \tag{58}$$

where ρ_m is the complex reflection coefficient at the module and φ is the one-way phase delay through the cable.

Note that $Z_{\text{in}}(0) = Z_m$, in agreement with Eq. (23). Then, obtaining the original reflection coefficient from Eq. (13), we have (Pozar, 2001, p. 35)

$$Z_{\text{in}}(\varphi) = \frac{1 + \left(\dfrac{Z_m - R_0}{Z_m + R_0}\right) e^{-j2\varphi}}{1 - \left(\dfrac{Z_m - R_0}{Z_m + R_0}\right) e^{-j2\varphi}} R_0 \tag{59}$$

$$= \frac{(Z_m + R_0)e^{+j\varphi} + (Z_m - R_0)e^{-j\varphi}}{(Z_m + R_0)e^{+j\varphi} - (Z_m - R_0)e^{-j\varphi}} R_0 \tag{60}$$

$$= \frac{Z_m(e^{+j\varphi} + e^{-j\varphi}) + R_0(e^{+j\varphi} - e^{-j\varphi})}{Z_m(e^{+j\varphi} - e^{-j\varphi}) + R_0(e^{+j\varphi} + e^{-j\varphi})} \tag{61}$$

$$= \frac{Z_m \cos\varphi + jR_0 \sin\varphi}{R_0 \cos\varphi + jZ_m \sin\varphi} R_0. \tag{62}$$

The impedance changes from $Z_{\text{in}}(0) = Z_m$ to R_0^2/Z_m at $\varphi = 90°$ and back to Z_m at $180°$.

If there is loss in the cable, $j\varphi$ in Eq. (57) must be replaced by $-h + jb$, as in Eq. (2.44).

F.5 SMITH CHART

The Smith chart (Pozar, 2001, pp. 42–47; Ramo et al., 1984, pp. 229–238; Gonzalez, 1984, pp. 43–49) is a plot of

$$\rho = \rho_m e^{-j\varphi} = u + jv, \tag{63}$$

where u is the abscissa and v is the ordinate. This complex variable ρ_m is related to the complex variable

$$Z_m/R_0 = R_m/R_0 + jX_m/R_0 \tag{64}$$

by Eq. (23), which describes a bilinear transformation that turns the real and imaginary axes of Z_m/R_0 into curved lines. The same relationship transforms the impedance Z into ρ at a distance corresponding to a phase delay of φ along the transmission line. This can be seen in Fig. F.2, which is a simplified Smith chart (showing only a few of its curves).

The Smith chart is widely used and is very valuable in representing impedance changes along a lossless transmission line. The heavy circle, which has been added in the center of this chart, represents a constant reflection magnitude of $|\rho| = 0.5$. Moving along this circle, we can read the value of Z/Z_0 as the line length, and thus φ, changes. We can begin at a point representing a module's input impedance and see how that impedance is transformed as the line is lengthened. Where the circle intersects $X/Z_0 = 0$, we can read the SWR according to Eq. (19).

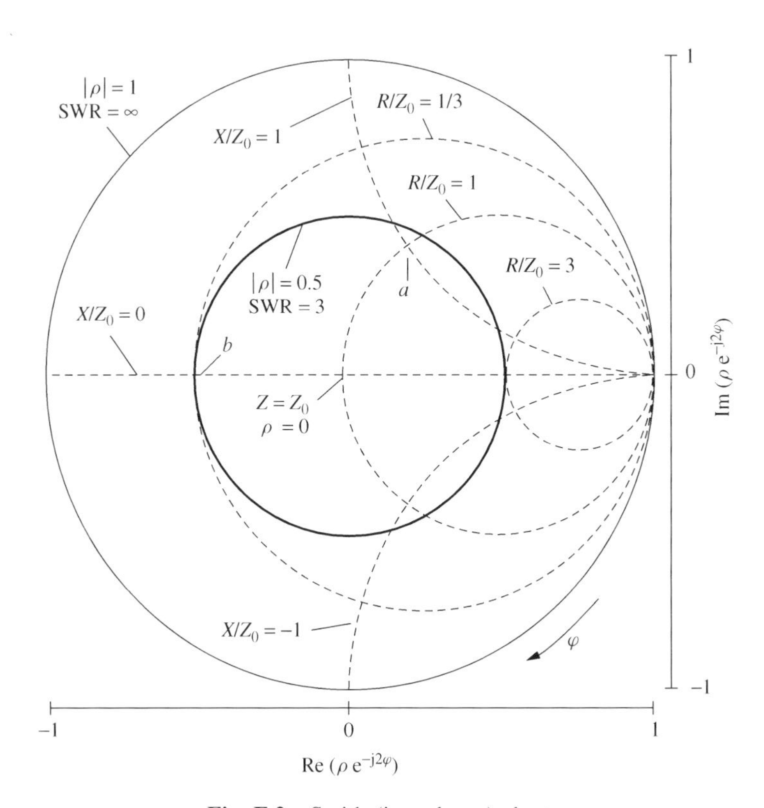

Fig. F.2 Smith (impedance) chart.

All of the possible impedances that could be seen at a terminal where the maximum SWR is 3 can be read inside the heavy circle (e.g., $Z/Z_0 = 1 + j$ at point a, 0.333 at b, or 1 at the center).

The same chart can also be used to represent $Y_{in}R_0$ by a change in the point where $\varphi = 0$, since the reciprocal of the ratio in Eq. (58) can be obtained by adding $\pi/2$ to φ. Both the value of $Y_{in}R_0$ and of Z_{in}/R_0 are often written on the same chart. Alternately, both $Y_{in}R_0$ and Z_{in}/R_0 can be plotted using the same origin for φ, forming two sets of curves, one the horizontal reflection of the other. Then the corresponding admittance and impedance can be read at any point on the chart.

APPENDIX G

TYPES OF POWER GAIN

Power gain is

$$g = p_{\text{out}}/p_{\text{in}}, \tag{1}$$

but there are various ways to define p_{out} and p_{in} (Petit and McWhorter, 1961, p. 2).

G.1 AVAILABLE GAIN

Figure G.1 illustrates available power gain. The input power p_{in} is the power that the source delivers to a load that is matched to that source. The output power p_{out} is the power the module delivers to a load that is matched to its output impedance when the module is driven by the source. The source impedance must be specified. (This term has also been used for what we are calling maximum available power gain (Jay, ed. 1977.)

Available gains can be multiplied to give the available gain of a cascade, as in Eq. (2.1), even though the variables [Eq. (2.2)] may not actually exist in the cascade. The intermediate variables u_j are powers that *would* be delivered to a matched load, whether such a load is present or not. This means that the available gain of a module depends on the output impedance of the preceding module and cannot be defined without that information. The same module will have different available gains in different cascades.

G.2 MAXIMUM AVAILABLE GAIN

Figure G.2 illustrates maximum available power gain. This is available power gain when the source impedance is matched to the module input impedance, rather

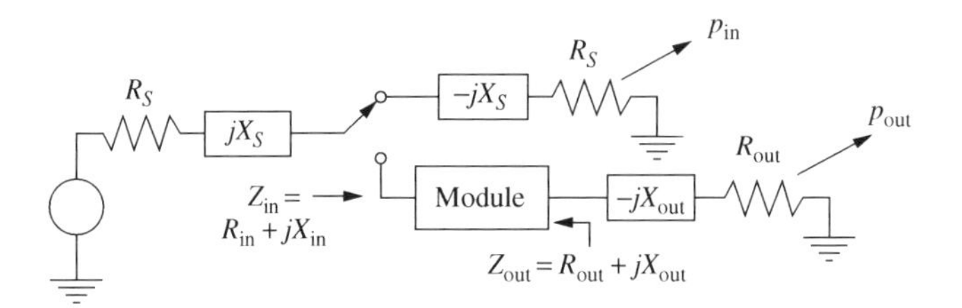

Fig. G.1 Available gain.

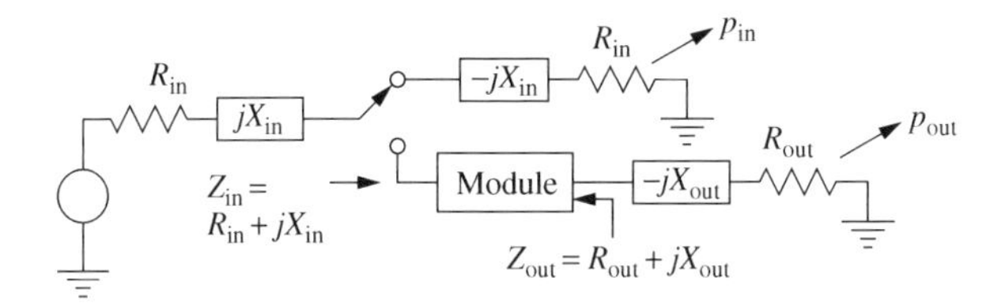

Fig. G.2 Maximum available gain.

than maintaining some specified value. This is sometimes called the completely matched power gain (Jay 1977). Since the source impedance is defined by the input impedance of the module, this gain can be defined independently of the driving source. However, it is unlikely to be directly useful in a cascade since each module would have to match the impedances at both of its interfaces for this gain to apply. Moreover, testing now requires an input matching network and, for the general case of a bilateral module, the two port impedances depend on each other, so finding the correct terminations could be difficult.

G.3 TRANSDUCER GAIN

Figure G.3 illustrates transducer power gain. The input power p_{in} is still the power that the source delivers to a load that is matched to the source. The output power p_{out} is the power the module delivers to a specified load when the module is driven by the source. Both source and load must be specified (50 Ω real is usual). From the definitions of the S and T parameters and the normalized waves

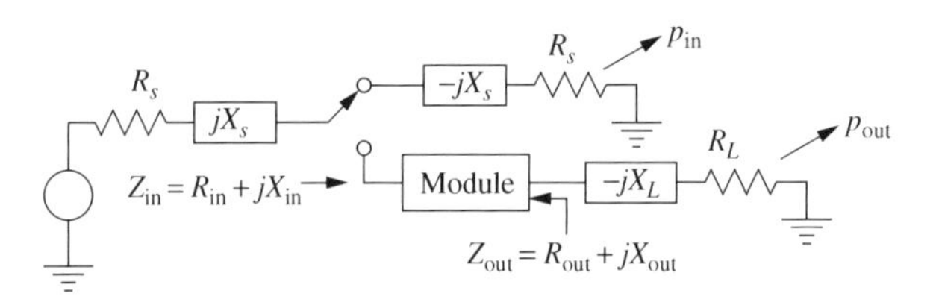

Fig. G.3 Transducer gain.

(Section 2.2) we can see that

$$|S_{21}|^2 = 1/|T_{11}|^2 = g, \tag{2}$$

where g is transducer power gain and the source and load impedances are real.

G.4 INSERTION GAIN

Figure G.4 illustrates insertion power gain. The input power p_{in} is the power that the source delivers to the load that is specified for the module. The output power p_{out} is the power the module delivers to the specified load when the module is inserted between the source and the specified load. This gain is the ratio of power delivered to the module load by the module when it is driven by the source to the power delivered to that load when the module is removed and the source drives it directly. As with transducer gain, the source and load impedances must be specified but the two gains can have different values because the input power is defined differently in the two cases.

G.5 ACTUAL GAIN

Figure G.5 illustrates the actual power gain. The input power p_{in} is the power that the source delivers to the module. The output power p_{out} is the power the module delivers to a specified load when the module is driven by the source. This is the ratio of the power out of the module to the power into it. The source impedance need not be specified. We just need to know how much power got into the module, not how it got there.

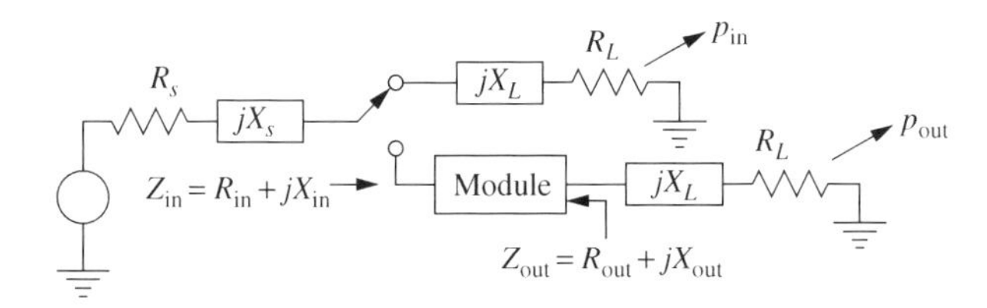

Fig. G.4 Insertion gain.

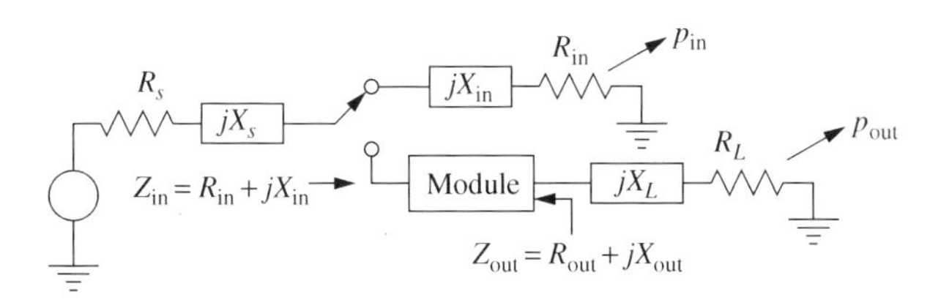

Fig. G.5 Actual gain.

APPENDIX H

FORMULAS RELATING TO IMs AND HARMONICS

This appendix gives various formulas expressing the relationships developed in Chapter 4. Capitalized variables represent values in dB. Output variables are related to the corresponding input variables by the (frequency independent) power gain.

H.1 SECOND HARMONICS

The second-harmonic output power is

$$p_{\text{out},H2} = \frac{p_{\text{out},F}^2}{p_{\text{OIP2},H}} \tag{1}$$

or

$$P_{\text{out},H2} = 2P_{\text{out},F} - P_{\text{OIP2},H}. \tag{2}$$

The equivalent second-harmonic input power is

$$p_{\text{in},H2} = \frac{p_{\text{in},F}^2}{p_{\text{IIP2},H}} \tag{3}$$

or

$$P_{\text{in},H2} = 2P_{\text{in},F} - P_{\text{IIP2},H}. \tag{4}$$

The ratio of the second-harmonic power to the fundamental is

$$\frac{p_{\text{out},H2}}{p_{\text{out},F}} = \frac{p_{\text{out},F}}{p_{\text{OIP2},H}} \tag{5}$$

$$= \frac{p_{\text{in},H2}}{p_{\text{in},F}} = \frac{p_{\text{in},F}}{p_{\text{IIP2},H}} \tag{6}$$

or

$$P_{\text{out},H2} - P_{\text{out},F} = P_{\text{out},F} - P_{\text{OIP2},H} \tag{7}$$

$$= P_{\text{in},H2} - P_{\text{in},F} = P_{\text{in},F} - P_{\text{IIP2},H}. \tag{8}$$

H.2 SECOND-ORDER IMs

The second-order output IM is

$$p_{\text{out,IM2}} = \frac{p_{\text{out},F1}\, p_{\text{out},F2}}{p_{\text{OIP2,IM}}} \tag{9}$$

or

$$P_{\text{out,IM2}} = P_{\text{out},F1} + P_{\text{out},F2} - P_{\text{OIP2,IM}}. \tag{10}$$

The equivalent second-order IM input power is

$$p_{\text{in,IM2}} = \frac{p_{\text{in},F1}\, p_{\text{in},F2}}{p_{\text{IIP2,IM}}} \tag{11}$$

or

$$P_{\text{in,IM2}} = P_{\text{in},F1} + P_{\text{in},F2} - P_{\text{IIP2,IM}}. \tag{12}$$

The ratio of the second-order IM power to the power in fundamental number 1 is

$$\frac{p_{\text{out,IM2}}}{p_{\text{out},F1}} = \frac{p_{\text{out},F2}}{p_{\text{OIP2,IM}}} \tag{13}$$

$$= \frac{p_{\text{in,IM2}}}{p_{\text{in},F1}} = \frac{p_{\text{in},F2}}{p_{\text{IIP2,IM}}} \tag{14}$$

or

$$P_{\text{out,IM2}} - P_{\text{out},F1} = P_{\text{out},F2} - P_{\text{OIP2,IM}} \tag{15}$$

$$= P_{\text{in,IM2}} - P_{\text{in},F1} = P_{\text{in},F2} - P_{\text{IIP2,IM}}. \tag{16}$$

H.3 THIRD HARMONICS

The third-harmonic output power is

$$p_{\text{out},H3} = \frac{p^3_{\text{out},F}}{p^2_{\text{OIP3},H}} \tag{17}$$

or

$$P_{\text{out},H3} = 3P_{\text{out},F} - 2P_{\text{OIP3},H}. \tag{18}$$

The equivalent third-harmonic input power is

$$p_{\text{in},H3} = \frac{p_{\text{in},F}^3}{p_{\text{IIP3},H}^2} \tag{19}$$

or

$$P_{\text{in},H3} = 3P_{\text{in},F} - 2P_{\text{IIP3},H}. \tag{20}$$

The ratio of the third-harmonic power to the fundamental is

$$\frac{p_{\text{out},H3}}{p_{\text{out},F}} = \left(\frac{p_{\text{out},F}}{p_{\text{OIP3},H}}\right)^2 \tag{21}$$

$$= \frac{p_{\text{in},H3}}{p_{\text{in},F}} = \left(\frac{p_{\text{in},F}}{p_{\text{IIP3},H}}\right)^2 \tag{22}$$

or

$$P_{\text{out},H3} - P_{\text{out},F} = 2(P_{\text{out},F} - P_{\text{OIP3},H}) \tag{23}$$

$$= P_{\text{in},H3} - P_{\text{in},F} = 2(P_{\text{in},F} - P_{\text{IIP3},H}). \tag{24}$$

H.4 THIRD-ORDER IMs

The third-order output IM adjacent to the fundamental $F1$, or to its harmonic, has power

$$p_{\text{out,IM3}}(\pm 2f_1 \pm f_2) = \frac{p_{\text{out},F1}^2 p_{\text{out},F2}}{p_{\text{OIP3,IM}}^2} \tag{25}$$

or

$$P_{\text{out,IM3}}(\pm 2f_1 \pm f_2) = 2P_{\text{out},F1} + P_{\text{out},F2} - 2P_{\text{OIP3,IM}}. \tag{26}$$

The equivalent third-order input IM is

$$p_{\text{in,IM3}}(\pm 2f_1 \pm f_2) = \frac{p_{\text{in},F1}^2 p_{\text{in},F2}}{p_{\text{IIP3,IM}}^2} \tag{27}$$

or

$$P_{\text{in,IM3}}(\pm 2f_1 \pm f_2) = 2P_{\text{in},F1} + P_{\text{in},F2} - 2P_{\text{IIP3,IM}}. \tag{28}$$

The ratio of this third-order IM to the fundamental $p_{\text{out},F1}$ adjacent to it is

$$\frac{p_{\text{out,IM3}}(\pm 2f_1 \pm f_2)}{p_{\text{out},F1}} = \frac{p_{\text{out},F1} p_{\text{out},F2}}{p_{\text{OIP3,IM}}^2} \tag{29}$$

$$= \frac{p_{\text{in,IM3}}(\pm 2f_1 \pm f_2)}{p_{\text{in},F1}} = \frac{p_{\text{in},F1} p_{\text{in},F2}}{p_{\text{IIP3,IM}}^2} \tag{30}$$

or

$$P_{\mathrm{out,IM3}}(\pm 2f_1 \pm f_2) - P_{\mathrm{out},F1} = P_{\mathrm{out},F1} + P_{\mathrm{out},F2} - 2P_{\mathrm{OIP3,IM}} \tag{31}$$

$$= P_{\mathrm{in,IM3}}(\pm 2f_1 \pm f_2) - P_{\mathrm{in},F1} = P_{\mathrm{in},F1} + P_{\mathrm{in},F2} - 2P_{\mathrm{IIP3,IM}}. \tag{32}$$

H.5 DEFINITIONS OF TERMS

IM is intermodulation product.
IP is intercept point.
$p_{\mathrm{IIP2},H}$ is the power at the second-order harmonic input IP, $\mathrm{IIP2}_H$.
$p_{\mathrm{IIP2,IM}}$ is the power at the second-order IM input IP, $\mathrm{IIP2}_{\mathrm{IM}}$.
$p_{\mathrm{IIP3},H}$ is the power at the third-order harmonic input IP, $\mathrm{IIP3}_H$.
$p_{\mathrm{IIP3,IM}}$ is power at the third-order IM input IP, $\mathrm{IIP3}_{\mathrm{IM}}$.
$p_{\mathrm{in},F}$ is the input power at the fundamental frequency.
$p_{\mathrm{in},Fi}$ is the input power in fundamental number i.
$p_{\mathrm{in,H2}}$ is the equivalent second-harmonic input power.
$p_{\mathrm{in,H3}}$ is the equivalent third-harmonic input power.
$p_{\mathrm{in,IM2}}$ is the equivalent second-order IM input power.
$p_{\mathrm{in,IM3}}$ is the equivalent third-order IM input power.
$p_{\mathrm{OIP2},H}$ is the power at the second-order harmonic output IP, $\mathrm{OIP2}_H$.
$p_{\mathrm{OIP2,IM}}$ is the power at the second-order IM output IP, $\mathrm{OIP2}_{\mathrm{IM}}$.
$p_{\mathrm{OIP3},H}$ is the power at the third-order harmonic output IP, $\mathrm{OIP3}_{\mathrm{IM}}$.
$p_{\mathrm{OIP3,IM}}$ is the power at the third-order IM output IP, $\mathrm{OIP3}_{\mathrm{IM}}$.
$p_{\mathrm{out},F}$ is the output power at the fundamental frequency.
$p_{\mathrm{out},Fi}$ is the output power in fundamental number i.
$p_{\mathrm{out},H2}$ is the second-harmonic output power.
$p_{\mathrm{out},H3}$ is the third-harmonic output power.
$p_{\mathrm{out,IM2}}$ is the second-order IM output power.
$p_{\mathrm{out,IM3}}$ is the third-order IM output power.

APPENDIX I

CHANGING THE STANDARD IMPEDANCE

Here we show how S or Z parameters for a module can be modified to represent different standard (interface) impedances than those used during measurement. For example, the S parameters of a module might be obtained in a 50-Ω measurement system, but we might connect one or both terminals to a 100-Ω cable in a system. Then we would need the S or T parameters for a standard impedance of 100 Ω at one or both ports.

One way to determine the module parameters for new standard impedances is to retest it with the new impedances. However, we can also compute the results of doing this.

I.1 GENERAL CASE

Figure I.1 shows a module M_a, whose parameters are defined for a characteristic impedance R_{1a} at the input port and R_{2a} at the output. The module is connected by cables of those standard impedances to transition modules T_1 and T_2. At the other sides of these transitions the system has the new standard impedances, R_{1b} and R_{2b}, respectively. The transition modules represent junctions between cables of the old and new impedances. The cables of impedance R_{1a} and R_{2a} are vanishingly short; so they introduce no phase shift or loss, but they allow us to visualize a place where the waves were defined for the old standard impedances. Module M_b has the new standard interface impedances and corresponding new T parameters,

$$\mathbf{T}_b = \mathbf{T}_1 \mathbf{T}_a \mathbf{T}_2, \tag{1}$$

where $\mathbf{T}_a$ is the T matrix of the module in its original representation and $\mathbf{T}_1$ and $\mathbf{T}_2$ are the T matrices for the transitions T_1 and T_2, respectively. The components

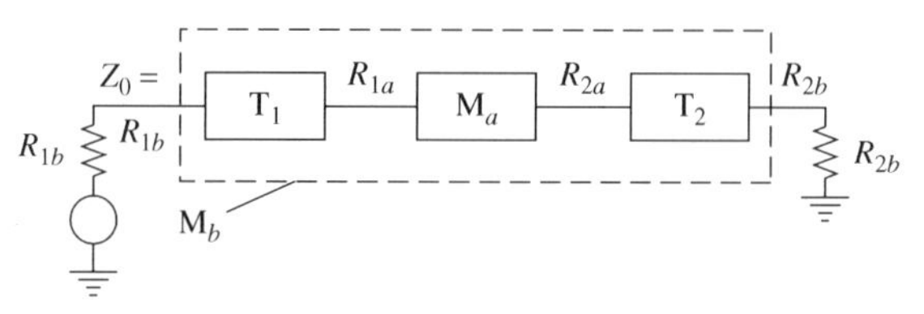

Fig. I.1 Module with originally measured parameters (M_{12a}) and with new standard impedances (M_{12b}).

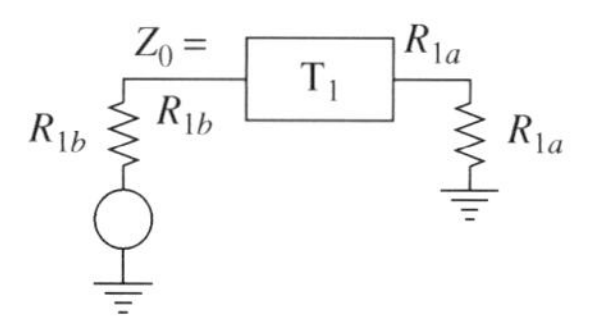

Fig. I.2 Input transition module in test.

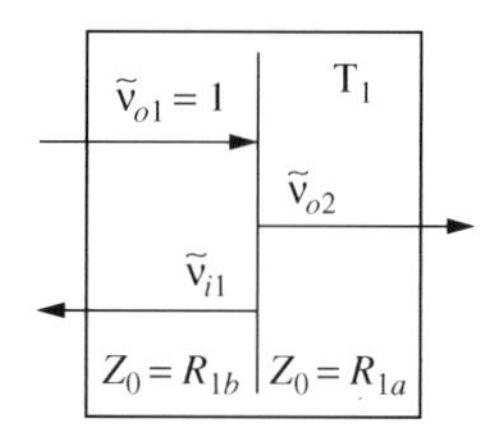

Fig. I.3 Input transition in test.

of $\mathbf{T}_a$ can be obtained from the original S parameters by Eq. (2.29) and the new S parameters can be found from $\mathbf{T}_b$ by Eq. (2.31). All that is needed for the transformation is the transition matrices.

The testing of module T_1 is illustrated in Fig. I.2. We choose the forward wave at the interface within the module (Fig. I.3) to be

$$v_{o1} = \frac{1}{\sqrt{R_{1b}}}, \tag{2}$$

corresponding to an rms voltage of $\tilde{v}_{o1} = 1$ (units arbitrary). By Eq. (13) in Appendix F, the reflected voltage is

$$\tilde{v}_{i1} = \frac{R_{1a} - R_{1b}}{R_{1a} + R_{1b}}. \tag{3}$$

The voltage at the interface equals the output wave voltage on the right of the transition and also the sum of the two waves on the left:

$$\tilde{v}_{o2} = 1 + \tilde{v}_{i1} = 2\frac{R_{1a}}{R_{1a} + R_{1b}}, \tag{4}$$

where Eq. (3) was substituted for $\tilde{v}_{i1}$. The corresponding normalized waves are

$$v_{i1} = \frac{1}{\sqrt{R_{1b}}} \frac{R_{1a} - R_{1b}}{R_{1a} + R_{1b}} \tag{5}$$

and

$$v_{o2} = \frac{2}{\sqrt{R_{1a}}} \frac{R_{1a}}{R_{1a} + R_{1b}}. \tag{6}$$

Parameter S_{11} can be obtained from the ratio between Eqs. (5) and (2) while S_{21} is the ratio between Eqs. (6) and (2). Measuring with the driving source at the other port gives S_{22} and S_{12}, which have similar forms. From all of this we can write the S matrix for T_1,

$$\mathbf{S}_{T1} = \begin{bmatrix} \left(\dfrac{R_{1a} - R_{1b}}{R_{1a} + R_{1b}}\right) & \left(\dfrac{2\sqrt{R_{1a}R_{1b}}}{R_{1a} + R_{1b}}\right) \\ \left(\dfrac{2\sqrt{R_{1a}R_{1b}}}{R_{1a} + R_{1b}}\right) & \left(\dfrac{R_{1b} - R_{1a}}{R_{1a} + R_{1b}}\right) \end{bmatrix}. \tag{7}$$

From Eq. (2.29) the equivalent T matrix is

$$\mathbf{T}_{T1} = \frac{1}{2\sqrt{R_{1a}R_{1b}}} \begin{bmatrix} (R_{1a} + R_{1b}) & (R_{1a} - R_{1b}) \\ (R_{1a} - R_{1b}) & (R_{1a} + R_{1b}) \end{bmatrix}. \tag{8}$$

Similarly, we can write, for T_2,

$$\mathbf{S}_{T2} = \begin{bmatrix} \left(\dfrac{R_{2b} - R_{2a}}{R_{2a} + R_{2b}}\right) & \left(\dfrac{2\sqrt{R_{2a}R_{2b}}}{R_{2a} + R_{2b}}\right) \\ \left(\dfrac{2\sqrt{R_{2a}R_{2b}}}{R_{2a} + R_{2b}}\right) & \left(\dfrac{R_{2a} - R_{2b}}{R_{2a} + R_{2b}}\right) \end{bmatrix} \tag{9}$$

and

$$\mathbf{T}_{T2} = \frac{1}{2\sqrt{R_{2a}R_{2b}}} \begin{bmatrix} (R_{2a} + R_{2b}) & (R_{2b} - R_{2a}) \\ (R_{2b} - R_{2a}) & (R_{2a} + R_{2b}) \end{bmatrix}. \tag{10}$$

To summarize, we convert the T matrix for the module with the old standard impedances to a T matrix with new standard impedances by multiplying by transition matrices $\mathbf{T}_{T1}$ before and $\mathbf{T}_{T2}$ behind. We use Eqs. (2.29) to change from S to T parameters initially and (2.31) and to change back to S parameters after the multiplication. If only one interface impedance is changed, the other corresponding T matrix will be missing from Eq. (1).

I.2 UNILATERAL MODULE

We can obtain some properties of the S parameters of transformed (by changes in standard impedances) unilateral modules. The transformed module is still unilateral so $S_{12b} = S_{12a} = 0$. If this is not apparent, it can be seen from Table S.1 in Appendix S. Limits on the SWR of a unilateral module transformed at either end can be obtained from Section F.3. For example, the maximum SWR of a module M_a with a transformation at the input is the product of the SWR of T_1 and that of M_a, where SWR is obtained for each module from ρ by Eq. (16) in Appendix F and $\rho = S_{11}$. The same is true at the output (but $\rho = S_{22}$ there). On the untransformed side the reflection is unchanged ($S_{nna} = S_{nnb}$) as a consequence of unilaterality, which can be seen also from Table S.1. As a result, if impedance change occurs at both ends, the limits of the resulting SWRs are the same as if the transformations had occurred separately (i.e., as described above).

We can obtain the transformed forward transfer ratio for a change of standard impedance at the module input from Table S1 and Eq. (7) as

$$S_{21b}|_{\Delta\text{in}} = S_{21a}\frac{2\sqrt{x}}{x + 1 + S_{11a}(x - 1)}, \tag{11}$$

where

$$x \triangleq \frac{R_{1a}}{R_{1b}}, \tag{12}$$

and, for a change at the output, as

$$S_{21b}|_{\Delta\text{out}} = S_{21a}\frac{2\sqrt{y}}{y + 1 + S_{22a}(y - 1)}, \tag{13}$$

where

$$y \triangleq \frac{R_{2a}}{R_{2b}}. \tag{14}$$

Since S_{22} is not changed by a transformation at the input (nor is S_{11} with a change at the output), we can use Eqs. (11) and (13) together to give the results of a change in standard impedance at both ends $\{S_{21b}|_{\Delta\text{in}}$ in Eq. (11) becomes S_{21a} for Eq. (13)$\}$:

$$S_{21b} = S_{21a}\frac{4\sqrt{xy}}{[x + 1 + S_{11a}(x - 1)][y + 1 + S_{22a}(y - 1)]}. \tag{15}$$

APPENDIX L

POWER DELIVERED TO THE LOAD

This development is referenced in Section 2.3.2.2. We will neglect cable loss here.

The ratio of the forward power at the cable output to the power injected by the source is, based on Eq. (2.40),

$$\frac{p_{o,j+1}}{p_{ojT}} = |a_{\mathrm{cbl},j}|^2. \tag{1}$$

The limits of this ratio are given by Eqs. (2.55) and (2.56), which, for the lossless case, give

$$\frac{1}{(1+|a_{\mathrm{RT}}|)^2} \le \frac{p_{o,j+1}}{p_{ojT}} \le \frac{1}{(1-|a_{\mathrm{RT}}|)^2}. \tag{2}$$

This can also be written in terms of the reflection coefficients at the cable ends as

$$\frac{1}{(1+|S_{11,j+1}S_{22,j-1}|)^2} \le \frac{p_{o,j+1}}{p_{ojT}} \le \frac{1}{(1-|S_{11,j+1}S_{22,j-1}|)^2}. \tag{3}$$

The power reflected at the load module input is $p_{o,j+1}$ multiplied by the reflection $|S_{11,j+1}|^2$, so the power absorbed by the load is what remains:

$$\frac{1-|S_{11,j+1}|^2}{(1+|S_{11,j+1}S_{22,j-1}|)^2} \le \frac{p_{\mathrm{absorbed},j+1}}{p_{ojT}} \le \frac{1-|S_{11,j+1}|^2}{(1-|S_{11,j+1}S_{22,j-1}|)^2}. \tag{4}$$

APPENDIX M

MATRIX MULTIPLICATION

First we will multiply a vector by a matrix. The matrix is

$$\mathbf{M}_a = \begin{bmatrix} a_{11} & a_{12} & a_{13} \\ a_{21} & a_{22} & a_{23} \\ a_{31} & a_{32} & a_{33} \end{bmatrix} \tag{1}$$

and the vector is

$$\mathbf{V}_1 = \begin{bmatrix} v_1 \\ v_2 \\ v_3 \end{bmatrix}. \tag{2}$$

We have given, to the matrix elements, subscripts that correspond to their row and column numbers, respectively. The product, $\mathbf{V}_2 = \mathbf{M}_a \times \mathbf{V}_1$, is

$$\mathbf{V}_2 = \begin{bmatrix} a_{11} & a_{12} & a_{13} \\ a_{21} & a_{22} & a_{23} \\ a_{31} & a_{32} & a_{33} \end{bmatrix} \begin{bmatrix} v_1 \\ v_2 \\ v_3 \end{bmatrix} = \begin{bmatrix} (a_{11}v_1 + a_{12}v_2 + a_{13}v_3) \\ (a_{21}v_1 + a_{22}v_2 + a_{23}v_3) \\ (a_{31}v_1 + a_{32}v_2 + a_{33}v_3) \end{bmatrix}. \tag{3}$$

Each summation has been placed in parentheses to emphasize that a single number is formed by the summation.

We can envision this result as being obtained as follows. The first row of $\mathbf{M}_a$ is rotated a quarter turn clockwise and the resulting column is placed against $\mathbf{V}_1$. Then adjacent elements (e.g., a_{11} and v_1) are multiplied. The three products so obtained are then added to give the first element in $\mathbf{V}_2$. In general, the element of $\mathbf{V}_2$ in its rth row and cth column is obtained by multiplying the rth row of $\mathbf{M}_a$ by the cth column of $\mathbf{V}_1$ in this manner. In this case, since there is only one column in $\mathbf{V}_1$, $\mathbf{V}_2$ turns out to be a one-column vector.

Two matrices are multiplied following the same rules. Let

$$\mathbf{M}_b = \begin{bmatrix} b_{11} & b_{12} & b_{13} \\ b_{21} & b_{22} & b_{23} \\ b_{31} & b_{32} & b_{33} \end{bmatrix}. \tag{4}$$

The product, $\mathbf{M}_c = \mathbf{M}_a \times \mathbf{M}_b$, is

$$\mathbf{M}_c \equiv \begin{bmatrix} c_{11} & c_{12} & c_{13} \\ c_{21} & c_{22} & c_{23} \\ c_{31} & c_{32} & c_{33} \end{bmatrix} = \begin{bmatrix} a_{11} & a_{12} & a_{13} \\ a_{21} & a_{22} & a_{23} \\ a_{31} & a_{32} & a_{33} \end{bmatrix} \begin{bmatrix} b_{11} & b_{12} & b_{13} \\ b_{21} & b_{22} & b_{23} \\ b_{31} & b_{32} & b_{33} \end{bmatrix} \tag{5}$$

$$= \begin{bmatrix} (a_{11}b_{11} + a_{12}b_{21} + a_{13}b_{31}) & (a_{11}b_{12} + a_{12}b_{22} + a_{13}b_{32}) & (a_{11}b_{13} + a_{12}b_{23} + a_{13}b_{33}) \\ (a_{21}b_{11} + a_{22}b_{21} + a_{23}b_{31}) & (a_{21}b_{12} + a_{22}b_{22} + a_{23}b_{32}) & (a_{21}b_{13} + a_{22}b_{23} + a_{23}b_{33}) \\ (a_{31}b_{11} + a_{32}b_{21} + a_{33}b_{31}) & (a_{31}b_{12} + a_{32}b_{22} + a_{33}b_{32}) & (a_{31}b_{13} + a_{32}b_{23} + a_{33}b_{33}) \end{bmatrix}. \tag{6}$$

The elements of $\mathbf{M}_c$ can be formed in the same manner as were those of $\mathbf{V}_2$. For example, c_{23}, the element of $\mathbf{M}_c$ in the second row and third column, can be obtained by rotating the second row of $\mathbf{M}_a$ clockwise a quarter turn, placing it against the third column of $\mathbf{M}_b$, multiplying adjacent pairs and adding the three results.

We can see from Eq. (6) that, in general,

$$\mathbf{M}_a \times \mathbf{M}_b \neq \mathbf{M}_b \times \mathbf{M}_a \tag{7}$$

because one side of this inequality differs from the other in that the a's and b's in Eq. (6) would be interchanged, but the subscripts would not and none of the elements of $\mathbf{M}_c$ would remain the same under such a change. (That is, the symbols would be different. It is possible that the numerical values might turn out to be the same.)

APPENDIX N

NOISE FACTORS—STANDARD AND THEORETICAL

This appendix contains details of noise factor relationships discussed in Chapter 3.

Figure N.1*a* represents a module with a noise source. The noise source, $a_k' v_{nk}$, is shown at the output, but we can also think of an equivalent source at the input, as in Fig. N.1*b*. These models are equivalent to a standard Z-parameter representation (Linvill and Gibbons, 1961, p. 211) with added noise source, as shown in Fig. N.1*c*. The source v_{nk} is placed so it will not interact with the source driving impedance (or the load). All of the figures represent isolated noise sources in that respect.

We ignore the feedback (reverse transmission) represented by Z_{12k}. Its value is zero for unilateral modules but, even in bilateral modules, it would appear to have no effect on spot noise figure. If some value of S/N_{out} occurs when $Z_{12k} = 0$, when $Z_{12k} \neq 0$ the same ratio between signal and noise will occur in the fed-back signal and in any reflection thereof. Therefore, the feedback will have the same effect on the signal as on noise at the same frequency and their ratio will not be altered.

N.1 THEORETICAL NOISE FACTOR

Although v_{nk} does not here change with source impedance, its contribution to noise factor will. Theoretically, when the module is tested, it should have the same source impedance as it will see in the cascade, that is, the output impedance of the previous stages, $Z_{22(k-)}$, as shown in Fig. 3.1. (If stage $k-1$ is unilateral, $k-$ can be replaced by $k-1$.) We define $e_{\text{added},s}$ to be a voltage in the source (Fig. N.2) that would produce v_{nk} across R_{11k}. We can see that it is related to v_{nk} by

$$v_{nk} = e_{\text{added},s} \hat{\gamma}, \tag{1}$$

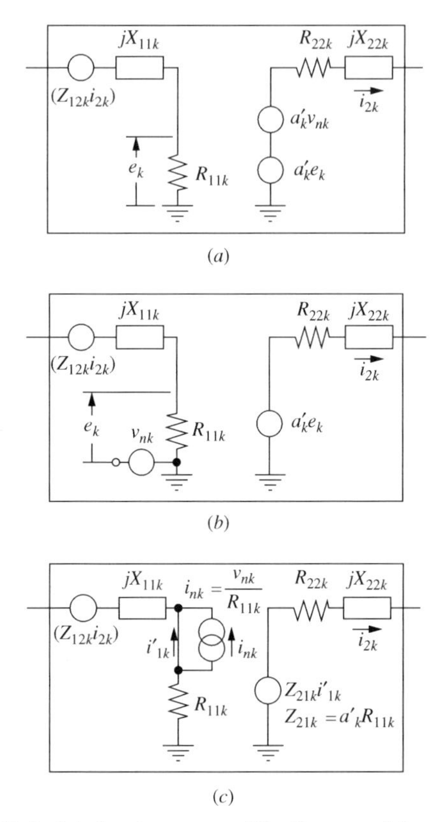

Fig. N.1 Module with isolated noise source. The three models are equivalent. The representation at (c) is the standard Z-parameter model with the noise source added.

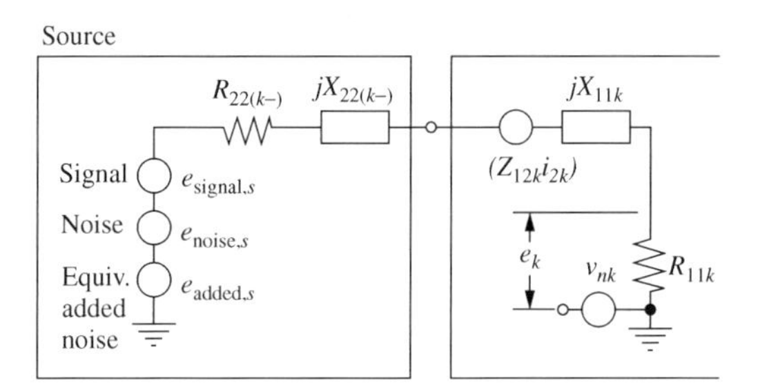

Fig. N.2 Source with equivalent added noise source.

where

$$\hat{\gamma} = \frac{R_{11k}}{R_{11k} + R_{22(k-)} + j(X_{11k} + X_{22(k-)})}. \tag{2}$$

Thus, having $e_{\text{added},s}$ in the source is equivalent to having v_{nk} in the module in that either would produce the observed module output noise. When tested into a matched load (Fig. 3.1), the theoretical noise source would produce a voltage $e_{\text{added},s}/2$ across the load resistor where, from Eq. (1),

$$e_{\text{added},s} = \frac{v_{nk}}{\hat{\gamma}}. \tag{3}$$

For this ideal definition of noise factor, Eq. (3.2) becomes

$$\hat{f}_k = \frac{P_{\text{signal},k}/(\overline{k}T_0B)}{g_{ak}P_{\text{signal},k} \Big/ \left\{ g_{ak}\left(\overline{k}T_0B + \dfrac{|v_{nk}/\hat{\gamma}|^2}{4R_{22(k-)}}\right)\right\}} \tag{4}$$

$$= 1 + \frac{v_{nk}^2}{|2\hat{\gamma}|^2 R_{22(k-)}(\overline{k}T_0B)}, \tag{5}$$

where g_{ak} is available power gain for module k and v_{nk} is the effective (rms) value of the noise voltage.

We can get $\hat{f}_k$ either by computing the signal and noise powers delivered at the output of the module or by comparing the total noise power theoretically dissipated in the matched source to the available thermal noise there.

The theoretical noise factor is valuable because it can be used in a cascade where such noise factors are divided by the preceding available gain (see Appendix G) to accurately determine the equivalent available noise in the source and, thus, the system noise factor. However, in our standard cascades of unilateral modules (Section 2.3.3), we use transducer gain, rather than available gain, and neither gain is known precisely due to reflections in the cascade. Moreover, the theoretical noise factor is not normally measured, due to the requirement to customize the test for each particular source impedance $R_{22(k-)}$ that will be seen in use. (Cases where it is given will be discussed in Section N.7.) What is commonly measured, fortunately, is consistent with our standard cascade.

N.2 STANDARD NOISE FACTOR

When the module is tested with standard interface impedances, as in Fig. N.3, the equivalents of Eqs. (2)–(5) are (assuming negligible cable losses in the setup)

$$\gamma = \frac{R_{11k}}{R_{11k} + R_0 + jX_{11k}}, \tag{6}$$

$$e_{\text{added},s} = \frac{v_{nk}}{\gamma}, \tag{7}$$

and

$$f_k = 1 + \frac{v_{nk}^2}{|2\gamma|^2 R_0(\overline{k}T_0B)}. \tag{8}$$

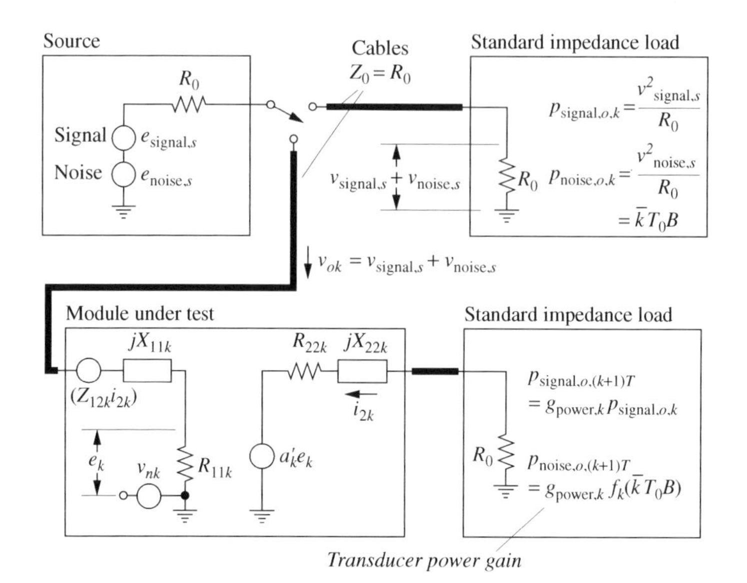

Fig. N.3 Noise figure test, typical (cable loss neglected or compensated).

Therefore, the relationship between the theoretical noise factor and the tested noise factor is given by

$$\frac{\hat{f}_k - 1}{f_k - 1} = \frac{|\gamma|^2 R_0}{|\hat{\gamma}|^2 R_{22(k-)}} \tag{9}$$

$$= \frac{\{[R_{11k} + R_{22(k-)}]^2 + [X_{11k} + X_{22(k-)}]^2\}/R_{22(k-)}}{\{[R_{11k} + R_0]^2 + X_{11k}^2\}/R_0} \tag{10}$$

$$= \frac{|Z_{11k} + Z_{22(k-)}|^2/R_{22(k-)}}{|Z_{11k} + R_0|^2/R_0}. \tag{11}$$

This shows how the theoretical noise factor $\hat{f}_k$ differs from the commonly measured noise factor f_k. Either may be larger, and one approximates the other to the degree that the preceding module has standard output impedance R_0.

N.3 STANDARD MODULES AND STANDARD NOISE FACTOR

We will now show that $f_k - 1$ represents the noise power, incident on a unilateral module, that would produce the same output noise level that is actually produced by the module, normalized to $\bar{k}T_0B$. (This differs from $(\hat{f}_k - 1)$, which represents *available* noise power.)

Assuming unilaterality, the reflection coefficient at the module input is [Appendix F, Eq. (13)]

$$S_{11k} = \frac{R_{11k} + jX_{11k} - R_0}{R_{11k} + jX_{11k} + R_0} \tag{12}$$

so the ratio of the noise power p_{nk} dissipated in the module to the incident forward noise power p_{nok} is

$$\frac{p_{nk}}{p_{nok}} = (1 - |S_{11}|^2) = 1 - \frac{(R_{11k} - R_0)^2 + X_{11k}^2}{(R_{11k} + R_0)^2 + X_{11k}^2} \tag{13}$$

$$= \frac{4R_{11k}R_0}{(R_{11k} + R_0)^2 + X_{11}^2} = |2\gamma|^2 \frac{R_0}{R_{11k}}. \tag{14}$$

If p_{nok} is the incident noise power that produces the same noise at the module output as does the internal source v_{nk}, then it would produce noise power

$$p_{nk} = \frac{v_{nk}^2}{R_{11k}} \tag{15}$$

in R_{11k}. From Eq. (14), its level would be

$$p_{nok} = \frac{v_{nk}^2}{R_{11k}} \frac{1}{|2\gamma|^2} \frac{R_{11k}}{R_0} = \frac{v_{nk}^2}{R_0} \frac{1}{|2\gamma|^2}. \tag{16}$$

The ratio of this incident wave power to thermal noise power is

$$\frac{p_{nok}}{\bar{k}T_0B} = \frac{v_{nk}^2}{|2\gamma|^2 R_0(\bar{k}T_0B)}, \tag{17}$$

which, from Eq. (8), is related to the tested noise factor by

$$\frac{p_{nok}}{\bar{k}T_0B} = f_k - 1. \tag{18}$$

Thus $f_k - 1$ does represent the incident wave power that would produce the added noise. Dividing it by the ratio of forward power, at the module input, to available power from the cascade source gives the equivalent available source noise for use in cascade analysis (all relative to $\bar{k}T_0B$).

N.4 MODULE NOISE FACTOR IN A STANDARD CASCADE

We use Fig. 3.2 to illustrate the equivalence between the noise factor measured in the module test and the noise factor used in obtaining the correct system noise level. Switch position 1 is used to measure source signal and noise power and, theoretically (it cannot be done experimentally), to measure the equivalent added noise power from the source (Fig. N.2).

Switch position 3 sends the same forward power to the module as was sent to (and absorbed in) the matched load in position 1. Multiplying by the module's transducer gain, we obtain the power, $p_{o(k+1)} = |\tilde{v}_{o(k+1)}|^2/R_0 = |v_{o(k+1)}|^2$, in the test load. By comparing the S/N in $p_{o(k+1)}$ to that in p_{ok}, we determine what relative increase in the noise power in p_{ok} would substitute for the module's internal noise [Eq. (18)].

This equivalent increase in p_{ok} is the same when the module is in the (standard) cascade. With the switch in position 2, this equivalent increase in incident noise power is related to the equivalent available noise power in the cascade source by the gain from the incident wave at the first module to the incident wave at module k. The output termination is different in the cascade than during test but that, presumably, does not affect the ratio between signal and noise, nor does it affect the module input, since modules in a standard cascade are unilateral. Thus the usually tested noise factor is the correct value to use in our standard cascade analysis (Section 2.3.3), where the gains are the ratios of forward waves (Fig. 2.5).

As further evidence of the validity of this approach, we can show (Section N.8) that the ratio of the power gain used with f_k (gain to the forward wave in the preceding cable) to that used with $\hat{f}_k$ (available gain at that cable's output) is given by Eq. (11). Therefore, the effective increase in noise factor referred to the cascade input, obtained by division of $f - 1$ by cascade gain, is the same using either $\hat{f}_k$ and available gain or using f_k and the gain to the forward wave in the interconnect cable. Thus the increase in cascade noise factor using either the theoretical or standard-cascade method is the same.

N.5 HOW CAN THIS BE?

How can this measure f be better than the theoretical noise factor $\hat{f}$? Because it matches our analysis, which has already given up precision in exchange for an estimate of mean values and variations therefrom. If we knew all of the module parameters and had the theoretical noise factors, we could use them to obtain precise results.

N.6 NOISE FACTOR OF AN INTERCONNECT

An interconnect is an attenuator with phase shift. Assume it is at a temperature of T_0. We then know that the noise factor of an attenuator with attenuation $1/g_2$ in a matched cascade is its attenuation. This is because the noise is the same at the output of the attenuator as at its input, both values being the noise available from R_0, but the signal power is lower by the attenuation at the output. Therefore, the noise factor is

$$f_2 = (S/N)_{\text{in}}/(S/N)_{\text{out}} = 1/g_2. \tag{19}$$

We know this without considering the details of the construction of the attenuator, whether it is a π or a T network, for example, or what values of resistors are used.

N.6.1 Noise Factor with Mismatch

We will use the circuit shown in Fig. N.4 to find f_2 when the interconnect is driven from a mismatch. It is not necessary to terminate the circuit in the characteristic impedance R_0 of the interconnect — the output S/N will be the same regardless of the load — but terminating the line in R_0 helps us to incorporate g_2 into the expression for f_2, as is done in a matched cascade. We assume initially that the interconnect is at temperature T_0.

The noise power introduced into the interconnect from the source is

$$N_{\text{in}} = 4N_T R_s \left(\frac{R_0}{|Z_s + R_0|} \right)^2 \frac{1}{R_0} = \frac{4N_T R_s R_0}{|Z_s + R_0|^2}, \tag{20}$$

where

$$N_T = \bar{k} T_0 B. \tag{21}$$

The noise power at its output is similar but relates to the impedance Z_{out} seen (looking back into the attenuator) there:

$$N_{\text{out}} = \frac{4N_T R_{\text{out}} R_0}{|Z_{\text{out}} + R_0|^2}. \tag{22}$$

The noise factor is

$$f_2 = \frac{(S/N)_{\text{in}}}{(S/N)_{\text{out}}} = \frac{S_{\text{in}}}{S_{\text{out}}} \frac{N_{\text{out}}}{N_{\text{in}}} \tag{23}$$

$$= \frac{1}{g_2} \frac{R_{\text{out}}}{R_s} \frac{|Z_s + R_0|^2}{|Z_{\text{out}} + R_0|^2}. \tag{24}$$

This is the factor by which mean-square noise voltage v_n^2 in Fig. N.4 would have to be multiplied to produce the observed $(S/N)_{\text{out}}$ from the interconnect if the interconnect were noiseless. The part due to the interconnect (removing the source noise) will be $v_n^2(f_2 - 1)$.

When the interconnect noise is referred further back the chain, what added noise, driving the module with output impedance Z_s (Source in Fig. N.4), would

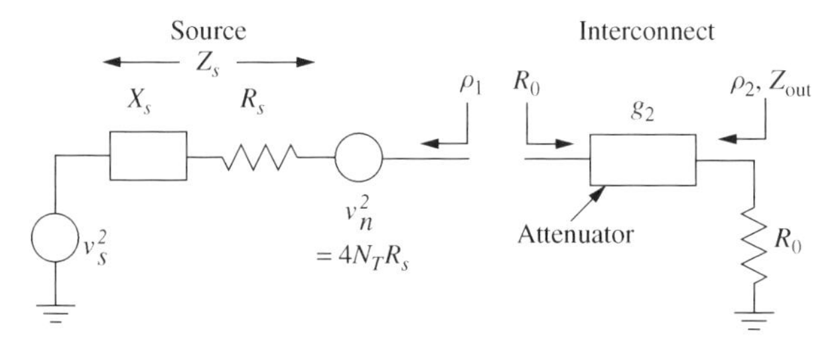

Fig. N.4 Interconnect driven from mismatch.

be equivalent to $v_n^2(f_2 - 1)$ in series with v_n^2 and Z_s? It would have a value $(f_2 - 1)N_T/g_{a1}$, where g_{a1} is the available gain of the source module. We know this because the available gain is measured into R_s so v_n^2 delivers $v_n^2/(4R_s) = N_T$. Therefore, $(f_2 - 1)v_n^2$ must deliver $(f_2 - 1)N_T$. However, the gain we are using for the module is its transducer gain into R_0. We need an effective noise factor f_{eff} for the interconnect that will give us the correct equivalent cascade input noise when we follow the normal procedure of dividing $(f_{\text{eff}} - 1)$ by the transducer gain, g_{t1}. Expressing this mathematically, we require

$$(f_2 - 1)N_T/g_{a1} = (f_{\text{eff}} - 1)N_T/g_{t1} \tag{25}$$

or

$$f_{\text{cbl}} \equiv f_{\text{eff}} = 1 + \frac{g_{t1}}{g_{a1}}(f_2 - 1) \tag{26}$$

$$= 1 + \frac{g_{t1}}{g_{a1}}\left(\frac{1}{g_2}\frac{R_{\text{out}}}{R_s}\frac{|Z_s + R_0|^2}{|Z_{\text{out}} + R_0|^2} - 1\right). \tag{27}$$

N.6.2 In More Usable Terms

Now we must perform the tedious task of writing Eq. (27) in more useable terms, using the variables we have available to our system analysis. Fortunately, we will find that the expression will simplify in the process.

Let the rms voltage v_s (Fig. N.4) be produced as a result of an input p_{in} to the source module. The available gain of the source module is the ratio of the power into a matched load to p_{in}. Half of the source voltage is applied across the real part of the matched load so the source module's available power gain is

$$g_{a1} = \frac{v_s^2}{4R_s p_{\text{in}}}. \tag{28}$$

Its transducer power gain is the ratio of the power into R_0 to p_{in}:

$$g_{t1} = \frac{v_s^2}{p_{\text{in}}}\frac{R_0}{|Z_s + R_0|^2}. \tag{29}$$

Therefore, the ratio of gains for the source module is

$$\frac{g_{t1}}{g_{a1}} = \frac{4R_s R_0}{|Z_s + R_0|^2}. \tag{30}$$

The source output impedance can be written, using Eq. (23) in Appendix F, as

$$Z_s = \frac{1 + \rho_1}{1 - \rho_1}R_0. \tag{31}$$

The denominator in Eq. (30) is

$$|Z_s + R_0|^2 = R_0^2 \left|1 + \frac{1+\rho_1}{1-\rho_1}\right|^2 = R_0^2 \left|\frac{2}{1-\rho_1}\right|^2 = \frac{4R_0^2}{(1-\rho_1)(1-\rho_1^*)} \tag{32}$$

$$= \frac{4R_0^2}{1 - 2|\rho_1|\cos\varphi_1 + |\rho_1|^2}, \tag{33}$$

where φ_1 is the phase of ρ_1. R_s, the real part of Z_s, can be written, from Eq. (31), as

$$R_s = \operatorname{Re} Z_s = \operatorname{Re}\left(R_0 \frac{1+\rho_1}{1-\rho_1}\right) = R_0 \operatorname{Re}\left(\frac{1+\rho_1}{1-\rho_1}\frac{1-\rho_1^*}{1-\rho_1^*}\right) \tag{34}$$

$$= R_0 \left(\frac{1-|\rho_1|^2}{1 - 2|\rho_1|\cos\varphi_1 + |\rho_1|^2}\right). \tag{35}$$

Substituting Eqs. (33) and (35) into (30), we obtain

$$\frac{g_{t1}}{g_{a1}} = 1 - |\rho_1|^2, \tag{36}$$

which is essentially the same Eq. (3.36).

This can also be obtained by comparing eqs. (3.2.2) and (3.2.4) in Gonzalez (1984, p. 92), using $\Gamma_L = 0$ because g_t is into R_0.

When there is attenuation, Eq. (58) in Appendix F becomes

$$Z_{\text{in}}(-d) = \frac{1 + \rho e^{-2(\alpha + j\varphi)}}{1 - \rho e^{-2(\alpha + j\varphi)}} R_0, \tag{37}$$

where α is attenuation (in nepers) and φ is phase shift in the interconnect (both one-way). Writing this in terms of our current variables, it becomes

$$Z_{\text{out}} = \frac{1 + \rho_1 g_2 e^{-j2\varphi_{\text{cbl}}}}{1 - \rho_1 g_2 e^{-j2\varphi_{\text{cbl}}}} R_0. \tag{38}$$

Following the procedure that produced Eq. (33), we obtain

$$|Z_{\text{out}} + R_0|^2 = \frac{4R_0^2}{1 - 2|\rho_1| g_2 \cos\theta + |\rho_1|^2 g_2^2}, \tag{39}$$

where

$$\theta = \varphi_1 - 2\varphi_{\text{cbl}}, \tag{40}$$

and $\varphi_{\rm cbl}$ is the one-way phase shift through the interconnect. Similarly, we can obtain $R_{\rm out}$ in the same manner in which we obtained R_s:

$$R_{\rm out} = R_0 \left(\frac{1 - |\rho_1|^2 g_2^2}{1 - 2|\rho_1| g_2 \cos\theta + |\rho_1|^2 g_2^2} \right). \tag{41}$$

Inserting Eqs. (33), (35), (36), (39), and (41) into Eq. (27), we obtain

$$f_{\rm cbl} = 1 + (1 - |\rho_1|^2) \left[\frac{1}{g_2} \frac{R_0 \left(\dfrac{1 - |\rho_1|^2 g_2^2}{1 - 2|\rho_1| g_2 \cos\theta + |\rho_1|^2 g_2^2} \right)}{R_0 \left(\dfrac{1 - |\rho_1|^2}{1 - 2|\rho_1| \cos\varphi_1 + |\rho_1|^2} \right)} \right.$$

$$\left. \times \frac{\dfrac{4R_0^2}{1 - 2|\rho_1| \cos\varphi_1 + |\rho_1|^2}}{\dfrac{4R_0^2}{1 - 2|\rho_1| g_2 \cos\theta + |\rho_1|^2 g_2^2}} - 1 \right] \tag{42}$$

$$= 1 + (1 - |\rho_1|^2) \left(\frac{1/g_2 - |\rho_1|^2 g_2}{1 - |\rho_1|^2} - 1 \right) \tag{43}$$

$$= 1/g_2 + |\rho_1|^2 (1 - g_2). \tag{3.27}$$

We can see that the noise factor reduces to the appropriate value in the absence of a mismatch (i.e., $\rho_1 = 0$).

In terms of SWR, we can write Eq. (3.27) as

$$f_{\rm cbl} = \frac{1}{g_2} + \left(\frac{\text{SWR}_1 - 1}{\text{SWR}_1 + 1} \right)^2 (1 - g_2). \tag{3.28}$$

In accordance with Section 3.5, the noise factor when the temperature of the interconnect differs from T_0 is

$$f_{\rm cbl}(T \neq T_0) = 1 + (f_{\rm cbl} - 1) T / T_0. \tag{3.29}$$

N.6.3 Verification

Figure N.5 represents a small system that we can use to verify Eq. (3.27). Figure N.6 shows a spreadsheet that compares results from circuit analysis to results using Eq. (3.27). The upper part of the spreadsheet contains calculations for a π attenuator matched to 1 Ω, as shown in Fig. N.5. The figure shows results for a 6-dB attenuator and several source and load resistances, including a 1-Ω match. The lower part of the spreadsheet computes the system noise factor using Eq. (3.27) for the noise factor of the attenuator and using the transducer gain of

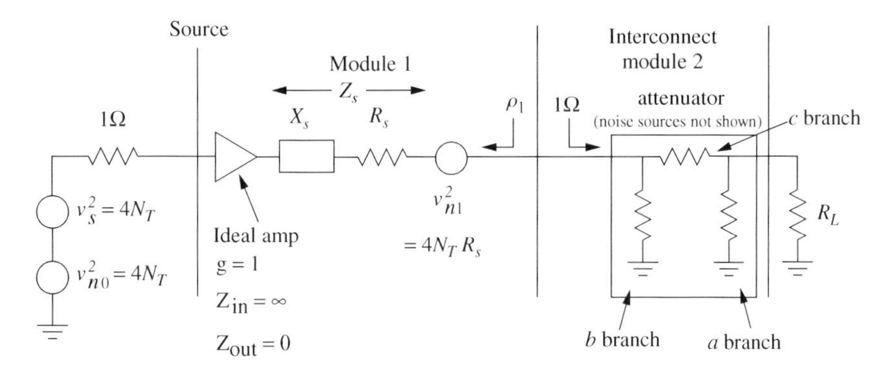

Fig. N.5 System with attenuator, $R_0 = 1\ \Omega$ (v values are rms).

	A	B	C	D	E	F	G	H	I
1	−6 dB pad driven by R = 1 through ideal amp with source output resistance.								
2	Powers normalized to *kT*o*B*.								
3	DETAILED NOISE CALCULATIONS								
4	signal v^2 in source	4		4		4		4	
5	shunt R(b,a)	3.0095		3.0095		3.0095		3.0095	
6	v^2	12.038		12.038		12.038		12.038	
7	series R(c)	0.74704		0.74704		0.74704		0.74704	
8	v^2	2.98816		2.98816		2.98816		2.98816	
9	source R (R_s)	1		2		0.5		3	
10	v^2	8		12		6		16	
11	load R (R_L)	1		3		0.5		0.5	
12	shunt out R	0.75059234		1.50237125		0.42876478		0.42876478	
13	shunt in R	0.75059234		1.20151712		0.42876478		1.50237125	
14	seen by *a* branch	0.59962082		1.18128804		0.35081795		0.40907144	
15	seen by *b* branch	0.59962082		1.05869313		0.35081795		0.84472683	
16	*a*-to-out volt ratio	0.16614041		0.28187731		0.10440022		0.11966152	
17	v^2 from *a* at out		0.33228053		0.95647707		0.13120706		0.17237066
18	*b*-to-out volt ratio	0.08326725		0.17381087		0.03807021		0.07992137	
19	v^2 from *b* at out		0.08346468		0.36367061		0.01744717		0.07689182
20	*c*-to-out volt ratio	0.33386002		0.43535278		0.26721483		0.16009582	
21	v^2 from *c* at out		0.33306783		0.56635206		0.21336587		0.07658855
22	source-to-out v ratio	0.25059278		0.26154191		0.22914462		0.08017445	
23	v^2 from *source* at out		0.50237392		0.82085003		0.31504353		0.10284708
24									
25	total noise v^2 at out		1.25118696		2.70734977		0.67706363		0.42869811
26	signal v^2 at out		0.25118696		0.27361668		0.21002902		0.02571177
27	S/N in:		1		1		1		1
28	S/N out:		0.20075893		0.1010644		0.31020573		0.0599764
29	f =	**4.98110**		**9.89468**		**3.22367**		**16.67322**	
30	CALCULATIONS FROM DEVELOPED FORMULA, Eq. (3.27)								
31	g2 =	0.251189 =	−6 dB						
32	rho, input reflection:	0		0.33333333		-0.33333333		0.5	
33	Eq. (3.27): fcbl =	3.98107171		4.06427297		4.06427297		4.16827454	
34	f1 =	2		3		1.5		4	
35	g1 =	1		0.4444444		1.77777778		0.25	
36	(fcb1−1) =	2.98107171		3.06427297		3.06427297		3.16827454	
37	(fcb1−1)/g1 =	2.98107171		6.89461418		1.72365354		12.6730982	
38	**f1 + (fcb1−1)/g1 = f =**	**4.98107**		**9.89461**		**3.22365**		**16.67310**	
39									
40	relative difference:	−5.4E−06		−6.8E−06		−4.2E−06		−7.5E−06	
41	accuracy of shunt Rs	1.7E−05							
42	accuracy of series R	6.7E−05							
43									
44	Resistor values take from								
45	E.C. Jordan, ed., Reference Data for Engineers, Radio, Electronics, Computers, and Communications, Seventh Ed.								
46	(Indianapolis, IN: Howard W. Sams & Co., 1986) p. 11–5.								

Fig. N.6 Comparison of formula to detailed noise calculations, 6-dB attenuator with various sources and loads.

the preceding module. The noise factors computed by the two methods are very close, the relative differences being on the order of inaccuracies in the resistor values due their truncation to four significant digits. Two other pages of the same spreadsheet show similar results for 3- and 20-dB attenuators, and another page shows the correspondence for the 6-dB attenuator with arbitrary temperature.

In these spreadsheets, a change in the value of load resistor (line 11) produces changes in many other cells, but the system noise figure, which incorporates these changing values, remains unchanged. The net effect is zero because the output signal and noise are affected similarly so their ratio does not change.

N.6.4 Comparison with Theoretical Value

Figure N.7 shows a module with an output reflection connected to a cable, which is terminated in its characteristic impedance R_0. The module is matched at its input. The noise factor for a cable with an input mismatch must produce the same effective increase in the source noise whether its value is that for a standard cascade or the theoretical value, each used within its corresponding system. We will show that our value of noise factor for the standard cascade implies the cable noise temperature given by Pozar's (2001) expression for use in a theoretical cascade.

Equations (3.27) and (3.29) for the standard system give a noise factor of

$$f_{\text{cbl}} = 1 + [1/g_2 + |\rho_1|^2(1 - g_2) - 1]T/T_0. \tag{44}$$

This is referred to the source by dividing by the module's transducer gain g_t, so the equivalent source noise power is

$$n_s = \overline{k}T_0 \frac{f_{\text{cbl}} - 1}{g_t} = \overline{k}T \frac{1/g_2 + |\rho_1|^2(1 - g_2) - 1}{g_t}. \tag{45}$$

Using eqs. (6.15) and (6.16) in Pozar (2001), the ratio of transducer gain to available gain for the module, matched as shown, is

$$\frac{g_t}{g_a} = 1 - |\rho_1|^2, \tag{46}$$

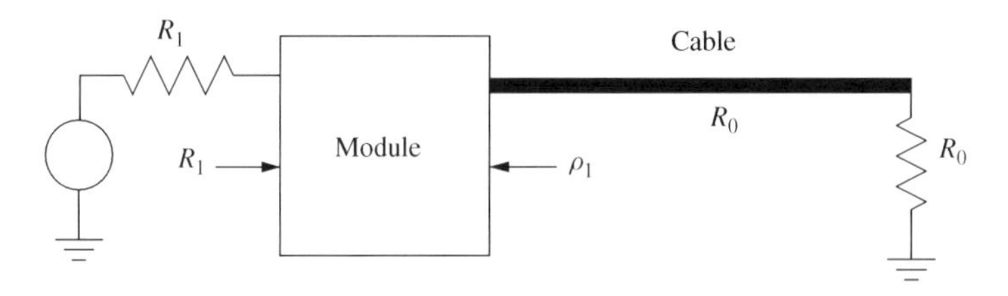

Fig. N.7 Cable with reflection at input.

which is the same as Eq. (36), so we can write Eq. (45) as

$$n_s = \bar{k}T \frac{1/g_2 + |\rho_1|^2(1 - g_2) - 1}{g_a(1 - |\rho_1|^2)}. \tag{47}$$

In the theoretical configuration, this noise is related to the module noise by available gain:

$$n_s = \frac{\bar{k}T_e}{g_a}, \tag{48}$$

where T_e is the noise temperature of the cable (to match Pozar's terminology). From Eqs. (48) and (47), this noise temperature is

$$T_e = \frac{g_a n_s}{\bar{k}} = \frac{1/g_2 + |\rho_1|^2(1 - g_2) - 1}{(1 - |\rho_1|^2)} T. \tag{49}$$

If we replace g_2 by $1/L$, as in Pozar, this is

$$T_e = \frac{L + |\rho_1|^2(1 - 1/L) - 1}{1 - |\rho_1|^2} T \tag{50}$$

$$= \frac{L^2 + |\rho_1|^2(L - 1) - L}{L(1 - |\rho_1|^2)} T \tag{51}$$

$$= \frac{(L - 1)(L + |\rho_1|^2)}{L(1 - |\rho_1|^2)} T, \tag{52}$$

which is equivalent to Pozar's (2001) Eq. (3.87).

N.7 EFFECT OF SOURCE IMPEDANCE

In the general case, the noise factor can be affected by source impedance differently than it is with the isolated noise source we have been considering. Haus et al. (1960b) have represented the general case by using a noise current generator and a noise voltage generator (Fig. 3.6). This results in a noise factor given by their Eq. (27) as

$$\hat{f} = 1 + \frac{G_u}{G_s} + \frac{R_n}{G_s}[(G_s + G_\gamma)^2 + (B_s + B_\gamma)^2], \tag{53}$$

where $Y_s = G_s + jB_s$ is source admittance, $Y_\gamma = G_\gamma + jB_\gamma$ is called the correlation admittance, R_n is called noise resistance, and G_u is a conductance representing noise that is uncorrelated between the two sources. We will rewrite this in terms of impedance so we can compare it to our expression for $\hat{f}$ for the isolated source:

$$\hat{f} = 1 + \frac{1}{G_s}(G_u + R_n|Y_s + Y_\gamma|^2) \tag{54}$$

$$= 1 + \frac{|Z_s|^2}{R_s}\left(G_u + R_n\frac{|Z_\gamma + Z_s|^2}{|Z_\gamma Z_s|^2}\right) \tag{55}$$

$$= 1 + \frac{1}{R_s}\left(|Z_s|^2 G_u + R_n\frac{|Z_\gamma + Z_s|^2}{|Z_\gamma|^2}\right). \tag{56}$$

By substituting the equivalent Z_{22k-} for Z_s and comparing this to Eqs. (5) and (2), we find that the isolated source can be represented by Eq. (53) if[1]

$$G_u = 0, \tag{57}$$

$$Z_\gamma = Z_{11k}, \tag{58}$$

and

$$R_n = \frac{v_{nk}^2}{4kT_0B}\frac{|Z_{11k}|^2}{R_{11k}^2}. \tag{59}$$

Equation (56) can be further manipulated to give Eq. (3.33), which can be represented by various circles on a Smith chart, depending on the value of $\hat{f}$ (see Section 3.8). Note that this form and this relationship between $\hat{f}$ and source impedance, which was developed to represent the dependence of noise factor on source impedance, applies even to the isolated source that we have analyzed here. Also note, however, how much simpler is our use of f to represent an isolated noise source in the standard cascade.

N.8 RATIO OF POWER GAINS

Here we determine the ratio of the power gain that is used to refer $\hat{f}$ to the cascade input to that which is used to refer f there. This is used in Section N.4.

Assume a Thevenin voltage source with mean-square voltage equal to $4R_{22k-}$ watts driving the output impedance Z_{22k-} that is seen by module k at its input. Half of this voltage will drop across the resistor in a matched load so the power dissipated in that resistor will be

$$p_a = \tilde{v}^2/R_{22k-} = 1\text{ W}. \tag{60}$$

If this source drives module k, the squared voltage appearing across the input to module k, when it is connected to the source, is

$$\tilde{v}_k^2 = 4R_{22k-}\text{ watts}\frac{|Z_{11k}|^2}{|Z_{11k} + Z_{22k-}|^2}. \tag{61}$$

The forward wave voltage $\tilde{v}_f$ in the cable is related to this voltage by

$$\tilde{v}_k = \tilde{v}_f|1 + S_{11k}| = \tilde{v}_f \left|1 + \frac{Z_{11k} - R_0}{Z_{11k} + R_0}\right| = 2\tilde{v}_f \left|\frac{Z_{11k}}{Z_{11k} + R_0}\right|. \quad (62)$$

Therefore, the square of the forward voltage is

$$\tilde{v}_f^2 = \frac{|Z_{11k} + R_0|^2}{|Z_{11k} + Z_{22k-}|^2} R_{22k-} \text{ watts}, \quad (63)$$

and the corresponding power is

$$p_f = \frac{|Z_{11k} + R_0|^2}{|Z_{11k} + Z_{22k-}|^2} \frac{R_{22k-}}{R_0} \text{ watts}. \quad (64)$$

The ratio of powers and, therefore of power gains, are therefore

$$\frac{g_a}{g_f} = \frac{p_a}{p_f} = \frac{|Z_{11k} + Z_{22k-}|^2/R_{22k-}}{|Z_{11k} + R_0|^2/R_0}. \quad (65)$$

This equals Eq. (11), as claimed in Section N.4.

ENDNOTE

[1]Equation (57) indicates that v_n and i_n in Fig. 3.6 are completely correlated when they represent the isolated noise source in Fig. 3.2. This is verified by circuit analysis, which shows that e_k will be produced across R_{11k}, independently of the driving source impedance, if $i_n = e_k/R_{11k}$ and $v_n = e_k Z_{11k}/R_{11k}$. But then $v_n = i_n Z_{11k} = i_n Z_\gamma$, so v_n and i_n are completely correlated.

APPENDIX P

IM PRODUCTS IN MIXERS

This development supports Section 7.3 and is also pertinent to Chapter 4.

In mixing, a relatively weak RF signal,

$$v_a = A \cos \varphi_a(t), \tag{1}$$

where

$$\varphi(t) = \omega t + \theta, \tag{2}$$

is basically raised to the nth power and shifted in phase by $m\varphi_b(t)$, resulting in a term given by Eq. (7.8). It follows that, if v_a should consist of two sinusoids, the results in the mixer output would also appear in this simple form, so that an input signal

$$v = v_1 + v_2 = k_1 \cos \varphi_1(t) + k_2 \cos \varphi_2(t) \tag{3}$$

would produce an output proportional to

$$v^n = (v_1 + v_2)^n \tag{4}$$

with phases additionally shifted by $m\varphi_b(t)$. (This is, in general, a fixed phase shift plus a frequency shift.) Our objective is to determine the amplitudes of the IMs, resulting from this process, relative to the linear terms. These can be determined by expanding Eq. (4). This expansion is well known to be

$$v^n = (v_1 + v_2)^n = \sum_{i=0}^{n} c(n, i) v_1^i v_2^{n-i}, \tag{5}$$

where $c(n, i)$ is the binomial coefficient, the values of which are given in Table P.1.

TABLE P.1 Binomial Coefficients

	i										
n	0	1	2	3	4	5	6	7	8	9	10
0	1										
1	1	1									
2	1	2	1								
3	1	3	3	1							
4	1	4	6	4	1						
5	1	5	10	10	5	1					
6	1	6	15	20	15	6	1				
7	1	7	21	35	35	21	7	1			
8	1	8	28	56	70	56	28	8	1		
9	1	9	36	84	126	126	84	36	9	1	
10	1	10	45	120	210	252	210	120	45	10	1

From Burrington (1954, p. 272).

Expanding, we obtain

$$v^n(\varphi(t)) = \sum_{i=0}^{n} c(n, i) k_1^i k_2^{n-i} \cos^i \varphi_1(t) \cos^{n-i} \varphi_2(t). \tag{6}$$

When the signals both have amplitude A, this becomes

$$v^n(\varphi(t)) = \sum_{i=0}^{n} c(n, i) A^n \cos^i \varphi_1(t) \cos^{n-i} \varphi_2(t). \tag{7}$$

This process applies directly to the IMs of Chapter 4, which are not frequency converted. But it also applies to IMs in mixers if we account for the addition of $m\varphi_b(t)$ to the phase, in which case we need $v^n[\varphi(t) + m\varphi_b(t)]$. However, since we are interested in the relative amplitude of the terms, we will not bother to show $m\varphi_b(t)$ explicitly.

The nth-order spurs are produced by the terms with $i = 0$ or $i = n$,

$$v_n = A^n \cos^n \varphi_j(t), \tag{8}$$

where j is 1 or 2. This term can be expanded to give

$$v_n = A^n \left\{ \frac{1 + \cos[2\varphi_j(t)]}{2} \right\} \cos^{n-2} \varphi_j(t) \tag{9}$$

$$= A^n \left\{ \frac{3\cos\varphi_j(t) + \cos[3\varphi_j(t)]}{2^2} \right\} \cos^{n-3} \varphi_j(t). \tag{10}$$

Repeating this process $n-3$ more times gives

$$v_n = \cdots + A^n \frac{\cos[n\varphi_j(t)]}{2^{n-1}}. \tag{11}$$

This last term is the nth-order spur with amplitude given in the spur-level table. The terms not shown are at lower frequencies. This frequency ω_j also occurs in other terms in Eq. (4.1) but with amplitudes raised to higher powers. As a result, at sufficiently low signal levels the term shown in Eq. (11) dominates.

For n odd, we can see from Table P.1 that the strongest IMs have $i = (n \pm 1)/2$. These are also the most troublesome producing frequencies near $\omega_1(t)$ and $\omega_2(t)$. We can expand the corresponding term in Eq. (7), by applying the process that we used to obtain Eq. (11) to each of its two cosines, to give

$$v_{n,(n\pm1)/2}(\varphi(t)) = A^n c[n,(n-1)/2] \cos^{(n-1)/2}\varphi_1(t)\cos^{(n+1)/2}\varphi_2(t) \tag{12}$$

$$= \cdots + A^n c[n,(n-1)/2]\frac{\cos\left[(n-1)\varphi_1(t)/2\right]\cos\left[(n+1)\varphi_2(t)/2\right]}{2^{n-2}} \tag{13}$$

$$\approx A^n c[n,(n-1)/2]\frac{\left\{\begin{array}{l}\cos([(n-1)\varphi_1(t)+(n+1)\varphi_2(t)]/2)\\ +\cos([(n-1)\varphi_1(t)-(n+1)\varphi_2(t)]/2)\end{array}\right\}}{2^{n-1}} \tag{14}$$

$$= A^n \frac{c[n,(n-1)/2]}{2^{n-1}}\left\{\begin{array}{l}\cos(\{\varphi_2(t)-\varphi_1(t)+n[\varphi_2(t)+\varphi_1(t)]\}/2)\\ +\cos(\{\varphi_1(t)+\varphi_2(t)+n[\varphi_2(t)-\varphi_1(t)]\}/2)\end{array}\right\}. \tag{15}$$

The amplitude of these IMs, relative to the nth-order spur from Eq. (11) is the binomial coefficient

$$r = c\left(n, \frac{n-1}{2}\right). \tag{16}$$

For $n=3$, Eq. (15) is

$$v_{3,(1\text{ or }2)} = A^3\frac{3}{4}\left\{\cos\begin{pmatrix}0.5[\varphi_2(t)-\varphi_1(t)]\\ +1.5[\varphi_2(t)+\varphi_1(t)]\end{pmatrix} + \cos\begin{pmatrix}0.5[\varphi_1(t)+\varphi_2(t)]\\ +1.5[\varphi_2(t)-\varphi_1(t)]\end{pmatrix}\right\} \tag{17}$$

$$= A^3\frac{3}{4}\{\cos[\varphi_1(t)+2\varphi_2(t)] + \cos\{\varphi_2(t)+[\varphi_2(t)-\varphi_1(t)]\}\}, \tag{18}$$

representing the IMs f and c or IMs d and g in Fig. 4.6, depending on whether φ_2 is a parameter of a or b.

With n even, the largest IM will be for $i = n/2$ and the equivalent of Eq. (15) will be

$$v_{n,n/2}(\varphi(t)) \approx A^n \frac{c\left(n, \frac{n}{2}\right)}{2^{n-1}}\left\{\cos\left(\frac{n}{2}[\varphi_2(t)+\varphi_1(t)]\right) + \cos\left(\frac{n}{2}[\varphi_2(t)-\varphi_1(t)]\right)\right\}. \tag{19}$$

The amplitude, relative to the spur is

$$r = c\left(n, \frac{n}{2}\right). \tag{20}$$

We can represent both Eq. (16) and (20) by saying that the ratio of the largest nth-order IMs to the nth order spur is

$$r = c[n, \text{int}(n/2)]. \tag{21}$$

For $n = 2$, Eq. (19) is represented by c and e in Fig. 4.2.

The RF is shifted by the LO phase and frequency $\varphi_b(t)$ to produce the IF. After the IMs are created, they are shifted by the same amount to accompany the signal through the conversion. The $1 \times n$ spur is the converted nth harmonic of the RF, and the IM level is related to that spur (or harmonic) level by r or $R = 20 \text{ dB} \log_{10}(r)$. The $1 \times n$ spur level, relative to the desired signal, can be obtained as the negative of the corresponding value in the spur table; R is added to that to give the level of the largest IM relative to the signal.

These IM levels are related to the signal level so we need not specify whether the levels are before or after conversion (assuming the IMs receive the same conversion loss as the signal). When we express the process in terms of IPs, we must specify whether we are referring to output levels or equivalent input levels.

APPENDIX S

COMPOSITE *S* PARAMETERS

We develop here the S parameters for a composite of two modules in terms of the parameters of the individual modules and show the effects of unilaterality in one of the individual modules. Results are referenced in Chapter 2.

Equations (2.6) and (2.7) are

$$v_{i,j} = S_{11j} v_{o,j} + S_{12j} v_{i,j+1}, \tag{2.6}$$

$$v_{o,j+1} = S_{21j} v_{o,j} + S_{22j} v_{i,j+1}. \tag{2.7}$$

Written for $j = 1$ and $j = 2$, they are

$$v_{i,1} = S_{111} v_{o,1} + S_{121} v_{i,2}, \tag{1}$$

$$v_{o,2} = S_{211} v_{o,1} + S_{221} v_{i,2}, \tag{2}$$

$$v_{i,2} = S_{112} v_{o,2} + S_{122} v_{i,3}, \tag{3}$$

and

$$v_{o,3} = S_{212} v_{o,2} + S_{222} v_{i,3}. \tag{4}$$

Rearranging Eq. (1), we obtain

$$v_{i,2} = \frac{v_{i,1} - S_{111} v_{o,1}}{S_{121}}. \tag{5}$$

Rearranging Eq. (4), we obtain

$$v_{o,2} = \frac{v_{o,3} - S_{222} v_{i,3}}{S_{212}}. \tag{6}$$

Equations (2) and (6) can be combined to give

$$\frac{v_{o,3} - S_{222}v_{i,3}}{S_{212}} = S_{211}v_{o,1} + S_{221}v_{i,2}, \tag{7}$$

which can be combined with Eq. (5) to give

$$\frac{v_{o,3} - S_{222}v_{i,3}}{S_{212}} = S_{211}v_{o,1} + S_{221}\frac{v_{i,1} - S_{111}v_{o,1}}{S_{121}} \tag{8}$$

$$= \left(S_{211} - \frac{S_{111}S_{221}}{S_{121}}\right)v_{o,1} + \frac{S_{221}}{S_{121}}v_{i,1}. \tag{9}$$

This can be rearranged to give

$$v_{o,3} = S_{212}\left[\left(S_{211} - \frac{S_{111}S_{221}}{S_{121}}\right)v_{o,1} + \frac{S_{221}}{S_{121}}v_{i,1}\right] + S_{222}v_{i,3}. \tag{10}$$

Equations (3) and (5) can be combined to give

$$\frac{v_{i,1} - S_{111}v_{o,1}}{S_{121}} = S_{112}v_{o,2} + S_{122}v_{i,3}, \tag{11}$$

which can be combined with Eq. (6) to give

$$\frac{v_{i,1} - S_{111}v_{o,1}}{S_{121}} = S_{112}\frac{v_{o,3} - S_{222}v_{i,3}}{S_{212}} + S_{122}v_{i,3} \tag{12}$$

$$= \frac{S_{112}}{S_{212}}v_{o,3} + \left(S_{122} - \frac{S_{112}S_{222}}{S_{212}}\right)v_{i,3}. \tag{13}$$

This can be combined with Eq. (10) to give

$$\frac{v_{i,1} - S_{111}v_{o,1}}{S_{121}} = \frac{S_{112}}{S_{212}}\left\{S_{212}\left[\begin{matrix}\left(S_{211} - \dfrac{S_{111}S_{221}}{S_{121}}\right)v_{o,1} \\ +\dfrac{S_{221}}{S_{121}}v_{i,1}\end{matrix}\right] + S_{222}v_{i,3}\right\}$$

$$+ \left(S_{122} - \frac{S_{112}S_{222}}{S_{212}}\right)v_{i,3} \tag{14}$$

$$= S_{112}\left\{\left(S_{211} - \frac{S_{111}S_{221}}{S_{121}}\right)v_{o,1} + \frac{S_{221}}{S_{121}}v_{i,1}\right\} + S_{122}v_{i,3}, \tag{15}$$

which can be rearranged to give

$$\frac{v_{i,1}}{S_{121}}(1 - S_{112}S_{221}) = \left(\frac{S_{111}}{S_{121}} + S_{112}S_{211} - \frac{S_{111}S_{112}S_{221}}{S_{121}}\right)v_{o,1} + S_{122}v_{i,3}, \tag{16}$$

and

$$v_{i,1} = \left(S_{111} + \frac{S_{112}S_{121}S_{211}}{1 - S_{112}S_{221}} \right) v_{o,1} + \frac{S_{121}S_{122}}{1 - S_{112}S_{221}} v_{i,3}. \tag{17}$$

Since port 3 is the output of the composite, this can be written

$$v_{i,1,\text{comp}} = S_{11,\text{comp}} v_{o,1,\text{comp}} + S_{12,\text{comp}} v_{i,2,\text{comp}}, \tag{18}$$

where

$$S_{11,\text{comp}} = S_{111} + \frac{S_{112}S_{121}S_{211}}{1 - S_{112}S_{221}}, \tag{19}$$

and

$$S_{12,\text{comp}} = \frac{S_{121}S_{122}}{1 - S_{112}S_{221}}. \tag{20}$$

Substituting Eq. (17) into Eq. (10), we obtain

$$v_{o,3} = S_{212} \left[\begin{array}{l} \left(S_{211} - \dfrac{S_{111}S_{221}}{S_{121}} \right) v_{o,1} \\ + \dfrac{S_{221}}{S_{121}} \left\{ \begin{array}{l} \left(S_{111} + \dfrac{S_{112}S_{121}S_{211}}{1 - S_{112}S_{221}} \right) v_{o,1} \\ + \dfrac{S_{121}S_{122}}{1 - S_{112}S_{221}} v_{i,3} \end{array} \right\} \end{array} \right] + S_{222} v_{i,3} \tag{21}$$

$$= S_{212} \left\{ \begin{array}{l} S_{211} - \dfrac{S_{111}S_{221}}{S_{121}} \\ + \dfrac{S_{111}S_{221}}{S_{121}} + \dfrac{S_{112}S_{221}S_{211}}{1 - S_{112}S_{221}} \end{array} \right\} v_{o,1} + \left\{ \begin{array}{l} S_{222} \\ + \dfrac{S_{122}S_{212}S_{221}}{1 - S_{112}S_{221}} \end{array} \right\} v_{i,3} \tag{22}$$

$$= \frac{S_{212}S_{211}}{1 - S_{112}S_{221}} v_{o,1} + \left\{ S_{222} + \frac{S_{122}S_{212}S_{221}}{1 - S_{112}S_{221}} \right\} v_{i,3}. \tag{23}$$

Since port 3 is the output of the composite, this can be written

$$v_{o,2,\text{comp}} = S_{21,\text{comp}} v_{o,1,\text{comp}} + S_{22,\text{comp}} v_{i,2,\text{comp}}, \tag{24}$$

where

$$S_{21,\text{comp}} = \frac{S_{212}S_{211}}{1 - S_{112}S_{221}}, \tag{25}$$

and

$$S_{22,\text{comp}} = S_{222} + \frac{S_{122}S_{212}S_{221}}{1 - S_{112}S_{221}}. \tag{26}$$

We can now write Eqs. (19), (20), (25), and (26) for the particular conditions of the second module being unilateral ($S_{122} = 0$) and for the first module being unilateral ($S_{121} = 0$). See Table S.1.

TABLE S.1 *S* Parameters for Composite of Two Modules

	General	Module 1 Unilateral ($S_{121} = 0$)	Module 2 Unilateral ($S_{122} = 0$)
$S_{11,\text{comp}}$	$S_{111} + \dfrac{S_{112}S_{121}S_{211}}{1 - S_{112}S_{221}}$	S_{111}	$S_{111} + \dfrac{S_{112}S_{121}S_{211}}{1 - S_{112}S_{221}}$
$S_{12,\text{comp}}$	$\dfrac{S_{121}S_{122}}{1 - S_{112}S_{221}}$	0	0
$S_{21,\text{comp}}$	$\dfrac{S_{212}S_{211}}{1 - S_{112}S_{221}}$	$\dfrac{S_{212}S_{211}}{1 - S_{112}S_{221}}$	$\dfrac{S_{212}S_{211}}{1 - S_{112}S_{221}}$
$S_{22,\text{comp}}$	$S_{222} + \dfrac{S_{122}S_{212}S_{221}}{1 - S_{112}S_{221}}$	$S_{222} + \dfrac{S_{122}S_{212}S_{221}}{1 - S_{112}S_{221}}$	S_{222}

APPENDIX T

THIRD-ORDER TERMS AT INPUT FREQUENCY

This development supports Section 5.1.4.2.

In the presence of additional frequencies, the first part of Eq. (4.20) becomes

$$\frac{a_3}{4}\left\{\begin{array}{l}(3A^3+6AB^2+6AC^2+6AD^2+\cdots)\cos\varphi_a(t)\\ \quad+(3B^3+6BA^2+6BC^2+6BD^2+\cdots)\cos\varphi_b(t)\\ \quad+(3C^3+6CA^2+6CB^2+6CD^2+\cdots)\cos\varphi_c(t)+\cdots\end{array}\right\}. \tag{1}$$

All of the squared amplitudes are from two terms at the same frequency whose frequencies subtracted to give zero (DC).

Consider the input power spectrum in Fig. 5.2 to be divided into n equally wide contiguous bands so the mean-square voltage equivalent of the power p_i in each is

$$|\tilde{v}_i|^2 = \frac{pR}{n}, \tag{2}$$

where p is the total input power, R is the resistance in which it is dissipated, and $\tilde{v}_i$ has a magnitude and phase like $\tilde{v}_x$ in Eq. (2.15). We will let these bands shrink enough so the signal represented by the power in each band is essentially constant over the time of interest. Then the term in Eq. (1) at the input frequency f_i would have

$$\sqrt{2}\tilde{v}_{3i} = \frac{a_3}{4}\left\{3\left(\frac{2pR}{n}\right)^{3/2} + 6\sum_{i=2}^{n}\left(\frac{2pR}{n}\right)^{3/2}\right\} e^{j\theta_i}, \tag{3}$$

where $i = a, b, c \ldots$ in Eq. (1).

As n becomes large, and, as a result, $\tilde{v}_i$ becomes small, the first term becomes negligible and the rms voltage at frequency f_i becomes

$$\tilde{v}_{3i} \approx \frac{1}{\sqrt{2}} \frac{a_3}{4} 6n \left(\frac{2pR}{n} \right)^{3/2} e^{j\theta_i} = 3a_3 \frac{(pR)^{3/2}}{\sqrt{n}} e^{j\theta_i}. \tag{4}$$

As in Eq. (4.20), the term with this rms value is coherent with the term at the same frequency from the first-order response:

$$\tilde{v}_{1i} \approx \frac{a_1}{\sqrt{2}} \left(\frac{2pR}{n} \right)^{1/2} e^{j\theta} = a_1 \frac{(pR)^{1/2}}{\sqrt{n}} e^{j\theta}. \tag{5}$$

The fundamental output power due to the sum of these voltages is

$$p_i = \frac{p_{1T}}{n} = \frac{|\tilde{v}_{1i} + \tilde{v}_{3i}|^2}{R} = \frac{|a_1 (pR)^{1/2} + 3a_3 (pR)^{3/2}|^2}{nR} \tag{6}$$

$$= \frac{|a_1^2 (pR) + 6a_1 a_3 (pR)^2 + 9a_3^2 (pR)^3|}{nR} \tag{7}$$

$$= a_1^2 \left| \frac{p}{n} + 6 \frac{a_3}{a_1} Rp \left(\frac{p}{n} \right) + 9 \left(\frac{a_3}{a_1} \right)^2 R^2 p^2 \left(\frac{p}{n} \right) \right| \tag{8}$$

$$= \frac{p_1}{n} \left| 1 + 6 \frac{a_3}{a_1} Rp + 9 \left(\frac{a_3}{a_1} \right)^2 R^2 p^2 \right|, \tag{9}$$

where p_{1T} is total power at the fundamental, and p_1 $(= |a_1|^2 p)$ is the power due to the first-order term.

We can divide both sides above by the bandwidth represented by each p_i, B/n, thereby changing p_1/n to the PSD of the first-order response, S_1. Then, using Eq. (4.27), the total two-sided PSD in Fig. 5.7 is

$$\frac{S_{1T}}{2} = \frac{S_1}{2} \left[1 + 6 \operatorname{sign}\left(\frac{a_3}{a_1} \right) \left(\frac{2}{3 p_{\text{IIP3,IM}}} \right) p + 9 \left(\frac{2}{3 p_{\text{IIP3,IM}}} \right)^2 p^2 \right] \tag{10}$$

$$= \frac{S_1}{2} \left| 1 + 4 \operatorname{sign}\left(\frac{a_3}{a_1} \right) \left(\frac{p}{p_{\text{IIP3,IM}}} \right) + 4 \left(\frac{p}{p_{\text{IIP3,IM}}} \right)^2 \right|. \tag{11}$$

Thus ε in Fig. 5.7 is

$$\varepsilon = 4 \left(\frac{S_1}{2} \right) \left[\operatorname{sign}\left(\frac{a_3}{a_1} \right) \left(\frac{p}{p_{\text{IIP3,IM}}} \right) + \left(\frac{p}{p_{\text{IIP3,IM}}} \right)^2 \right]. \tag{5.20}$$

APPENDIX V

SENSITIVITIES AND VARIANCE OF NOISE FIGURE

This appendix is referenced several places in Chapter 3.

From Eq. (3.14) we know that the cascade noise factor f_{cas} is a function of module and cable noise factors and gains. From Eqs. (3.28) and (2.46), we know that the cable noise factors and gains are both functions of the module SWRs. We can write this complex dependence symbolically as

$$F_{\text{cas}} = F(f_i, g_i, \text{SWR}_i). \tag{1}$$

We can write a Taylor series for small changes in F_{cas} retaining only the first derivatives, based on an assumption of small changes in the individual variables:

$$dF_{\text{cas}} = \sum_i \left[\frac{\partial F_{\text{cas}}}{\partial f_i} df_i + \frac{\partial F_{\text{cas}}}{\partial g_i} dg_i + \frac{\partial F_{\text{cas}}}{\partial \text{SWR}_i} d\text{SWR}_i \right]. \tag{2}$$

Since dF_{cas} is composed of these small changes, the variance of F_{cas} will be the sum of the variances of each of the elements of the summation:

$$\sigma^2_{F_{\text{cas}}} = \sum_i [\hat{S}^2_{fi}\sigma^2_{fi} + \hat{S}^2_{gi}\sigma^2_{gi} + \hat{S}^2_{\text{SWR}i}\sigma^2_{\text{SWR}i}], \tag{3.30}$$

where

$$\hat{S}_{xk} = \frac{\partial F_{\text{cas}}}{\partial x_k} \tag{3.78}$$

is the sensitivity of F_{cas} to the parameter x_k.

We can write dF_{cas} more simply for fixed g_i and SWR_i, allowing only variations df_j in module noise factors. Based on Eq. (3.14), the variation in f_{cas} would then be

$$df_{cas}(df_j) = \frac{\partial f_{cas}}{\partial f_1} df_1 + \frac{\partial f_{cas}}{\partial f_3} df_3 + \frac{\partial f_{cas}}{\partial f_5} df_5 + \cdots, \tag{3}$$

where even-numbered elements are cables and are not included.

The noise figure is related to the noise factor by

$$F_{cas} = 10 \text{ dB} \log_{10} f_{cas} = 10 \text{ dB} \log_{10}(e) \ln(f_{cas}) = 4.343 \text{ dB} \ln(f_{cas}), \tag{4}$$

so a change in the noise figure can be related to a change in the noise factor by

$$dF_{cas} = \frac{dF_{cas}}{df_{cas}} df_{cas} = \frac{4.343 \text{ dB}}{f_{cas}} df_{cas} = 4.343 \text{ dB} \times 10^{-F_{cas}/10 \text{ dB}} df_{cas}. \tag{5}$$

An inverse relationship can also be written:

$$0.23 \times 10^{F_j/10 \text{ dB}} \frac{dF_j}{\text{dB}} = df_j. \tag{6}$$

Using Eqs. (3) and (3.14), we can write (5) as

$$dF_{cas}(df_j) = 4.343 \text{ dB} \times 10^{-F_{cas}/10 \text{ dB}} \left\{ df_1 + \frac{df_3}{g_1 g_2} + \cdots \right\}. \tag{7}$$

Using Eq. (6), this is

$$dF_{cas}(dF_j) = 10^{-F_{cas}/10 \text{ dB}} \left\{ \begin{array}{l} 10^{F_1/10\text{dB}} dF_1 \\ \quad +10^{(F_3-G_1-G_2)/10 \text{ dB}} dF_3 + \cdots \end{array} \right\}. \tag{3.79}$$

This gives approximately the change in F_{cas} as a function of small changes in the module noise figures about the operating point described by the values of the G_j's and F_j's. For small enough changes, it can be used to give the variance of F_{cas} as a function of the variances of the noise figures of individual modules:

$$\sigma^2_{F_{cas}}(dF_j) = 10^{-F_{cas}/5 \text{ dB}} \left\{ \begin{array}{l} 10^{F_1/5 \text{ dB}} \sigma^2_{F1} \\ \quad +10^{(F_3-G_1-G_2)/5 \text{ dB}} \sigma^2_{F_3} \\ \quad +10^{\left(F_5 - \sum_{j=1}^{4} G_j\right) / 5 \text{ dB}} \sigma^2_{F5} + \cdots \end{array} \right\}. \tag{8}$$

This could be expanded to many stages, following the pattern indicated here, or we can reduce it to two stages and sequentially use stage 1 as the composite of all previous stages as we move from input to output of a cascade. For example,

$$\sigma^2_{F_{cas3}} = 10^{-F_{cas3}/5 \text{ dB}} \left\{ \begin{array}{l} 10^{F_1/5 \text{ dB}} \sigma^2_{F1} \\ \quad +10^{(F_3-G_1-G_2)/5 \text{ dB}} \sigma^2_{F3} \end{array} \right\} \tag{9}$$

and

$$\sigma^2_{F_{\text{cas5}}} = 10^{-F_{\text{cas5}}/5\ \text{dB}} \left\{ \begin{array}{l} 10^{F_{\text{cas3}}/5\ \text{dB}} \sigma^2_{F\text{cas3}} \\ +10^{\left(F_5 - \sum_{j=1}^{4} G_j\right) / 5\ \text{dB}} \sigma^2_{F5} \end{array} \right\}, \tag{10}$$

where $F_{\text{cas}\,j}$ is F for the cascade through stage j.

Equation (10) is the same as Eq. (8). We can also write this as

$$\sigma^2_{F_{\text{cas}.n}} = 10^{-F_{\text{cas}.n}/5\ \text{dB}} \left\{ \begin{array}{l} 10^{F_{\text{cas}(n-1)}/5\ \text{dB}} \sigma^2_{F\text{cas}(n-1)} \\ +10^{\left(F_{\text{cas}.n} - \sum_{j=1}^{n-1} G_j\right) / 5\ \text{dB}} \sigma^2_{Fn} \end{array} \right\}, \tag{3.31}$$

depending upon σ^2_{Fn} being zero for all interconnects. Then every other application of Eq. (3.31), those that apply to interconnects, will produce a change in cascade noise figure variance only because of the change in cascade noise figure.

APPENDIX X

CROSSOVER SPURS

Table X.1 is a list of all crossover spurs for the range of m and n shown in its upper left. The ratios of RF to IF and of RF to LO, which apply to the two types of normalizations that we have considered, are listed for the three desired-signal

TABLE X.1 Crossover Spurs

set max n = 5
giving max m
10
10

desired m(LO) =	1		
desired n(RF) =	−1		
	spur		
RF/IF	m	n	RF/LO
0.2	0	5	0.1666667
0.25	0	4	0.2
0.3333333	0	3	0.25
0.3333333	2	−5	0.25
0.5	−1	5	0.3333333
0.5	0	2	0.3333333
0.5	2	−4	0.3333333
0.6666667	−1	4	0.4
1	−2	5	0.5
1	−1	3	0.5
1	0	1	0.5
1	2	−3	0.5
1	3	−5	0.5
1.5	−2	4	0.6
2	−3	5	0.6666667
2	−1	2	0.6666667
2	3	−4	0.6666667
3	−2	3	0.75
3	4	−5	0.75
4	−3	4	0.8
5	−4	5	0.8333333

TABLE X.1 Crossover Spurs (*continued*)

desired m(LO) = 1				desired m(LO) = −1			
desired n(RF) = 1				desired n(RF) = 1			
	spur				spur		
RF/IF	m	n	RF/LO	RF/IF	m	n	RF/LO
0.1428571	2	−5	0.1666667	1.1	10	0	11
0.1666667	2	−4	0.2	1.1111111	9	0	10
0.2	0	5	0.25	1.125	−10	2	9
0.2	2	−3	0.25	1.125	8	0	9
0.25	0	4	0.3333333	1.1428571	−9	2	8
0.25	2	−2	0.3333333	1.1428571	7	0	8
0.25	3	−5	0.3333333	1.1666667	−8	2	7
0.2857143	3	−4	0.4	1.1666667	6	0	7
0.3333333	−1	5	0.5	1.2	−7	2	6
0.3333333	0	3	0.5	1.2	5	0	6
0.3333333	2	−1	0.5	1.2222222	10	−1	5.5
0.3333333	3	−3	0.5	1.25	−6	2	5
0.3333333	4	−5	0.5	1.25	4	0	5
0.375	4	−4	0.6	1.25	9	−1	5
0.4	−1	4	0.6666667	1.2857143	−10	3	4.5
0.4	3	−2	0.6666667	1.2857143	8	−1	4.5
0.4	5	−5	0.6666667	1.3333333	−9	3	4
0.4285714	−2	5	0.75	1.3333333	−5	2	4
0.4285714	4	−3	0.75	1.3333333	3	0	4
0.4444444	5	−4	0.8	1.3333333	7	−1	4
0.4545455	6	−5	0.8333333	1.375	10	−2	3.6666667
0.5	−3	5	1	1.4	−8	3	3.5
0.5	−2	4	1	1.4	6	−1	3.5
0.5	−1	3	1	1.4285714	9	−2	3.3333333
0.5	0	2	1	1.5	−10	4	3
0.5	2	0	1	1.5	−7	3	3
0.5	3	−1	1	1.5	−4	2	3
0.5	4	−2	1	1.5	2	0	3
0.5	5	−3	1	1.5	5	−1	3
0.5	6	−4	1	1.5	8	−2	3
0.5	7	−5	1	1.5714286	10	−3	2.75
0.5384615	8	−5	1.1666667	1.6	−9	4	2.6666667
0.5454545	7	−4	1.2	1.6	7	−2	2.6666667
0.5555556	−4	5	1.25	1.6666667	−6	3	2.5
0.5555556	6	−3	1.25	1.6666667	4	−1	2.5
0.5714286	−3	4	1.3333333	1.6666667	9	−3	2.5
0.5714286	5	−2	1.3333333	1.75	−8	4	2.3333333
0.5714286	9	−5	1.3333333	1.75	6	−2	2.3333333
0.5833333	8	−4	1.4	1.8	−10	5	2.25
0.6	−5	5	1.5	1.8	8	−3	2.25
0.6	−2	3	1.5	1.8333333	10	−4	2.2
0.6	4	−1	1.5	2	−9	5	2
0.6	7	−3	1.5	2	−7	4	2
0.6	10	−5	1.5	2	−5	3	2
0.6153846	9	−4	1.6	2	−3	2	2
0.625	−4	4	1.6666667	2	1	0	2
0.625	6	−2	1.6666667	2	3	−1	2
0.6363636	−6	5	1.75	2	5	−2	2
0.6363636	8	−3	1.75	2	7	−3	2
0.6428571	10	−4	1.8	2	9	−4	2
0.6666667	−7	5	2	2.2	10	−5	1.8333333

TABLE X.1 Crossover Spurs (*continued*)

desired m(LO) = 1				desired m(LO) = −1			
desired n(RF) = 1				desired n(RF) = 1			
	spur				spur		
RF/IF	m	n	RF/LO	RF/IF	m	n	RF/LO
0.6666667	−7	5	2	2.2	10	−5	1.8333333
0.6666667	−5	4	2	2.25	8	−4	1.8
0.6666667	−3	3	2	2.3333333	−8	5	1.75
0.6666667	−1	2	2	2.3333333	6	−3	1.75
0.6666667	3	0	2	2.5	−6	4	1.6666667
0.6666667	5	−1	2	2.5	4	−2	1.6666667
0.6666667	7	−2	2	2.5	9	−5	1.6666667
0.6666667	9	−3	2	2.6666667	7	−4	1.6
0.6923077	−8	5	2.25	3	−7	5	1.5
0.6923077	10	−3	2.25	3	−4	3	1.5
0.7	−6	4	2.3333333	3	2	−1	1.5
0.7	8	−2	2.3333333	3	5	−3	1.5
0.7142857	−9	5	2.5	3	8	−5	1.5
0.7142857	−4	3	2.5	3.5	6	−4	1.4
0.7142857	6	−1	2.5	4	−5	4	1.3333333
0.7272727	−7	4	2.6666667	4	3	−2	1.3333333
0.7272727	9	−2	2.6666667	4	7	−5	1.3333333
0.7333333	−10	5	2.75	5	−6	5	1.25
0.75	−8	4	3	5	4	−3	1.25
0.75	−5	3	3	6	5	−4	1.2
0.75	−2	2	3	7	6	−5	1.1666667
0.75	4	0	3				
0.75	7	−1	3				
0.75	10	−2	3				
0.7692308	−9	4	3.3333333				
0.7777778	−6	3	3.5				
0.7777778	8	−1	3.5				
0.7857143	−10	4	3.6666667				
0.8	−7	3	4				
0.8	−3	2	4				
0.8	5	0	4				
0.8	9	−1	4				
0.8181818	−8	3	4.5				
0.8181818	10	−1	4.5				
0.8333333	−9	3	5				
0.8333333	−4	2	5				
0.8333333	6	0	5				
0.8461538	−10	3	5.5				
0.8571429	−5	2	6				
0.8571429	7	0	6				
0.875	−6	2	7				
0.875	8	0	7				
0.8888889	−7	2	8				
0.8888889	9	0	8				
0.9	−8	2	9				
0.9	10	0	9				
0.9090909	−9	2	10				
0.9166667	−10	2	11				

curves, with m and n (± 1) for them shown at the top. The spurs have been sorted from lowest to highest frequency ratios or vise versa. Crossovers at zero or infinity are not shown.

Table X.1 has three major divisions, according to the sign of m and n that applies to the desired response. Within the divisions for the 1×1 or 1×-1

desired responses, the values of m and n for the crossover spurs, on any given line, are the same for RF/IF as for RF/LO. That is because, as the RF increases along these curves (Fig. 7.28), both RF/IF and RF/LO increase. Therefore, the sequence in which they cross spurs is the same. However, within the third segment, that for the -1×1 desired response, RF/IF increases with increasing RF but RF/LO decreases. For this reason, the ratio RF/LO has been sorted by decreasing value in this last segment so that both ratios on a line refer to the same m and n values.

APPENDIX Z

NONSTANDARD MODULES

Here we treat unilateral modules that are specified by their input and output impedances and by their transducer gains or their maximum available gains (Appendix G). Figure Z.1 shows two such modules, each represented by its input and output impedances (Z_{11}) and (Z_{22}) and a voltage generator that depends on the voltage across the input resistance (thus on the square root of the input power). We will see how to compute the gain of a cascade of such modules, using a spreadsheet as an aid, and how to find the S parameters for such modules and cascades.

Z.1 GAIN OF CASCADE OF MODULES RELATIVE TO TESTED GAIN

What is the gain of a cascade of modules that interfaces with impedances that are different than those used in obtaining their gains when they were tested with matched loads (maximum available gain), assuming negligible reverse transmission (Z_{12}, $S_{12} = 0$)?

The ratio of the voltage across the real part of a driven load to the voltage across the real part of the module's input is[1]

$$a_j \stackrel{\Delta}{=} \frac{e_{(j+1)}}{e_j} = a'_j \frac{R_{11(j+1)}}{Z_{22j} + Z_{11(j+1)}}. \tag{1}$$

The voltage at the cascade source is e_1. The load $Z_{11(j+1)}$ is the input to the next stage except that $Z_{11(N+1)}$ is the load for a cascade of N modules.

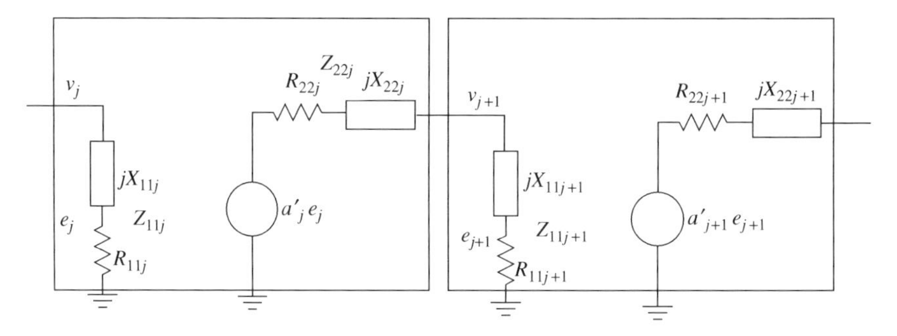

Fig. Z.1 The jth module in the cascade.

During module characterization (test), when the load is matched to (i.e., the complex conjugate of) the module's output impedance, Eq. (1) becomes

$$a_{Tj} \stackrel{\Delta}{=} \frac{e_{T(j+1)}}{e_{Tj}} = a'_j \frac{R_{22j}}{R_{22j} + R_{22j}} = \frac{a'_j}{2}. \tag{2}$$

(The imaginary parts in the denominator cancel since the impedances are conjugate.)

Since the gain is defined with respect to the voltage across R_{11}, the input match during test is not significant here. From Eq. (2) we can obtain the internal parameter a' in terms of the tested transfer function a_T. Substituting a' from Eq. (2) into Eq. (1), we obtain the transfer function in the cascade relative to the tested transfer function for the module:

$$a_j = a_{Tj} \frac{2R_{11(j+1)}}{Z_{22j} + Z_{11(j+1)}}. \tag{3}$$

The voltage transfer function for a cascade of N modules is then

$$a_{\text{cas}} = \left(\prod_{j=1}^{N} a_{Tj} \right) \left(\prod_{j=1}^{N} \frac{2R_{11(j+1)}}{Z_{22j} + Z_{11(j+1)}} \right). \tag{4}$$

Note that the first product is the transfer function the cascade would have if not for the differing impedances between test and use, and the second product is the modification due to the differing impedances.

The (actual) power gain $g_{\text{act},j}$ is the ratio of power absorbed in the load ($R_{11,j+1}$) to the power dissipated by the module input resistance ($R_{11,j}$). Due to the impedance matches at input and output, the power gain during test is the maximum available power gain $g_{\text{ma}T}$:

$$g_{\text{ma}Tj} = \left| \frac{e^2_{T(j+1)}/R_{22j}}{e^2_{Tj}/R_{11j}} \right| = |a_{\text{ma}Tj}|^2 \frac{R_{11j}}{R_{22j}}, \tag{5}$$

but the power gain of a module in the cascade is

$$g_{\text{act},j} = \left| \frac{e_{j+1}^2 / R_{11(j+1)}}{e_j^2 / R_{11j}} \right| \tag{6}$$

$$= |a_j|^2 \frac{R_{11j}}{R_{11(j+1)}} = |a_{\text{ma}Tj}|^2 \frac{4R_{11(j+1)}^2}{|Z_{22j} + Z_{11(j+1)}|^2} \frac{R_{11j}}{R_{11(j+1)}} \tag{7}$$

$$= g_{\text{ma}Tj} \frac{R_{22j}}{R_{11j}} \frac{4R_{11(j+1)}^2}{|Z_{22j} + Z_{11(j+1)}|^2} \frac{R_{11j}}{R_{11(j+1)}} = g_{\text{ma}Tj} \frac{4R_{22j} R_{11(j+1)}}{|Z_{22j} + Z_{11(j+1)}|^2}. \tag{8}$$

Here Eqs. (3) and (5) have been used. From this we obtain the actual power gain (Appendix G) for a cascade of N modules:

$$g_{\text{act,cas}} = \prod_{j=1}^{N} g_{\text{act},j} = \left(\prod_{j=1}^{N} g_{\text{ma}Tj} \right) \left(\prod_{j=1}^{N} \frac{4R_{22j} R_{11(j+1)}}{|Z_{22j} + Z_{11(j+1)}|^2} \right). \tag{9}$$

Here the subscript $N+1$ refers to the load and $g_{\text{ma}Tj}$ is the maximum available power gain measured when the module was characterized with a matched source and load. The actual power gain is the ratio of power delivered to the load to the power absorbed at the input of the cascade.

Example Z.1 Cascade Gain, Nonstandard Modules Figure Z.2 shows a spreadsheet that executes Eq. (9). The input and output impedances and the power gain in test ($g_{\text{ma}Ti}$) are listed for each module in lines 5–8 with the cascade's load

	A	B	C	D	E	F
1		Maximum				
2		Available				
3		Power Gain				
4		in test	R11	X11	R22	X22
5	Module 1	6.00 dB	200.00 Ω	100.00 Ω	300.00 Ω	150.00 Ω
6	Module 2	9.00 dB	1500.00 Ω	-250.00 Ω	1200.00 Ω	-200.00 Ω
7	Module 3	4.50 dB	1000.00 Ω	200.00 Ω	500.00 Ω	200.00 Ω
8	Module 4	22.00 dB	250.00 Ω	45.00 Ω	55.00 Ω	10.00 Ω
9	Load		100.00 Ω	25.00 Ω		
10		actual gain in use				
11		Module	CUMULATIVE			
12	Module 1	3.41 dB	3.41 dB			
13	Module 2	8.94 dB	12.36 dB			
14	Module 3	3.53 dB	15.88 dB			
15	Module 4	21.38 dB	37.27 dB			

Fig. Z.2 Spreadsheet for cascade of nonstandard modules.

given in line 9. The actual gain of each module in the cascade [Eq. (8)] is given in cells B12–B15 with the cumulative gain in cells C12–C15. Note that the sum of the tested gains is 41.5 dB, whereas the cascade gain is only 37.27 dB.

Z.2 FINDING MAXIMUM AVAILABLE GAIN OF A MODULE

We can obtain the value of $g_{\text{ma}Tj}$ from test data that was not obtained with matched source and load but rather in a transducer-gain test, with a signal generator and power meter (Fig. Z.3). Using Eqs. (6) and (8), we write

$$g_{\text{ma}Tj} = \left| \frac{e_{j+1}^2/R_{11(j+1)}}{e_j^2/R_{11j}} \right| \frac{|Z_{22j} + Z_{11(j+1)}|^2}{4R_{22j}R_{11(j+1)}}. \tag{10}$$

Here, $Z_{11(j+1)}$, including $R_{11(j+1)}$, is the test load. Assume that the power meter presents a real impedance equal to that of the connecting cable. Then $Z_{11(j+1)} = R_{11(j+1)}$. We recognize the first ratio as the ratio of power absorbed in the meter (assuming negligible cable loss) to that absorbed in the module, so we can write

$$g_{\text{ma}Tj} = \frac{p_{o,j+1}}{p_{o,j} - p_{i,j}} \frac{|Z_{22j} + R_{11(j+1)}|^2}{4R_{22j}R_{11(j+1)}} = \frac{p_{o,j+1}}{p_{o,j} - p_{i,j}} \frac{|1 + Z_{22j}/R_{11(j+1)}|^2}{4R_{22j}/R_{11(j+1)}}. \tag{11}$$

If we do not have a value for $p_{i,j}$, we can relate it to $p_{o,j}$ and to the transducer gain of the module, g_{tj}, by

$$g_{\text{ma}Tj} = \frac{p_{o,j+1}}{p_{o,j}(1 - |S_{11j}|^2)} \frac{|1 + Z_{22j}/R_{11(j+1)}|^2}{4R_{22j}/R_{11(j+1)}} \tag{12}$$

$$= g_{tj} \frac{|1 + Z_{11j}/R_{0j}|^2}{4R_{11j}/R_{0j}} \frac{|1 + Z_{22j}/R_{11(j+1)}|^2}{4R_{22j}/R_{11(j+1)}}. \tag{13}$$

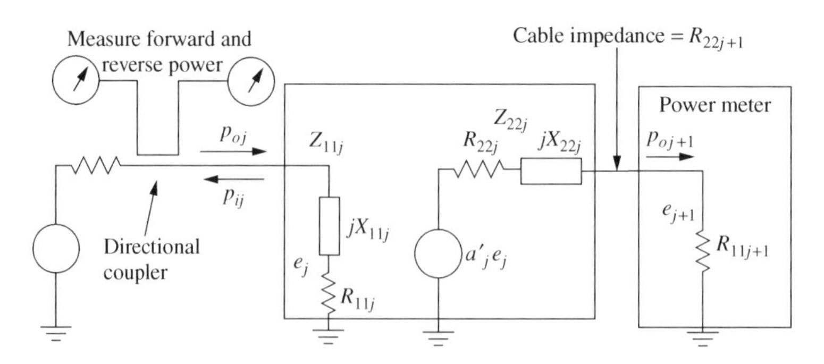

Fig. Z.3 Testing module for available gain.

Here R_{0j} is the characteristic impedance of the input cable (at port j) during test, and S_{11j} can be obtained in terms of impedances from Eq. (14) below.

Z.3 INTERCONNECTS

Interconnect impedances may be included as part of the input impedance $Z_{11,j+1}$ of the following module or the output impedance Z_{22j} of the preceding module. When a true transmission line is used, Section F.4 may be helpful in translating the input impedance of the following module to a value at the output of the preceding module. If the line is made part of the following module and is lossless, whatever power is absorbed into the combined input structure must be absorbed in $R_{11,j+1}$ of the following module. This power is part of the power gain equation [the numerator in Eq. (6)]. For use with voltage gain [Eq. (1)], e_{j+1}^2 can be obtained from the power by multiplying the power by $R_{11,j+1}$.

Z.4 EQUIVALENT S PARAMETERS

Here we will consider how to convert the description of a module in terms of nonstandard impedances into a description using S parameters for standard impedances. If the interfaces in a cascade are matched to various resistive values, the modules on either side of an interface being each matched to the same resistance with specified deviation therefrom, such a cascade can be treated as a standard cascade. The variation of standard impedance (e.g., 75 Ω, 120 Ω, etc.) from interface to interface does not invalidate that method. However, when we convert to an S-parameter description at an impedance significantly different than the actual interface impedance and apply the methods of Section 2.3, we may be throwing away significant information and, as a result, generating unnecessarily large uncertainties in overall performance. In other words, if the impedances of modules in a cascade are known in detail, rather than by their allowed deviation from a standard impedance, converting them to the latter type of description throws away useful information. It may be better to compute cascade gain as in Section Z.1 and then possibly describe the overall cascade by S parameters, as we will do here.

Figure Z.4 shows the module as it is during characterization. (Note that only $v_{\text{in},j}$, $v_{\text{out},j}$, and $v_{\text{out},j+1}$ are normalized variables here.) From this figure we can see (Section F.2)

$$S_{11j} = \frac{Z_{11j} - R_{0j}}{Z_{11j} + R_{0j}}, \tag{14}$$

$$S_{22j} = \frac{Z_{22j} - R_{0,j+1}}{Z_{22j} + R_{0,j+1}}, \tag{15}$$

$$S_{21j} = \frac{v_{\text{out},(j+1)}}{v_{\text{in},j}} = \frac{\tilde{v}_{\text{out},(j+1)}}{\tilde{v}_{\text{in},j}} \sqrt{\frac{R_{0j}}{R_{0,j+1}}}, \tag{16}$$

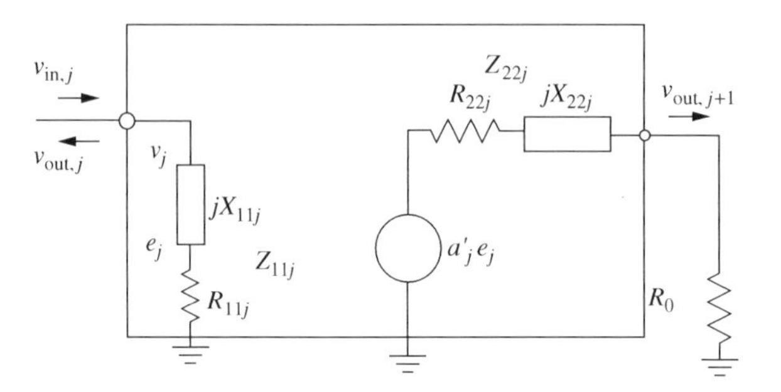

Fig. Z.4 The jth module in test.

$$= \frac{R_{0,j+1}}{Z_{22j} + R_{0,j+1}} a'_j \frac{e_j}{\tilde{v}_{in,j}} \sqrt{\frac{R_{0j}}{R_{0,j+1}}} \tag{17}$$

$$S_{12j} = 0. \tag{18}$$

To put S_{21j} in a usable form, we write e_j in terms of the waves used in defining S parameters by observing

$$\frac{Z_{11j}}{R_{11j}} e_j = \tilde{v}_{in,j} + \tilde{v}_{out,j} = \tilde{v}_{in,j}(1 + S_{11j}). \tag{19}$$

We then substitute e_j from (19) and a'_j from (2) into (17) to obtain

$$S_{21j} = \frac{\sqrt{R_{0j} R_{0,j+1}}}{Z_{22j} + R_{0,j+1}} 2a_{Tj} \frac{R_{11j}}{Z_{11j}} (1 + S_{11j}). \tag{20}$$

Then, substituting for S_{11j} from Eq. (14), we obtain

$$S_{21j} = 4a_{Tj} \frac{\sqrt{R_{0j} R_{0,j+1}} R_{11j}}{(Z_{11j} + R_{0j})(Z_{22j} + R_{0,j+1})}. \tag{21}$$

This expression, along with those of Eqs. (14), (15), and (18), allow S parameters to be written in terms of nonstandard-module parameters.

The second page of the workbook containing Fig. Z.2 shows how these conversions can be made using a spreadsheet. The spreadsheet is written for the usual case where $R_{0j} = R_{0,j+1} = R_0$. It uses the module parameters from the Fig. Z.2 spreadsheet and only the phase of a_{Tj} must be added.

Z.5 *S* PARAMETERS FOR CASCADE OF NONSTANDARD MODULES

We determine the S parameters for a cascade of nonstandard modules so we can use that cascade as an element in a cascade with standard modules having

R_0 interfaces. Because the modules are unilateral, the input impedance of the cascade is that of the first module from Eq. (14):

$$S_{11} = \frac{Z_{111} - R_0}{Z_{111} + R_0}, \tag{22}$$

and the output impedance is that of the last module in the cascade,

$$S_{22} = \frac{Z_{22N} - R_0}{Z_{22N} + R_0}, \tag{23}$$

where the last subscript refers to the number of the module in the cascade of nonstandard modules.

The forward transfer ratio is (Fig. Z.1)

$$S_{21} = \frac{v_{o(N+1)}}{v_{o1}} = \frac{v_1}{v_{o1}} \frac{e_1}{v_1} \frac{v_{o(N+1)T}}{e_1} \tag{24}$$

$$= (1 + S_{111}) \frac{R_{111}}{Z_{111}} a_{\text{cas}}|_{Z_{11(N+1)}=R_0} \tag{25}$$

$$= 2 \frac{R_{111}}{Z_{111} + R_0} a_{\text{cas}}|_{Z_{11(N+1)}=R_0}, \tag{26}$$

where $a_N|_{Z_{11(N+1)}=R_0}$ is given by Eq. (4) with $Z_{11(N+1)} = R_{11(N+1)} = R_0$ (i.e., with the nonstandard cascade properly terminated). [Equation (22) was used in obtaining Eq. (26)].

Due to our assumption that $S_{12j} \approx 0$ for these modules, $S_{12} = 0$ for the cascade also. Thus the cascade meets the unilaterality requirement for modules in Section 2.3.

Note, from Eqs. (22) and (23), that it is easy to determine variations in the reflection coefficients of the cascade from the variations in the individual modules. Also, from Eqs. (26) and (4), the effect of variations in a_{Tj} on S_{21} can be easily determined. Sensitivity analysis may be helpful in determining the effects of the various impedances in Eq. (4) on overall gain if that becomes important.

ENDNOTE

[1]The forward and reverse gain constants in standard Z parameters, Z_{21} and Z_{12}, respectively, are ratios of voltages to terminal currents. The internal gain factor used here is related to the corresponding standard Z parameters by $a'_j = Z_{21j}/R_{11j}$.

REFERENCES

Abromowitz, M., and I. Stegun (1964). *Handbook of Mathematical Functions.* Washington, DC: U.S. Gov't Printing Office.

Allan, D. W. (1966). "Statistics of Atomic Frequency Standards." *Proceedings of IEEE*, Vol. 54, No. 2, February, pp. 221–230.

Amphenol (1995). *Amphenol Reverse Polarity TNC and SMA Connectors.* Data sheet F122-RF/PDS034 Issue 1. Wallingford, CT.

Anaren (2000). *www.anaren.com/catalog.shtml*, on-line catalog for Anaren Microwave, Inc.

Arntz, B. (2000). "Second Order Effects in Feedforward Amplifiers." *Applied Microwaves and Wireless*, January, pp. 66–75.

Baier, S. (1996). "Noise Sources and Noise Calculations for Op Amps." *RF Design*, May, pp. 66–74.

Barkley, K. (2001). "Two-Tone IMD Measurement Techniques." *RF Design*, June, pp. 36–52.

Blachman, N. (1966). *Noise and Its Effect on Communication.* New York: McGraw-Hill, pp. 15, 89–92.

Bracewell, R. (1965). *The Fourier Transform and Its Applications.* New York: McGraw-Hill.

Bullock, S. R. (1995). *Transceiver System Design for Digital Communications.* Tucker, GA: Noble.

Burington, R. (1954). *Handbook of Mathematical Tables and Formulas.* Sandusky, OH: Handbook.

Cain, S. (1999). Composite Triple Beat Count Program. *http://slcain.home.mindspring.com/ctb.htm.*

Cheadle, D. (1973) "Selecting Mixers for Best Intermod Performance (Part 1 and Part 2)." Part 1 in *Microwaves*, November 1973, pp. 48–52 and part 2 in *Microwaves*, December 1973.

Cheadle, D. (1993). *RF and Microwave Designer's Handbook.* San Jose, CA: Stellex Microwave Systems (formerly Watkins-Johnson in Palo Alto, CA), 1993, pp. 484–494. This material also appears in Cheadle (1973).

Davenport, W. B., Jr., and W. L. Root (1958). *An Introduction to the Theory of Random Signals and Noise.* New York: McGraw-Hill, pp. 253–265.

Deats, B., and R. Hartman (1997). "Measuring the Passive-IM Performance of RF Cable Assemblies." *Microwaves and RF*, March, pp. 108–114.

Dechamps, G. A., and J. D. Dyson (1986). "Scattering Matrices" in Edward C. Jordan, ed. *Reference Data for Engineers: Radio, Electronic, Computer, and Communications*, 7th ed. Indianapolis, IN: Howard. W. Sams, pp. 31-3–31-4.

Dicke, R. H. (1948). "General Microwave Circuit Theorems" in C. G. Montgomery, R. H. Dicke, and E. M. Purcell, eds. *Principles of Microwave Circuits*, Vol. 8 of *Radiation Laboratory Series.* New York: McGraw-Hill, p. 150 (S_{12} and S_{21} are interchanged in some places).

Domino, W., N. Vakilian, and D. Agahi (2001). "Polynomial Model of Blocker Effects on LNA/Mixer Devices." *Applied Microwave & Wireless*, June, pp. 30–44.

Drakhlis, B. (2001). "Calculate Oscillator Jitter by Using Phase-Noise Analysis." *Microwaves & RF*, January, pp. 82–90, 157 and February, pp. 109–119.

Egan, W. F. (1981). "The Effects of Small Contaminating Signals in Nonlinear Elements Used in Frequency Synthesis and Conversion." *Proceedings of the IEEE*, Vol. 69, No. 7, July, pp. 279–811.

Egan, W. F. (1988). "An Efficient Algorithm to Compute Allan Variance from Spectral Density." *IEEE Transactions on Instrumentation and Measurement*, Vol. 37, No. 2, June, pp. 240–244.

Egan, W. F. (1998). *Phase-Lock Basics.* New York: Wiley.

Egan, W. F. (2000). *Frequency Synthesis by Phase Lock*, 2nd ed. New York: Wiley.

Egan, W. F. (2002). *Reflections and Mismatches in Interconnects (Appendix R). ftp://ftp.Wiley.com/public/sci_tech_med/rf_system.*

Fano, R. M., and A. W. Lawson (1948). "The Theory of Microwave Filters" in George L. Ragan, ed., *Microwave Transmission Circuits*, Vol. 9 of *Radiation Laboratory Series.* New York: McGraw-Hill, pp. 551–554.

Fong, A., R. Coackley, J. Dupre, M. Fischer, R. Pratt, and D. K. Rytting (1986). "Measurements and Analysis" in Edward C. Jordan, ed. *Reference Data for Engineers: Radio, Electronic, Computer, and Communications*, 7th ed. Indianapolis, IN: Howard. W. Sams, pp. 12-33–12-34.

Germanov, V. (1998). "Calculating the CSO/CTB Spectrums of CATV Amplifiers and Optical Receivers." *IEEE Transactions on Broadcasting*, Vol. 44, No. 3, September, pp. 363–370.

Goldman, S. (1948). *Frequency Analysis, Modulation and Noise*. New York: McGraw-Hill, pp. 172–175.

Gonzalez, G. (1984). *Microwave Transistor Amplifiers, Analysis and Design*. Englewood Cliffs, NJ: Prentice-Hall.

Hardy, J. (1979). *High Frequency Circuit Design*. Reston, VA: Preston.

Haus, H. A., W. R. Atkinson, G. M. Branch, W. B. Davenport, Jr., W. H. Fonger, W. A. Harris, S. W. Harrison, W. W. McLeod, E. K. Stodola, T. E. Talpey (1960a). "IRE Standards on Methods of Measuring Noise in Linear Twoports, 1959." *Proceedings of the IRE*, January, pp. 60–68.

Haus, H. A., W. R. Atkinson, G. M. Branch, W. B. Davenport, Jr., W. H. Fonger, W. A. Harris, S. W. Harrison, W. W. McLeod, E. K. Stodola, T. E. Talpey (1960b). "Representation of Noise in Linear Twoports." *Proceedings of the IRE*, January, pp. 69–74.

Henderson, B. C. (1983). "Reliably Predict Mixer IM Suppression." *Microwaves & RF*, November, pp. 63–70, 132.

Henderson, B. C. (1989). "Mixers in Microwave Systems." *MSN*. Part 1, October, pp. 64–74; Part 2, November, pp. 71–75.

*Henderson, B. C. (1993a). "Predicting Intermodulation Suppression in Double-Balanced Mixers." *97–98 RF and Microwave Designer's Handbook*. San Jose, CA: Stellex Microwave Systems (formerly Watkins-Johnson in Palo Alto, CA), pp. 495–501. See also [Henderson 1983].

*Henderson, B. C. (1993b). "Mixers: Part 1 Characteristics and Performance." *97–98 RF and Microwave Designer's Handbook*. San Jose, CA: Stellex Microwave Systems (formerly Watkins-Johnson in Palo Alto, CA), pp. 469–475.

*Henderson, B. C. (1993c). "Mixers: Part 2 Theory and Technology." *97–98 RF and Microwave Designer's Handbook*. San Jose, CA: Stellex Microwave Systems (formerly Watkins-Johnson in Palo Alto, CA), pp. 476–483.

Hellwig, H., D. Allan, P. Kartaschoff, J. Vanier, J. Vig, G. Winkler, and N. Yannoni (1988). *IEEE Std 1139–1988 Standard Definitions of Physical Quantities for Fundamental Frequency and Time Metrology* (New York, IEEE).

Heutmaker, M. S., J. R. Welch, and E. Wu (1997). "Using Digital Modulation to Measure and Model RF Amplifier Distortion." *Applied Microwave and Wireless*, March/April, pp. 34–39.

Hewlett-Packard (1983). *Fundamentals of RF and Microwave Noise Figure Measurements, Application Note 57-1*. Palo Alto, CA: Hewlett Packard Co. [Agilent Technologies], July.

Hewlett-Packard (1996). *S-Parameter Techniques, Application Note 95-1*. Palo Alto, CA: Hewlett Packard Co. [Agilent Technologies], *http://www.hp.com/go/tmappnotes*.

Howe, D. A. (1976). "Frequency Domain Stability Measurement: A Tutorial Introduction." *NBS Technical Note (U. S.)* 679 (Washington, D.C.: U.S. Government Printing Office), March.

Huh, J. W., I. S. Chang, and C. D. Kim (2001). "Spectrum Monitored Adaptive Feedforward Linearization." *Microwave Journal*, September, pp. 160–166.

Jay, F., ed. (1977). *IEEE Standard Dictionary of Electrical and Electronic Terms, IEEE Std 100-1977.* New York: IEEE, p. 45.

Johnson, K. (2002). "Optimizing Link Performance, Cost and Interchangeability by Predicting Residual BER: Part 1—Residual BER Overview and Phase Noise," July, pp. 20–30, and "...: Part 2—Nonlinearity and System Budgeting," September, pp. 96–131, *Microwave Journal.*

Jordan, E. C., ed. (1986). *Reference Data for Engineers: Radio, Electronic, Computer, and Communications*, 7th ed. Indianapolis, IN: Howard. W. Sams.

Kalb, R. M., and W. R. Bennett (1935). "Ferromagnetic Distortion of a Two-Frequency Wave," *Bell System Technical Journal*, Vol. 14, p. 322.

Katz, A. (1999). "SSPA Linearization." *Microwave Journal*, April, pp. 22–44. (Note: Figure 7 does not appear to represent true separation of the signals into quadrature components since it does not employ coherent detection as does Fig. 8.)

Klipper, H. (1965). "Sensitivity of Crystal Video Receivers with RF Pre-Amplification." *Microwave Journal*, August, pp. 85–92.

Kurokawa, K. (1965). "Power Waves and the Scattering Matrix." *IEEE Transactions on Microwave Theory and Techniques*, Vol. MTT-13, No. 2, March, pp. 194, 195.

Kyle, R. R. (1999). *Spurplot, Mixer Spurious-Response Analysis with Tunable Filtering, Software and User's Manual, Version 2.0.* Boston, MA: Artech House (for Windows 95 and NT).

Laico, J. P. (1956). "A Medium Power TWT for 6000 MHz Radio Relay." *Bell System Technical Journal*, Vol. 35, No. 6, November, pp. 1318–1346 (starts before p. 1318).

Latimer, K. E. (1935–36). "Intermodulation in Loaded Telephone Cables." *Electrical Communications*, Vol. 14, p. 275.

Leeson, D. B. (1966). "A Simple Model of Feedback Oscillator Noise Spectrum." *Proceedings of the IEEE*, Vol. 54, No. 2, February, pp. 329–330. (Note: The symbols S_ϕ and $S_{\dot\phi}$ are interchanged several times in this work.)

Lindsey, W. C., and M. K. Simon (1973). *Telecommunication Systems Engineering.* Englewood Cliffs, NJ; Prentice-Hall.

Linnvill, J. G., and J. F. Gibbons (1961). *Transistors and Active Circuits.* New York: McGraw-Hill.

MA-COM (2000). *www.macom.com*, on-line catalog and application notes for MA-COM, part of Tyco Electronics Corp.

Maas, S. A. (1993). *Microwave Mixers*, 2nd ed. Boston: Artech House.

Maas, S. A. (1995). "Third-Order Intermodulation Distortion in Cascaded Stages." *IEEE Microwave and Guided Wave Letters*, Vol. 5, No. 6, June, pp. 189–191.

Mashhour, A., W. Domino, and N. Beamish (2001). "On the Direct Conversion Receiver — A Tutorial." *Microwave Journal*, June, pp. 114–128.

McClaning, K., and T. Vito (2000). *Radio Receiver Design*. Atlanta, GA: Noble.

Myer, D. P. (1994). "A Multicarrier Feed-forward Amplifier Design." *Microwave Journal*, October, pp. 78–88.

Petit, J. M., and M. M. McWhorter (1961). *Electronic Amplifier Circuits*, New York: McGraw-Hill.

Pozar, D. M. (1990). *Microwave Engineering*. New York: Wiley.

Pozar, D. M. (1998). *Microwave Engineering*, 2nd ed. New York: Wiley.

Pozar, D. M. (2001). *Microwave and RF Wireless Systems*. New York: Wiley.

Ragan, G. L. (1948). "Elementary Line Theory" in George L. Ragan, ed., *Microwave Transmission Circuits*, Vol. 9 of *Radiation Laboratory Series*. New York: McGraw-Hill, p. 35.

Ramo, S., J. R. Whinnery, and T. Van Duzer (1984). *Fields and Waves in Communication Electronics*, 2nd ed. New York: Wiley.

Reuter, W. (2000). "Source and Synthesizer Phase Noise Requirements for QAM Radio Applications," on *www.cti-inc.com*, Whippany, NJ: Communications Techniques Inc., site ©1997–2000.

RF Micro-Devices (2001). Data sheet for RF2317 Linear CATV Amplifier, Rev. A15, March 16.

Rice, S. O. (1944, 1945). "Mathematical Analysis of Random Noise." *Bell System Technical Journal*, Vol. 23, July 1944, pp. 282–332; continued in Vol. 24, January 1945, pp. 46–156.

Robins, W. P. (1984). *Phase Noise in Signal Sources*. London: Peter Peregrinus for Institution of Electrical Engineers.

Roetter, A., and D. Belliveau (1997). "Single-Tone IMD Analysis via the Web: A Spur Chart Calculator Written in Java." *Microwave Journal*, Vol. 40, No. 11, November. The paper and the calculator program are available at *http://www.hittite.com*. Paths such as *published papers/mixers and converters, product support/mixer spur chart calculator*, or *company information/engineering tools/mixer spur chart calculator* may be helpful.

Rohde, U. L., and T. T. N. Bucher (1988). *Communications Receivers: Principles and Design*, New York: McGraw-Hill.

Schwartz, M., W. R. Bennett, and S. Stein (1966). *Communication Systems and Techniques*. New York: McGraw-Hill, pp. 107–114.

Seidel, H. (1971a). "A Feedforward Experiment Applied to an L-4 Carrier System Amplifier." *IEEE Transactions on Communications Technology*, Vol. Com-19, No. 3, June, pp. 320–325.

Seidel, H. (1971b). "A Microwave Feed-Forward Experiment." *Bell System Technical Journal*, November, pp. 2879–2916.

Seidel, H., H. R. Beurrier, and A. N. Friedman (1968). "Error-Controlled High Power Linear Amplifiers and VHF." *Bell System Technical Journal*, May–June, pp. 651–722.

Sevick, J. (1987). *Transmission Line Transformers*. Newington, CT: American Radio Relay League.

Snelling, E. C. (1988). *Soft Ferrites: Properties and Applications*. London: Butterworths, pp. 39–40.

Snyder, R. E. (1978). "Use Filter Models to Analyze Receiver IM." *Microwaves*, November, pp. 78–82.

Steffes, M. (1998). "Noise Analysis for High Speed Op Amps." *Burr-Brown Applications Bulletin* from Texas Instruments website. Modified 10/18/2000.

Stellex Catalog (1997). "Mixer Application Information." *RF and Microwave Designer's Handbook*. San Jose, CA: Stellex Microwave Systems (formerly Watkins-Johnson Company in Palo Alto, CA), pp. 465–468.

Thomas, J. L. (1995). *Cable Television Proof-of-Performance*. Upper Saddle River, NJ: Prentice-Hall.

Toolin, M. J. (2000). "A Simplified Approach to Determining AM/PM Conversion Coefficient in Microwave Low Noise Amplifiers and Systems." *Microwave Journal*, August, pp. 80–90.

Tsui, J. B. (1985). *Microwave Receivers and Related Components*. Los Altos, CA: Peninsula.

Tsui, J. B. (1995). *Digital Techniques for Wideband Receivers*. Boston: Artech House.

Vizmuller, P. (1995). *RF Design Guide—Systems, Circuits, and Equations*. Boston: Artech House.

Watkins-Johnson Catalog (1993). *RF and Microwave Designers Handbook, 1993–1994*. Palo Alto, CA: Watkins-Johnson.

Winder, S. (1993). "Single Tone Intermodulation Testing." *RF Design*, December, pp. 34–44. Steve Franke, on p. 16 of the March 1994 issue of the same magazine, identifies some errors in the development, which, however, do not invalidate results.

Wood, R. A. (2001a). *SpurFinder, v. 3.0*. R. A. Wood Associates Software Products, *http://www.rawood.com/software_products/index.html*.

Wood, R. A. (2001b). *TunerHelper v. 2.0*. R. A. Wood Associates Software Products, *http://www.rawood.com/software_products/index.html*.

Yang, Y., J. Yi, J. Nam, and B. Kim (2000). "Behavioral Modeling of High Power Amplifiers Based on Measured Two-Tone Transfer Characteristics." *Microwave Journal*, December, pp. 90–104.

Yola, D. C. (1961). "On Scattering Matrices Normalized to Complex Port Numbers." *Proceedings IRE*, Vol. 49, No. 7, July, p. 1221.

ENDNOTE

*First publication date of these works is uncertain due to the nature of the source. They may have appeared in versions of the catalog that are earlier than the date given in parentheses.

INDEX